Student Study Guide with Answers to Text Exercises

*Student Study Guide with
Answers to Text Exercises*

Introduction to Organic and Biological
Chemistry FOURTH EDITION

Stuart J. Baum

STATE UNIVERSITY OF NEW YORK, COLLEGE AT PLATTSBURGH

MACMILLAN PUBLISHING COMPANY
NEW YORK
COLLIER MACMILLAN PUBLISHERS
LONDON

Macmillan Publishing Company
866 Third Avenue, New York, New York 10022

Collier Macmillan Canada, Inc.

ISBN 0-02-306760-8

Printing: 3 4 5 6 7 8   Year: 1 2 3 4 5 6

ISBN 0-02-306760-8

Contents

Note to the Student

It is difficult for most students to learn chemistry simply by reading the text and attending lectures. Knowledge is gained by actively taking part in the learning process. You should get in the habit of studying and reviewing rather than cramming before major exams. That is, read the material in the text, go over your lecture notes, and write out your answers to the exercises at the end of each chapter. Then test your understanding by working through sample tests.

This study guide is divided into two sections. Part I contains answers to all the text exercises. You ought to do each exercise and then check your answer–not read the question and then look up the answer. If you don't attempt to answer the question yourself, the answer will have little meaning and will be quickly forgotten.

Part II contains numerous questions for self-study. They are comprised of multiple choice, fill in the blank, and true-false types. In each set the questions follow the order of coverage of the material in the text. The more questions you can answer successfully, the greater will be your confidence that you understand the subject. If you have trouble with these questions, you should seek tutoring. Although the answers are given at the end of Part II, again it is vital that you attempt the questions first, in writing, before you look at the answers.

Doing all the exercises in the text and the questions in this study guide is a lot of work. The payoff for you will be a better understanding of the subject–and a good grade in the course.

S. J. B.

PART I

Answers to Text Exercises

CHAPTER 1

Introduction to Organic Chemistry

1.1 Widespread belief in the vital force theory, which held that living organisms possessed some unexplained force that was necessary for the formation of organic substances. Wöhler's experiment was significant because it was the first time in history that an organic compound was prepared from inorganic substances. This experiment is considered to mark the beginning of modern organic chemistry.

1.2 The number of carbon-containing compounds appears to be unlimited because carbon has the ability to bond to other carbon atoms to form chains of various lengths. As successively more carbons are added to the chain, the possible number of ways in which these carbon atoms arrange themselves relative to one another increases astronomically, so that there can be several compounds with identical molecular composition but different chemical structures.

1.3 In general, organic compounds can be considered to exhibit those properties associated with covalent compounds because their bonding is usually covalent. Thus, most organic compounds have low melting points and low boiling points and are insoluble in water but soluble in the nonpolar organic solvents. One often hears of fires occurring in the organic chemistry laboratory because organic compounds are usually flammable. Solutions containing organic compounds do not conduct an electrical current. Because carbon can bond to other carbons to form long chains and form compounds with identical molecular formulas but different structural formulas, organic compounds are said to exhibit isomerism. Organic compounds exist at room temperature as gases, liquids, and solids. As a rule, the reactions of organic compounds are slow.

Inorganic compounds exhibit predominantly ionic bonding, which results in very different properties. The high melting and boiling points found whenever ionic bonds must be disrupted are evidenced by inorganic compounds. Because they are ionic, these compounds are soluble in water, but generally insoluble in the nonpolar solvents, and their solutions tend to conduct an electrical current. Inorganic compounds are generally nonflammable and exhibit isomerism only as a rare exception. Their reactions are generally quite rapid. These compounds usually exist in the solid phase at room temperature.

1.4 Tests to determine if a newly discovered compound is organic or inorganic:

1. Try to dissolve the compound in both water and an organic solvent (e.g., toluene or CCl_4). If the compound is insoluble in water but soluble in the organic solvent, it is probably organic. If the compound is soluble in water but insoluble in the organic solvent, it is probably inorganic.

2. Try to burn the substance (under carefully controlled conditions). If the substance is flammable, it is probably organic. A nonflammable substance is probably inorganic.

3. Determine the melting point and boiling point of the substance. Very high melting points and boiling points are more apt to be found among inorganic compounds. Low melting points and boiling points are indicative of an organic substance.

A combination of all of these tests would give more conclusive evidence as to the nature of the compound being considered.

1.5 Carbon is exceptional among the elements because it has the ability to form bonds with other carbon atoms to form carbon chains of various lengths. As successive carbons are added to the chain, the ways in which these carbons can be arranged relative to one another increase. This results in the phenomenon known as isomerism.

1.6 Structural formulas are more informative. They show the bonding arrangement of the atoms in the molecule.

1.7 (a) :F̈:F̈:   (b) :Ö::C::Ö: (c) S̈::Ö   (d) H:Ö: (e) H:N̈:H

(c, below) :Ö:

(d) H

(e) H

(f) H:C::Ö: (g) H:C:::N̈   (h) H:Ö:C̈l: (i) H:C̈:C̈l:

(f) H

(i, above) H   (i) H

(j) H:Ö:N̈::Ö: (k) H:C̈:C::C:H   (l) H:C̈:C̈:C̈:H   (m) :Cl: ·Cl:P:Cl: ·Ö:·Ö:

(k) H   H H H

(l) H H H   H H H

(n) H:Ö:C̈l::Ö: Ö: Ö: (o) Mg²⁺ :C̈l:⁻ :C̈l:⁻   (p) Li⁺ :B̈r:⁻

1.8 (a) polar (b) polar (c) polar (d) nonpolar (e) nonpolar (f) polar
(g) polar (h) nonpolar

1.9 The high melting point of lithium hydride is indicative of ionic bonding between lithium cations and hydrogen anions (hydride ions) in the crystal lattice. The low melting point of methane suggests that the molecule is maintained by covalent bonds and that intermolecular forces of attraction are extremely weak.

1.10 (a) A typical ionic compound has a high melting point and a high boiling point and is soluble in water.
(b) A typical covalent compound has relatively low melting and boiling points and is insoluble in water.

1.11 Organic substances: alcohol, ether, gasoline, sugar, cholesterol, etc. Inorganic substances: water, ammonia, sodium hydroxide, sulfuric acid, salt, etc.

1.12 (a) 109° (b) 180° (c) 109° (d) 109° (e) 120° (f) 109° (g) 120° (h) 120°
(i) 120° (j) 120° (k) 120° (l) 120°

1.13 (a) alcohol (b) alkyne (c) amine (d) ether (e) alkene (f) alkane
(g) ketone (h) ester (i) aldehyde (j) carboxylic acid (k) amide (l) amide

1.14 H—C(H)(H)—C(H)(H)—H   C(H)(H)=C(H)(H)   H—C≡C—H   H—C(H)(H)—C(=O)—H

All of the other compounds in the table have isomers.

1.15 (a) C_5H_{12} (b) C_5H_8 (c) C_6H_{12} (d) $C_3H_6Br_2$ (e) $C_6H_{14}O$ (f) C_2H_7N
(g) $C_6H_{12}O$ (h) C_4H_8O (i) $C_4H_{10}O$ (j) C_2H_5ON (k) $C_5H_{10}O_2$ (l) $C_3H_6O_2$

1.16 (a) $CH_3CH_2CH_2Br$ (b) $CH_3CH_2CH_2CH_2OH$ (c) $CH_3C≡CCH_2CH_3$ (d) $CH_3CH_2OCH_3$
(e) $CH_3CH_2NH_2$ (f) $CH_3CH_2CH_2CH=CHCH_3$

1.17 **(a)**

```
    H   H   H              H  OH  H              H   H        H
    |   |   |              |   |   |             |   |        |
H—C—C—C—OH          H—C—C—C—H          H—C—C—O—C—H
    |   |   |              |   |   |             |   |        |
    H   H   H              H   H   H             H   H        H
```

(b)

```
    H  Cl                  Cl  Cl
    |   |                   |   |
H—C—C—Cl            H—C—C—H
    |   |                   |   |
    H   H                   H   H
```

(c)

```
    H  Cl                  Cl  Cl
    |   |                   |   |
H—C—C—Cl            H—C—C—Cl
    |   |                   |   |
    H  Cl                   H   H
```

(d)

```
    H   H   O              H   O   H
    |   |   ||             |   ||  |
H—C—C—C—H          H—C—C—C—H
    |   |                  |       |
    H   H                  H       H
```

(e)

```
                           H   O                 H       O
                           |   ||                |       ||
HO—C=C—OH          HO—C—C—H          H—C—O—C—H
     |   |                 |                     |
     H   H                 H                     H
```

(f)

```
    H   H   H              H  Cl   H
    |   |   |              |   |   |
H—C—C—C—Cl         H—C—C—C—H
    |   |   |              |   |   |
    H   H   H              H   H   H
```

(g)

```
    H   H   Br             H   Br  H             H  Cl   H             H   H   H
    |   |   |              |   |   |             |   |   |             |   |   |
H—C—C—C—Cl         H—C—C—C—Cl         H—C—C—C—Br         H—C—C—C—H
    |   |   |              |   |   |             |   |   |             |   |   |
    H   H   H              H   H   H             H   H   H             Br  H  Cl
```

```
    H  Br   H
    |   |   |
H—C—C—C—H
    |   |   |
    H  Cl   H
```

(h)

```
    H   H   H   H          H   H   Br  H          H   H   H             H  Br   H
    |   |   |   |          |   |   |   |          |   |   |             |   |   |
H—C—C—C—C—Br     H—C—C—C—C—H     H—C—C—C—Br     H—C—C—C—H
    |   |   |   |          |   |   |   |          |   |   |             |   |   |
    H   H   H   H          H   H   H   H          H   C   H             H   C   H
                                                     / \                   / \
                                                    H   H  H             H   H  H
```

(i)

```
    H   H                  H  OH                  H       H             H           H
    |   |                  |   |                  |       |             |           |
HO—C—C—OH          H—C—C—OH          H—C—O—C—OH          H—C—O—O—C—H
    |   |                  |   |                  |       |             |           |
    H   H                  H   H                  H       H             H           H
```

```
    H   H
    |   |
H—C—C—O—O—H
    |   |
    H   H
```

```
      H  H              H  H  H
      |  |              |  |  |
(j) H—C—C—N—H      H—C—N—C—H
      |  |  |           |     |
      H  H  H           H     H
```

1.18 Formula pairs (b), (d), (f), (g), (h), (i), (j), and (k) are isomers.

1.19 No. Since the molecular formula of one homolog differs from that of the next by a $-CH_2-$ unit, and isomers of each other must have the same molecular formula, homologs, by definition, cannot be isomers.

1.20 (a) A positively charged ion, e.g., sodium ion, Na^+.
 (b) A negatively charged ion, e.g., chloride, Cl^-.
 (c) The forces or interactions that cause atoms to be held together as molecules or cause atoms, ions, and molecules to be held together as more complex aggregates, e.g., $Na^+ Cl^-$, CH_3OH.
 (d) An electrostatic attraction between a positively charged ion (cation) and a negatively charged ion (anion), e.g., Na^+Cl^-.
 (e) The attraction between two atoms as a result of sharing one or more outer-shell electrons by the two atoms, e.g., H–H.
 (f) One electron pair shared between two atoms, e.g., H–H.
 (g) Two electron pairs shared between two atoms, e.g., $CH_2=CH_2$.
 (h) When there is separation of positive and negative charges, e.g., H–Cl.
 (i) When there is equal sharing of the shared electrons, e.g., H–H.
 (j) The center-to-center distance between the two nuclei; e.g., the C–C bond distance in ethane is 1.54 Å.
 (k) The energy required to break the chemical bond; e.g., the bond energy of the C–C bond in ethane is 83 kcal/mole.
 (l) The angle formed by two covalent bonds; e.g., the H–C–H bond angle in methane is 109.5°.
 (m) Also called simplest formula, it gives the relative numbers of atoms in a molecule, e.g., CH_3 for ethane.
 (n) The chemical formula for a molecule that gives the actual numbers of atoms of each element in the molecule, e.g., ethane, C_2H_6.
 (o) A representation of how the individual atoms are bonded to each other in a molecule, e.g.,

```
      H  H
      |  |
   H—C—C—H
      |  |
      H  H
     ethane
```

 (p) Compounds having the same molecular formula but different structural formulas, e.g.,

```
   H  H  H  H                    H   H  H
   |  |  |  |                     \  | /
H—C—C—C—C—H    and        H   C   H
   |  |  |  |                  H—C—C—C—H
   H  H  H  H                     |  |  |
     butane                       H  H  H
                                 isobutane
```

 (q) A family of similar compounds, the members of which fulfill the following requirements: (1) all of the members of the series must contain the same elements and be able to be represented by one general formula; (2) the molecular formula of any given homolog in the series differs from the one below it and the one above it by a methylene ($-CH_2-$) group; (3) as the molecular weights of the homologs increase with addition of further methylene groups, there is a slight change in physical properties; (4) all of the homologs in a series exhibit similar chemical behavior. Example: methane, ethane, propane.

(r) Any particular arrangement of a few atoms that bestows characteristic properties on an organic molecule, e.g., $-OH$, $C=C$.

1.21 C_2H_6

1.22 (a) C_2H_5 **(b)** C_4H_{10} **(c)** $CH_3CH_2CH_2CH_3$ and $CH_3-\overset{\displaystyle H}{\underset{\displaystyle CH_3}{C}}-CH_3$

CHAPTER 2

The Saturated Hydrocarbons

2.1 $CH_3CH_2CH_2CH_2CH_2CH_2CH_3$

n-heptane

$$CH_3CH_2CH_2CH_2\overset{\overset{\displaystyle CH_3}{|}}{C}HCH_3$$

2-methylhexane

$$CH_3CH_2CH_2\overset{\overset{\displaystyle CH_3}{|}}{C}HCH_2CH_3$$

3-methylhexane

$$CH_3CH_2\overset{\overset{\displaystyle CH_3}{|}}{\underset{\underset{\displaystyle CH_3}{|}}{C}}CH_2CH_3$$

3,3-dimethylpentane

$$CH_3\overset{\overset{\displaystyle CH_3}{|}}{C}HCH_2\overset{\overset{\displaystyle CH_3}{|}}{C}HCH_3$$

2,4-dimethylpentane

$$CH_3\overset{\overset{\displaystyle CH_3}{|}}{C}H\underset{\underset{\displaystyle CH_3}{|}}{C}HCH_2CH_3$$

2,3-dimethylpentane

$$CH_3CH_2CH_2\overset{\overset{\displaystyle CH_3}{|}}{\underset{\underset{\displaystyle CH_3}{|}}{C}}CH_3$$

2,2-dimethylpentane

$$CH_3\overset{\overset{\displaystyle H_3C}{|}}{C}H\underset{\underset{\displaystyle CH_3}{|}}{\overset{\overset{\displaystyle CH_3}{|}}{C}}CH_3$$

2,2,3-trimethylbutane

$$CH_3CH_2\overset{\overset{\displaystyle CH_3}{|}}{\overset{\overset{\displaystyle CH_2}{|}}{C}}HCH_2CH_3$$

3-ethylpentane

2.2 (a) and (h); (b) and (j); (c) and (g); (d) and (i); (e) and (f)

2.3 (a) $CH_3CH_2\overset{\overset{\displaystyle CH_3}{|}}{C}HCH_2CH_3$

3-methylpentane

(b) $CH_3\overset{\overset{\displaystyle CH_3}{|}}{C}H\underset{\underset{\displaystyle CH_3}{|}}{C}HCH_3$

2,3-dimethylbutane

(c) $CH_3\underset{\underset{\displaystyle CH_2CH_3}{|}}{C}HCH_2CH_3$

3-methylpentane

(d) $CH_3\overset{\overset{\displaystyle CH_3}{|}}{C}H-\underset{\underset{\displaystyle CH_3}{|}}{\overset{\overset{\displaystyle CH_3}{|}}{C}}-\underset{\underset{\displaystyle CH_2CH_3}{|}}{\overset{\overset{\displaystyle CH_2CH_3}{|}}{C}}-CH_2CH_3$

4,4-diethyl-2,3,3-trimethylhexane

(e) $CH_3-\underset{\underset{\displaystyle CH_3}{|}}{\overset{\overset{\displaystyle CH_3}{|}}{C}}-CH_2CH_3$

2,2-dimethylbutane

(f) $CH_3\overset{\overset{\displaystyle CH_3}{|}}{C}HCH_2\overset{\overset{\displaystyle CH_3}{|}}{C}HCH_3$

2,4-dimethylpentane

(g) $CH_3CH_2\overset{\overset{\displaystyle CH_2CH_3}{|}}{C}HCH_2CH_2CH_3$

3-ethylhexane

(h) $CH_3CH_2CH_2CH_2\overset{\overset{\displaystyle CH_3}{|}}{C}H\underset{\underset{\displaystyle CH_2CH_2CH_3}{|}}{C}HCH_3$

4-isopropyloctane

(i) $CH_3\overset{\overset{\displaystyle CH_3}{|}}{C}HCH_2\underset{\underset{\displaystyle CH_3}{|}}{\overset{\overset{\displaystyle CH_3}{|}}{C}}-\overset{\overset{\displaystyle CH_2CH_3}{|}}{\overset{\overset{\displaystyle CH_2}{|}}{C}}H-\underset{\underset{\displaystyle CH_3}{|}}{\overset{\overset{\displaystyle CH_3}{|}}{C}}-CH_3$

3-ethyl-2,2,4,4,6-
pentamethylheptane

8

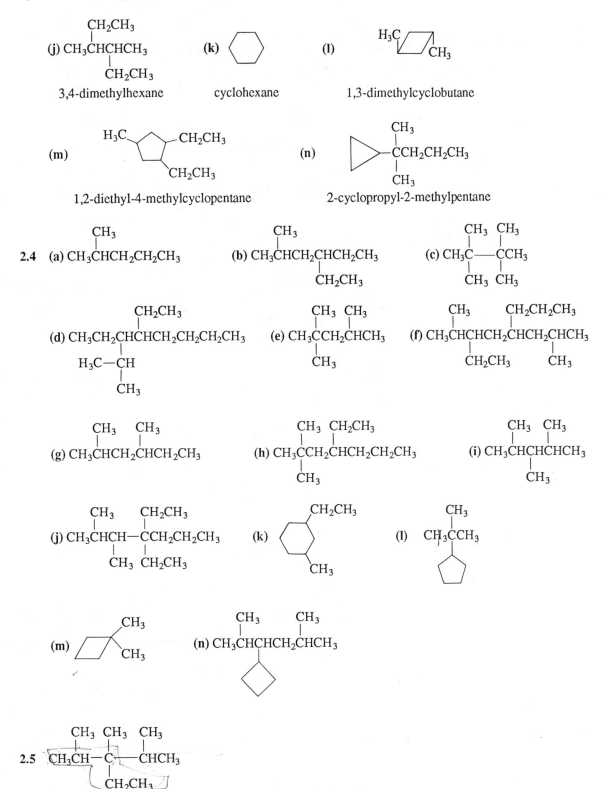

(j) 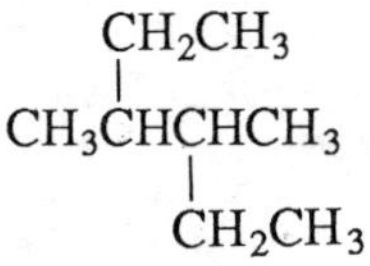

3,4-dimethylhexane

(k) 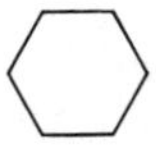

cyclohexane

(l) 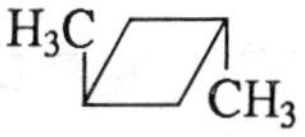

1,3-dimethylcyclobutane

(m) 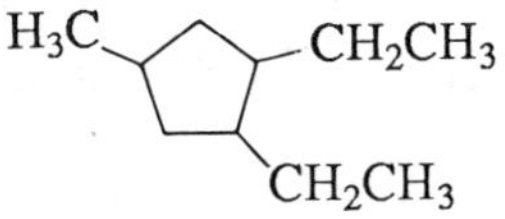

1,2-diethyl-4-methylcyclopentane

(n) 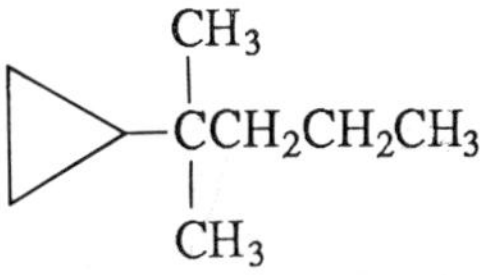

2-cyclopropyl-2-methylpentane

2.4 (a) 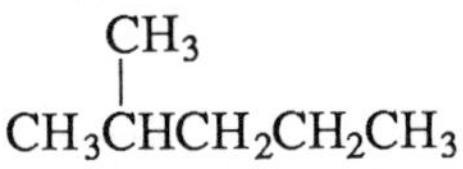

(b) 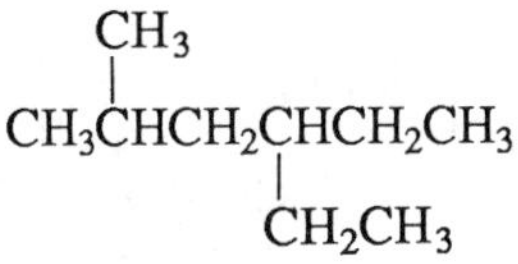

(c) 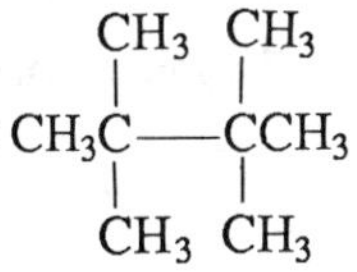

(d) 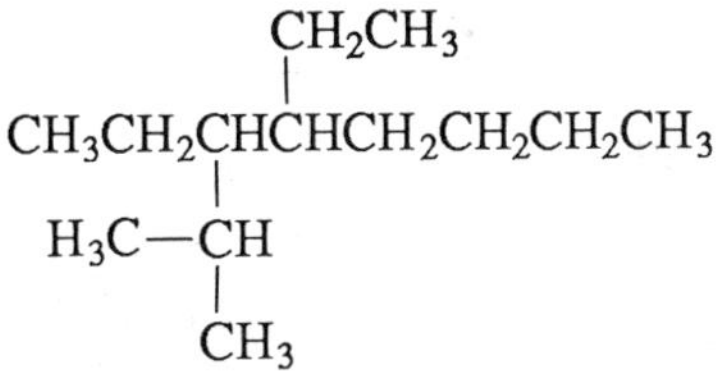

(e) 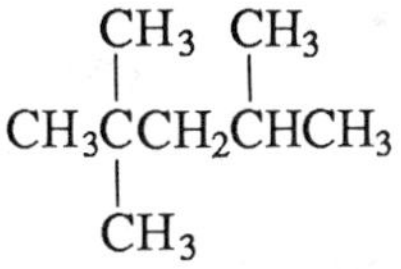

(f) 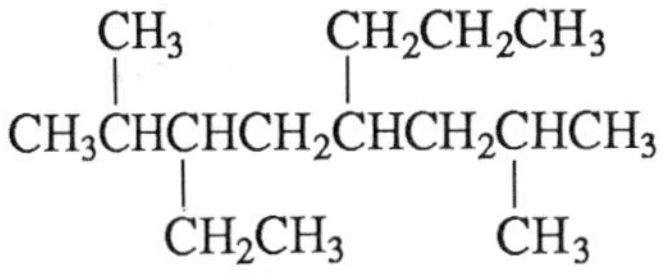

(g) 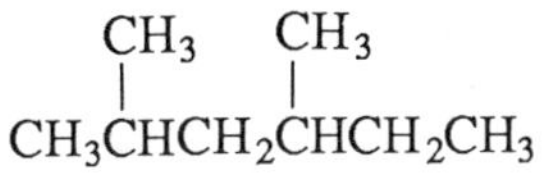

(h) 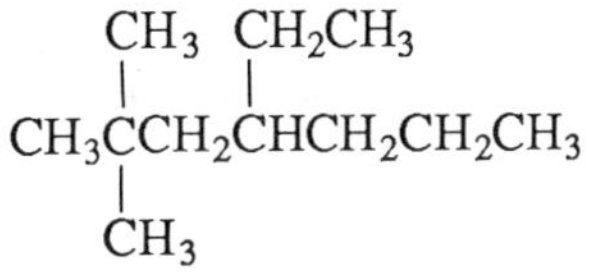

(i) 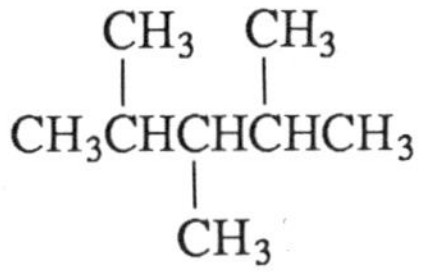

(j) 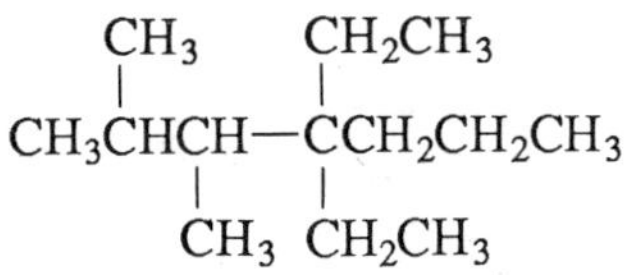

(k) 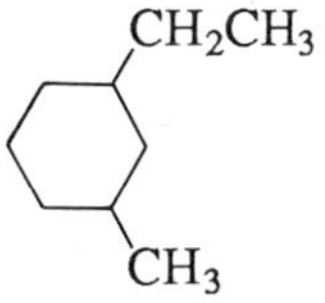

(l) 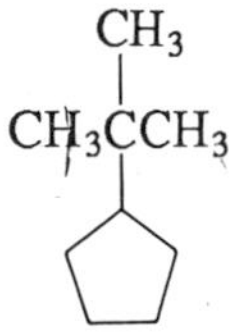

(m) 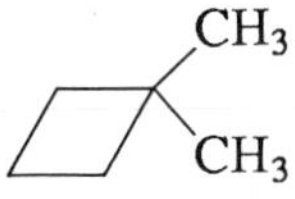

(n) 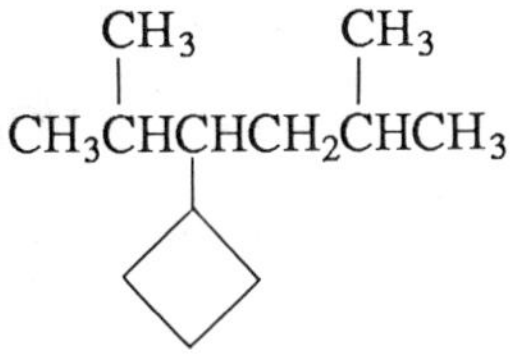

2.5 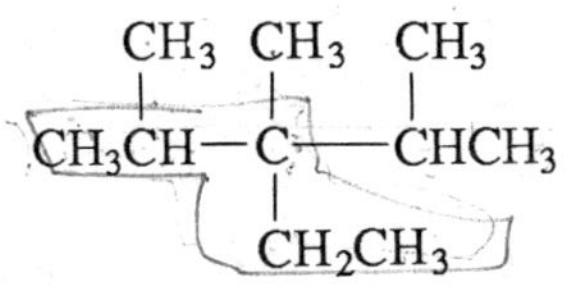

3-isopropyl-2,3-dimethylpentane

2.6 (a)

$$CH_3-\underset{\underset{CH_3}{|}}{\overset{\overset{CH_3}{|}}{C}}-CH_3$$

2,2-dimethylpropane

(b)

$$CH_3CH-\underset{\underset{CH_3}{|}}{\overset{\overset{CH_3\ CH_3}{|\ \ \ |}}{C}}-CH_3$$

2,2,3-trimethylbutane

(c)

$$CH_3\underset{\underset{CH_2CH_3}{|}}{CHCH_2}\underset{\overset{CH_2CH_3}{|}}{CH}CH_3$$

3,5-dimethylheptane

(d)

$$CH_3CH_2CHCHCHCH_3$$
$$\underset{CH_3}{|}\ \ \underset{CH_2}{|}$$
$$\underset{CH_2}{|}$$
$$\underset{CH_3}{|}$$

3,4,5-trimethyloctane

(e)

$$CH_3CH_2CH_2\underset{\underset{CH_3}{|}}{CH}\overset{\overset{CH_3CHCH_3}{|}}{CH}CH_2CH_3$$

4-isopropyl-3-methylheptane

(f)

$$CH_3CH_2\overset{\overset{CH_3}{|}}{CH}\underset{\underset{CH_2CH_3}{|}}{CH}CH_3$$

3-ethyl-2-methylpentane

(g)

$$CH_3CH_2\underset{\underset{CH_2CH_3}{|}}{\overset{\overset{CH_2CH_3}{|}}{C}}-CH_2-\underset{\underset{CH_3}{|}}{\overset{\overset{CH_3}{|}}{C}}CH_3$$

4,4-diethyl-2,2-dimethylhexane

(h)

$$CH_3\underset{\underset{CH_2CH_3}{|}}{\overset{\overset{CH_2CH_3}{|}}{C}}-CH_2-\underset{\underset{CH_3}{|}}{\overset{\overset{CH_3}{|}}{C}}CH_2CH_2CH_3$$

3-ethyl-3,5,5-trimethyloctane

(i)

$$CH_3\underset{\underset{CH_3}{|}}{CH}CH_2CH_2CH_2CH_2CH_2\overset{\overset{CH_3}{|}}{CH_2}$$

2-methylnonane

(j)

1,2-dimethylcyclobutane

(k)

1,4-diethylcyclohexane

(l)

isopropylcyclopentane

(m)

3-ethyl-1,1-dimethylcycloheptane

(n)

$$CH_3CH_2CHCH_3$$

2-cyclopropylbutane

2.7 methane and ethane

2.8 (a) pentane   (b) pentane   (c) $CH_3(CH_2)_4CH_3$ (d) isohexane   (e) cyclohexane
      (f) $CH_3(CH_2)_7CH_3$

2.9 (a) pentane   (b) neopentane   (c) $(CH_3)_2CHCH(CH_3)_2$ (d) isohexane   (e) cyclohexane
      (f) $CH_3(CH_2)_7CH_3$

2.10 (I) CH_3CH_2Br (II) CH_3CHBr_2 (III) CH_2BrCH_2Br (IV) CH_3CBr_3

(V) $CH_2BrCHBr_2$ (VI) $CHBr_2CHBr_2$ (VII) CH_2BrCBr_3 (VIII) $CHBr_2CBr_3$

(IX) CBr_3CBr_3

Isomers are (II) and (III); (IV) and (V); (VI) and (VII).

2.11 (a) 3; $CH_3CH_2CH_2CH_2CH_2Cl$, $CH_3CH_2CH_2CHClCH_3$ and $CH_3CH_2CHClCH_2CH_3$
  (b) 4 monochloro isomers of isopentane; 1 monochloro derivative of neopentane.

2.12 (a) petroleum (crude oil and natural gas)   (b) fuels, lubricating oils, greases, fertilizers, synthetic chemicals (e.g., plastics)

2.13 (a) alkyl group, CH_3CH_2- alkane, CH_3CH_3
  (b) alkyl anion, $CH_3:^-$ alkyl free radical, $CH_3\cdot$
  (c) common name, isopentane   IUPAC name, 2-methylbutane

  (d) n-propyl group, $CH_3CH_2CH_2-$ isopropyl group, $CH_3-\overset{\overset{CH_3}{|}}{\underset{\underset{H}{|}}{C}}-$

  (e) straight chain, $CH_3CH_2CH_2CH_3$ branched chain, $CH_3\overset{\overset{CH_3}{|}}{C}HCH_3$

  (f) acyclic, $CH_3CH_2CH_2CH_2CH_3$ alicyclic,

  (g) chair form,   boat form,

  (h) axial hydrogen,   equatorial hydrogen,

  (i) polar group, $-OH$ nonpolar group, CH_3CH_2-
  (j) antioxidant, BHT   antiknock agent, $(C_2H_5)_4Pb$

2.14 (a) Natural gas is composed of 80% methane, 10% ethane, and 10% mixture of higher alkanes; it has a density of 0.65 g/L. Bottled gas is composed chiefly of propane; it has a density of 1.6 g/L.
  (b) Physical properties can be described without reference to any other specific chemical substance: density, melting point, boiling point, etc. Chemical properties describe the reactions of a substance with other chemicals, e.g., halogenation, hydrogenation, oxidation.
  (c) Exothermic reaction is one in which heat is liberated. Endothermic reaction is one in which heat is used up.
  (d) Crude oil is the liquid component of petroleum. Natural gas is the gaseous component of petroleum.
  (e) In the free radical reaction mechanism, the chain-initiating step involves the cleavage of a diatomic halogen by ultraviolet light into two highly reactive species (free radicals), each containing an odd number of electrons. The chain-terminating step of the mechanism involves the combination of two free radicals to form a stable species.

2.15 An octane rating of 90 means that the gasoline has the same knocking properties as a mixture of 10% heptane and 90% isooctane. In general, the more branched the hydrocarbon chain, the higher the octane rating of that compound.

2.16 hexane, 2-methylhexane, 2-methylbutane, 2,2,4-trimethylpentane.

2.17 (a) no reaction   (b) $CO_2 + H_2O$ (c) ⬡—Br + HBr   (d) no reaction

       (e) no reaction   (f) no reaction

2.18 (a)

$$CH_3-\underset{\underset{CH_3}{|}}{\overset{\overset{CH_3}{|}}{C}}-CH_3 \;+\; Br_2 \;\xrightarrow{\text{light}}\; CH_3-\underset{\underset{CH_3}{|}}{\overset{\overset{CH_3}{|}}{C}}-CH_2Br$$

(b)

$$CH_3-\underset{\underset{H}{|}}{\overset{\overset{CH_3}{|}}{C}}-CH_2CH_3 \;+\; Br_2 \;\longrightarrow\;
\begin{cases}
CH_3-\underset{\underset{H}{|}}{\overset{\overset{CH_2Br}{|}}{C}}-CH_2CH_3 \\[2mm]
CH_3-\underset{\underset{Br}{|}}{\overset{\overset{CH_3}{|}}{C}}-CH_2CH_3 \\[2mm]
CH_3-\underset{\underset{H}{|}}{\overset{\overset{CH_3}{|}}{C}}-CH_2CH_2Br \\[2mm]
CH_3-\underset{\underset{H}{|}}{\overset{\overset{H_3C}{|}}{C}}-\underset{\underset{H}{|}}{\overset{\overset{Br}{|}}{C}}-CH_3
\end{cases}$$

2.19 $2\,C_8H_{18} + 25\,O_2 \longrightarrow 16\,CO_2 + 18\,H_2O$
1 gal = 4 qt = 3.8 L
mass of 3.8 L of gasoline = (0.69)(3800) = 2600 g
2600 g of octane = 2600/114 moles of octane = 23 moles of octane
23 moles of octane will yield 180 moles of CO_2 and 210 moles of H_2O
mass of CO_2 produced = (180)(44) = 7900 g
mass of H_2O produced = (210)(18) = 3800 g

2.20 $2\,C_2H_6 + 7\,O_2 \longrightarrow 4\,CO_2 + 6\,H_2O$
2.24 L of ethane = 0.1 mole of ethane
0.1 mole of ethane requires 7/2 (0.1) = 0.35 mole of oxygen
0.35 mole of oxygen = (0.35)(22.4) = 7.8 L of oxygen
7.8 L of oxygen is provided by 39 L of air

2.21 Simplest formula of hydrocarbon is CH_2.
molecular weight of hydrocarbon = (1.25)(22.4) = 28 g
Therefore, molecular formula is C_2H_4, and structural formula is $H-\underset{\underset{H}{|}}{C}=\underset{\underset{H}{|}}{C}-H$.

The Unsaturated Hydrocarbons

3.1 CH_2=$CHCH_2CH_2CH_2CH_3$ CH_3CH=$CHCH_2CH_2CH_3$ CH_3CH_2CH=$CHCH_2CH_3$
           1-hexene   2-hexene   3-hexene

$$CH_3$$
CH_2=$CCH_2CH_2CH_3$
2-methyl-1-pentene

$$CH_3$$
CH_3C=$CHCH_2CH_3$
2-methyl-2-pentene

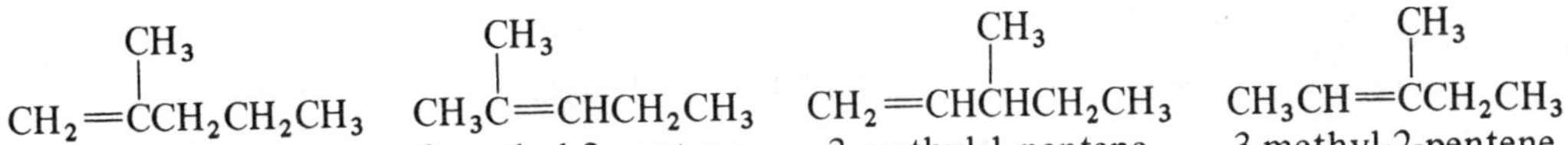

$$CH_3$$
CH_2=$CHCHCH_2CH_3$
3-methyl-1-pentene

$$CH_3$$
CH_3CH=CCH_2CH_3
3-methyl-2-pentene

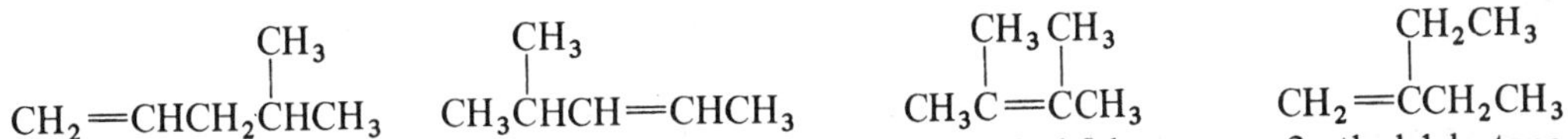

$$CH_3$$
CH_2=$CHCH_2CHCH_3$
4-methyl-1-pentene

$$CH_3$$
CH_3CHCH=$CHCH_3$
4-methyl-2-pentene

$$CH_3 \quad CH_3$$
CH_3C=CCH_3
2,3-dimethyl-2-butene

$$CH_2CH_3$$
CH_2=CCH_2CH_3
2-ethyl-1-butene

$$CH_3$$
CH_2=$CCHCH_3$
$$CH_3$$
2,3-dimethyl-1-butene

$$CH_3$$
CH_2=$CHCCH_3$
$$CH_3$$
3,3-dimethyl-1-butene

3.2

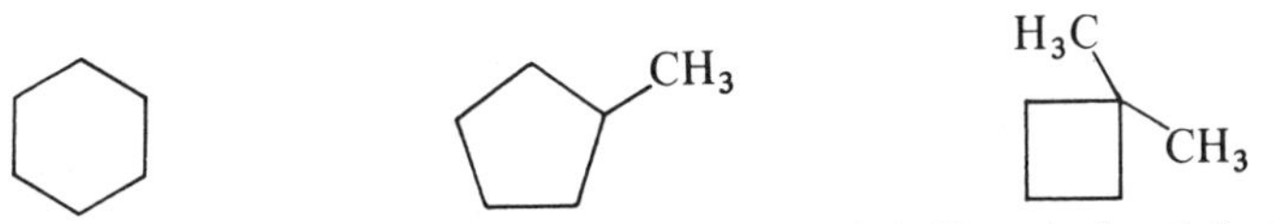

cyclohexane

$$CH_3$$
methylcyclopentane

$$H_3C$$
$$CH_3$$
1,1-dimethylcyclobutane

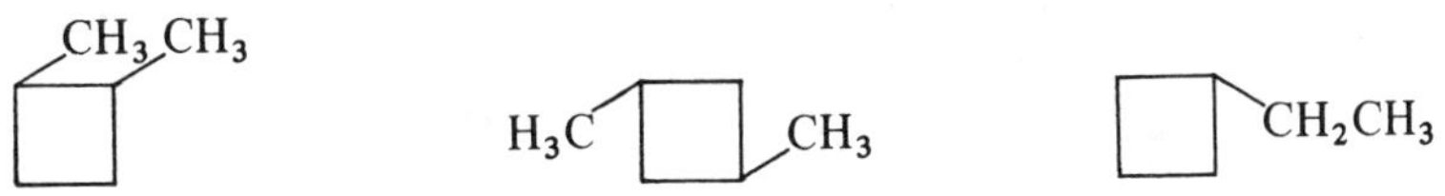

$$CH_3 \quad CH_3$$
1,2-dimethylcyclobutane

H_3C CH_3
1,3-dimethylcyclobutane

CH_2CH_3
ethylcyclobutane

$$CH_3$$
H_3C CH_3
1,2,3-trimethylcyclopropane

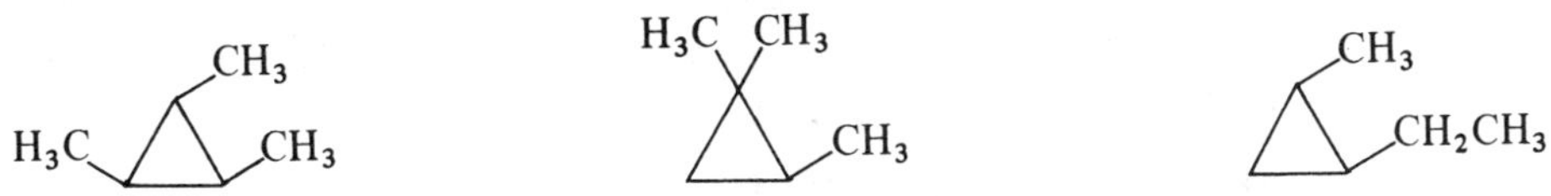

$$H_3C \quad CH_3$$
CH_3
1,1,2-trimethylcyclopropane

$$CH_3$$
CH_2CH_3
1-ethyl-2-methylcyclopropane

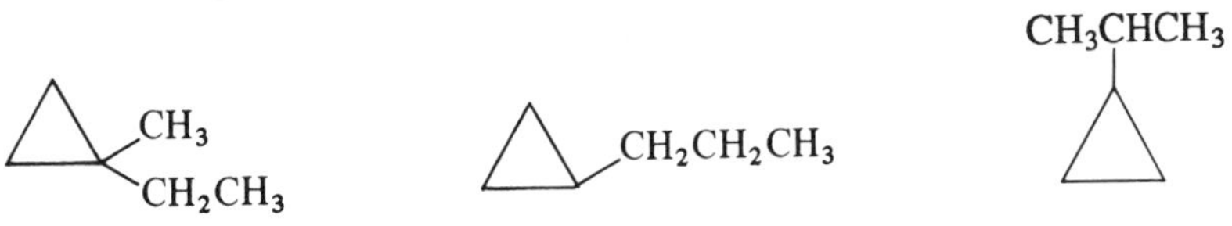

CH_3
CH_2CH_3
1-ethyl-1-methylcyclopropane

$CH_2CH_2CH_3$
propylcyclopropane

$$CH_3CHCH_3$$

isopropylcyclopropane

3.3 $CH \equiv CCH_2CH_2CH_3$ $CH_3C \equiv CCH_2CH_3$ $CH_3CH = C = CHCH_3$
1-pentyne   2-pentyne   2,3-pentadiene

$CH_2 = C = CHCH_2CH_3$ $CH_2 = CHCH = CHCH_3$ $CH_2 = CHCH_2CH = CH_2$
1,2-pentadiene   1,3-pentadiene   1,4-pentadiene

$\overset{\displaystyle CH_3}{\underset{|}{CH_2 = CCH = CH_2}}$ $\overset{\displaystyle CH_3}{\underset{|}{CH \equiv CCHCH_3}}$ $\overset{\displaystyle CH_3}{\underset{|}{CH_2 = C = CCH_3}}$
2-methyl-1,3-butadiene   3-methyl-1-butyne   3-methyl-1,2-butadiene

3.4 (a) 1-pentene   (b) $CH_3CH_2CH = \overset{\displaystyle CH_3}{\underset{|}{C}}CH_3$ (c) $CH_3\overset{\displaystyle CH_3}{\underset{|}{\underset{\displaystyle CH_3}{\underset{|}{C}}}}CH = \overset{\displaystyle CH_3}{\underset{|}{C}}CH_2CH_3$

2-methyl-2-pentene   2,2,4-trimethyl-3-hexene

(d) $CH_3\overset{\displaystyle CH_3}{\underset{|}{C}} = CH\overset{\displaystyle Br}{\underset{|}{C}}HCH_3$ (e) $CH_3\overset{\displaystyle Cl}{\underset{|}{C}}HCH_2CH = CHCH_2CH_3$ (f) 3-ethylcyclohexene

4-bromo-2-methyl-2-pentene   6-chloro-3-heptene

(g) 1,3-cyclopentadiene   (h) $CH_3CH_2\overset{\displaystyle Cl}{\underset{|}{C}}HCH = CH\overset{\displaystyle Cl}{\underset{|}{C}}HCl$ (i) 1,2-butadiene

1,1,4-trichloro-2-hexene

(j) $CH_3\overset{\displaystyle CH_3}{\underset{|}{C}}HC \equiv C\overset{\displaystyle CH_3}{\underset{|}{C}}HCH_3$ (k) $CH_3CH_2CH_2\overset{\displaystyle CH_2CH_3}{\underset{\displaystyle CH_2CH_3}{\underset{|}{C}}}-C \equiv CH$ (l) $CH_3CH_2CH_2\overset{\displaystyle}{\underset{\displaystyle CH_2CH_2CH_3}{\underset{|}{C}}}HC \equiv CCH_3$

2,5-dimethyl-3-hexyne   3,3-diethyl-1-hexyne   4-propyl-2-heptyne

(m) 4-cyclopropyl-1-iodo-1-butene   (n) 1-bromo-2-methylcyclobutene

3.5 (a) $CH_3CH_2CH = \overset{\displaystyle CH_3}{\underset{|}{C}}CH_3$ (b) $CH_2 = CHCH_2CH_2\overset{\displaystyle CH_3}{\underset{|}{C}}HCH_3$ (c) $CH_2 = CHCH = CH_2$

(d) $CH_2 = \overset{\displaystyle CH_2CH_3}{\underset{|}{C}}CH_2CH_3$ (e) $CH_3\overset{\displaystyle CH_3}{\underset{|}{C}} = CH\overset{\displaystyle CH_3}{\underset{|}{C}}HCH_2\overset{\displaystyle CH_3}{\underset{\displaystyle CH_3}{\underset{|}{C}}}CH_3$ (f) $CH_3\overset{\displaystyle CH_3}{\underset{|}{C}} = CH$ (g)

(h) $CH_3\overset{\displaystyle CH_3}{\underset{|}{C}}CH = \overset{\displaystyle CH_2CH_3}{\underset{|}{C}}CH_2CH_3$ (i) $CH_3\overset{\displaystyle CH_3}{\underset{\displaystyle CH_3}{\underset{|}{C}}}CH_2\overset{\displaystyle H}{\underset{|}{C}} = \overset{\displaystyle H}{\underset{|}{C}}CH_2CH_2CH_3$ (j) $CH_3CH = CHCH_2\overset{\displaystyle CH_2CH_3}{\underset{\displaystyle CH_3}{\underset{|}{C}}}CH_2CH_3$

(k) $CH \equiv CCHCH_2CHCH_3$ (with two CH_3 substituents)

(l) $CH_3C \equiv CCHCH_3$ (with CH_3 substituent)

(m) 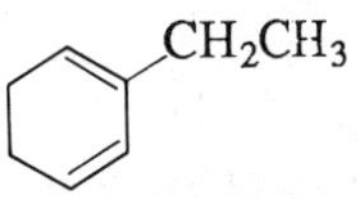 with CH_2CH_3

(n) 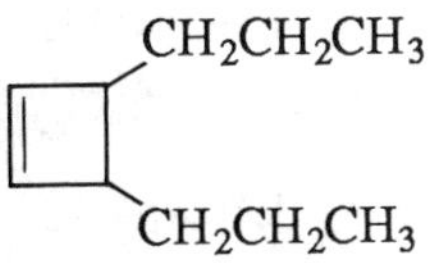 with $CH_2CH_2CH_3$ (top) and $CH_2CH_2CH_3$ (bottom)

3.6 (a) $CH_3CHCH_2CH=CHCH_2CH_3$ (with CH_3)

6-methyl-3-heptene

(b) $CH_3CHCH=CHCH_2CH_3$ (with CH_2CH_3)

5-methyl-3-heptene

(c) $CH_3C=CHCH_2CH=CH_2$ (with CH_3)

5-methyl-1,4-hexadiene

(d) $CH_3CCH=CHCH_3$ (with two CH_3)

4,4-dimethyl-2-pentene

(e) 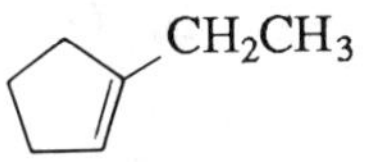 cyclopentene with CH_2CH_3

1-ethylcyclopentene

(f) $CH_3C \equiv CH$

1-propyne

(g) $CH_3C \equiv CCH_2$ (with I)

1-iodo-2-butyne

(h) $CH_3CHCH_2C \equiv CH$ (with Cl)

4-chloro-1-pentyne

(i)  cyclopentadiene

1,3-cyclopentadiene

(j) 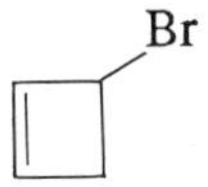 cyclobutene with Br

3-bromocyclobutene

3.7 (a) $CH_3CH_2CH=CHCH_3$ (b) $CH_2=C(CH_3)_2$ (c)  cyclopentene (d) $CH_2=CH_2$ (e) 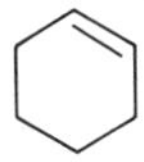 cyclohexene

(f) $(CH_3)_3CCH=C(CH_3)_2$ (g) $(CH_3)_2CBrCH_2Br$ (h) 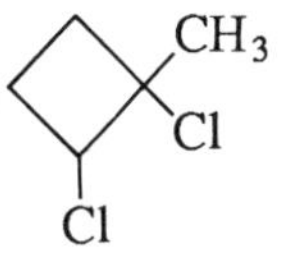 cyclobutane with CH_3, Cl, Cl (i) $CH_3CH(CH_3)CH_2CH_3$

(j) $CH_3CH_2CH_2CH_3$ (k) $(CH_3)_2CBrCH_2CH_3$ (l)  cyclohexane with CH_3, Cl (m) $(CH_3)_2C(OSO_3H)CH_3$

(n) $CH_3CH_2CH_2Cl(CH_3)CH_2CH_3$ (o) $(CH_3)_2COHCOH(CH_3)_2$ (p) 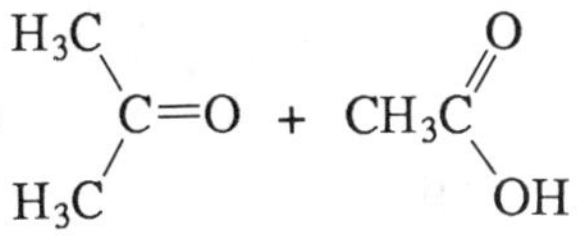 $(CH_3)_2C=O + CH_3C(=O)OH$

(q) $CH_3CH_2CBr=CHBr$ (r) $CH_3CH_2CH_2CH_3$ (s) $CH_3CCl=CH_2$ (t) 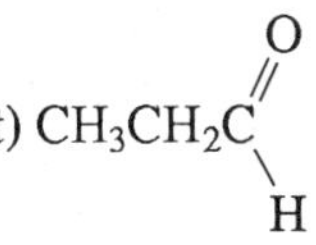 $CH_3CH_2C(=O)H$

3.8 (a) $CH_3CH_2CH_3$ (b) $CH_3CHBrCH_2Br$ (c) $CH_3CHClCH_3$ (d) $CH_3CH(OSO_3H)CH_3$

(e) $CH_3CHOHCH_2OH$ (f) 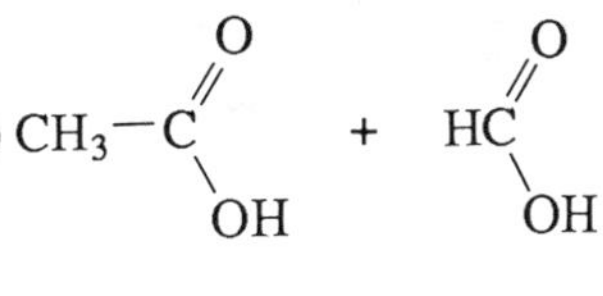 $CH_3C(=O)OH + HC(=O)OH$

3.9 **(a)** O_2, lighted match   **(b)** $2\,Br_2$ **(c)** H_2SO_4 **(d)** H_2, Ni   **(e)** HBr   **(f)** $2\,Cl_2$
       (g) cold, dilute $KMnO_4$ **(h)** hot, conc. $KMnO_4$

3.10 **(a)** Addition reaction: $CH_2{=}CH_2 \;+\; Cl_2 \longrightarrow CH_2ClCH_2Cl$

    Substitution reaction: $CH_4 \;+\; Cl_2 \xrightarrow{\text{light}} CH_3Cl + HCl$

   (b) Dehydration reaction: $CH_3CH_2OH \xrightarrow[\Delta]{H_2SO_4} CH_2{=}CH_2 \;+\; HOH$

    Dehydrohalogenation reaction: $CH_3CH_2Br \xrightarrow[\Delta]{KOH} CH_2{=}CH_2 \;+\; KBr \;+\; HOH$

   (c) Oxidation reaction: $CH_2{=}CH_2 \xrightarrow[KMnO_4]{\text{cold, dilute}} CH_2OHCH_2OH$

    Reduction reaction: $CH_2{=}CH_2 \;+\; H_2 \xrightarrow{Ni} CH_3CH_3$

3.11 **(a)** An ion or molecule that seeks a pair of electrons (electron-pair acceptor), e.g., H^+.
   (b) An ion in which there is a positive charge on carbon, e.g.,

$$CH_3-\overset{+}{\underset{\underset{\textstyle H}{|}}{C}}-CH_3$$

   (c) The process in which small molecules react with themselves to form very large molecules, e.g.,

$$n\,(CH_2{=}CH_2) \xrightarrow{\text{catalyst}} -\!(CH_2-CH_2)\!-_n$$

   (d) The process in which saturated hydrocarbons are heated to very high temperatures in the presence of a catalyst, e.g.,

$$CH_3CH_2CH_2CH_3 \xrightarrow[\text{catalyst}]{>400°C} \begin{cases} CH_2{=}CH_2 + CH_3CH_3 \\ CH_3CH{=}CH_2 + CH_4 \\ CH_3CH{=}CHCH_3 + H_2 \end{cases}$$

   (e) Atoms or groups of atoms that have high electronegativities such that they attract electrons, e.g., Cl^-, Br^-.
   (f) Atoms or groups of atoms that repel shared electrons, e.g., CH_3.
   (g) In the addition of unsymmetrical reagents to alkenes, the positive portion of the reagent adds to that carbon atom that already has the most hydrogen atoms.

$$CH_3-\overset{\overset{\textstyle CH_3}{|}}{C}{=}\overset{\underset{\textstyle H}{|}}{C}-CH_3 + HBr \longrightarrow CH_3-\overset{\overset{\textstyle CH_3}{|}}{\underset{\underset{\textstyle Br}{|}}{C}}-\overset{\overset{\textstyle H}{|}}{\underset{\underset{\textstyle H}{|}}{C}}-CH_3$$

   (h) A five-carbon unit found in many natural products,

$$CH_2{=}\overset{\underset{\textstyle CH_3}{|}}{C}-CH{=}CH_2$$

(i)   All terpenes should be formally divisible into isoprene units. See Figure 3.3 for examples.

(j)   Terpenes are compounds found in the oils of certain plants, e.g.,

camphor

(k)   The reaction of a compound with halogen molecules.

$$CH_3CH=CH_2 \;+\; Br_2 \longrightarrow CH_3CHBrCH_2Br$$

(l)   The reaction of a compound with hydrogen.

(m) Electrostatic factors in which electrons are either attracted to, or repelled by, an atom or group of atoms. Alkyl groups bound to a carbon atom donate electron density to the carbon, e.g.,

$$CH_3 \rightarrow \overset{|}{\underset{|}{C}}-$$

(n)   Compounds containing two hydroxyl groups, e.g., CH_2OHCH_2OH.

(o)   A small molecule (building block) that reacts with other small molecules to form a large molecule (polymer); e.g., $CH_2=CH_2$ is the monomer of the polymer called polyethylene.

(p)   The large molecule that results from the joining (polymerization) of many small molecules, e.g.,

polyethylene

(q)   A diene in which the double bonds are separated by one carbon–carbon single bond, e.g., $CH_2=CH-CH=CH_2$.

(r)   The vinyl group is the ethylene group minus one hydrogen: $-CH=CH_2$.

3.12 (a) The positively charged ion, in this case H^+.

(b) At the carbon that contains the most hydrogens.

(c)

3.13 Add a few drops of a dilute potassium permanganate solution to each sample. Pentane does not react, the purple color remains. 1-Pentene reacts with $KMnO_4$, and we observe a color change from purple to brown. A second test is to add a few drops of bromine to each sample. With pentane there is no reaction, and the red-brown color of the bromine remains. Pentene reacts with the bromine, and the color disappears.

3.14 $(CH_3)_3C^+ \;>\; CH_3CH_2\overset{+}{C}HCH_3 \;>\; CH_3CH_2CH_2\overset{+}{C}H_2$

$3° \;>\; 2° \;>\; 1°$

3.15 Single bonds > double bonds > triple bonds (see Table 1.2).

3.16 It is stored and shipped in an acetone solution within pressurized cylinders.

3.17 Chief uses of acetylene: (1) starting material for production of vinyl acetate, (2) fuel in oxyacetylene torches, (3) in production of acrylonitrile, (4) in production of vinyl acetylene, (5) in production of metal acetylides. Industrial preparation: from calcium carbide and water, or by the cracking of methane.

3.18 Similar to ethene: (1) addition of symmetric reagents,

$$CH_3C{\equiv}CH + Br_2 \longrightarrow CH_3\underset{\underset{\displaystyle Br}{|}}{\overset{\overset{\displaystyle Br}{|}}{C}}{=}CH$$

(2) addition of unsymmetric reagents,

$$CH_3C{\equiv}CH + HCl \longrightarrow CH_3\underset{\underset{\displaystyle Cl}{|}}{C}{=}CH_2$$

Unique to acetylene:

(1) $HC{\equiv}CH + HCN \xrightarrow{\ catalyst\ } H_2C{=}\underset{\underset{\displaystyle CN}{|}}{CH}$

(2) $HC{\equiv}CH + CH_3{-}C\overset{\displaystyle O}{\underset{\displaystyle OH}{\diagdown}} \longrightarrow CH_2{=}CH{-}O{-}\overset{\overset{\displaystyle O}{\|}}{C}{-}CH_3$

3.19 See Table 3.4.

3.20

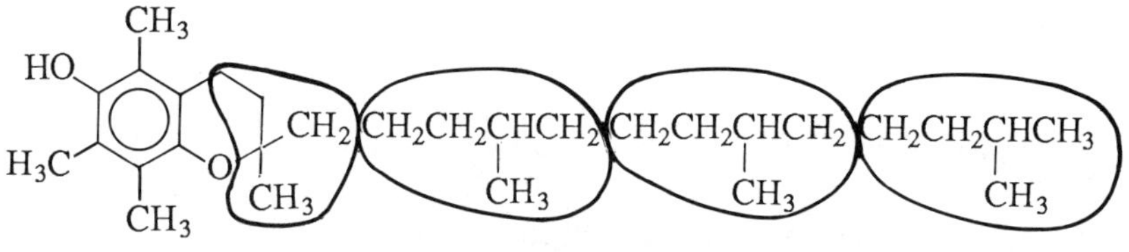

Vitamin E

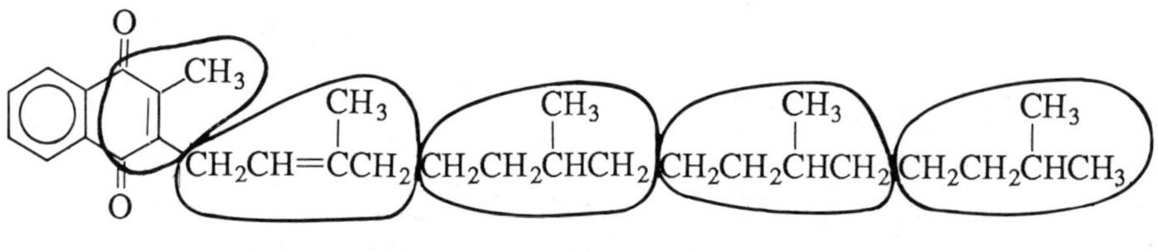

Vitamin K

3.21

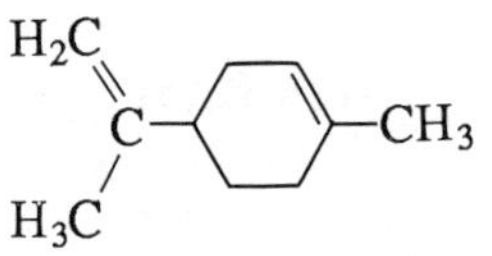

3.22 (a) 56 amu; C_4H_8.

(b) $CH_3CH=CHCH_3$ $CH_2=CHCH_2CH_3$ $(CH_3)_2C=CH_2$

3.23 Each of the following four compounds will react with 2 moles of Cl_2 to form a compound with the formula $C_4H_6Cl_4$: $HC\equiv CCH_2CH_3$ $CH_3C\equiv CCH_3$ $CH_2=CHCH=CH_2$ $CH_2=C=CHCH_3$. The information given is insufficient to determine the exact structure of the compound.

3.24 $HC\equiv CCHCH_3$, $CH_2=CCH=CH_2$,   or $CH_3C=C=CH_2$
           $\quad\quad\;|$ $\quad\quad\;|$ $\quad\quad|$
           $\quad\quad CH_3$ $\quad\quad CH_3$ $\quad CH_3$

3.25

3.26 $X = CH_3C=CHCH_3$ $Y = CH_3CHCH=CH_2$ $Z = CH_2=CCH_2CH_3$
          $\quad\quad\quad\;|$ $\quad\quad\quad\;\;|$ $\quad\quad\quad\;|$
          $\quad\quad\quad CH_3$ $\quad\quad\quad CH_3$ $\quad\quad\quad CH_3$

3.27 $CH\equiv CH + 2\,HBr \longrightarrow CH_3CHBr_2$

13 g of acetylene $= \dfrac{13}{26} = 0.50$ mole of acetylene

0.50 mole of acetylene requires 1.0 mole of HBr $= 81$ g

3.28 $CH_2=CH_2 + Br_2 \longrightarrow CH_2BrCH_2Br$

% yield $= \dfrac{150}{188}(100) = 80\%$

CHAPTER 4

The Aromatic Hydrocarbons

4.1 Presence (with a few exceptions) of at least one benzene ring, chemical stability of the aromatic ring structure, compounds exhibit resonance.

4.2 A compound is said to exhibit resonance if two or more forms can be drawn that differ only in the position of electrons. The true structure is a composite or a weighted average of the structures of the contributing forms.

4.3 (a)

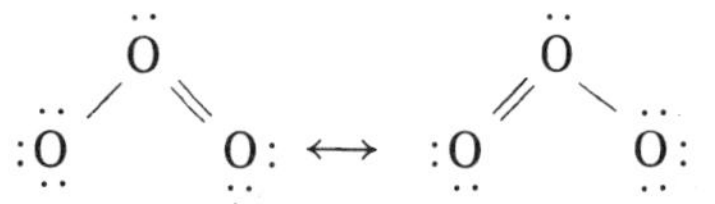

(b)

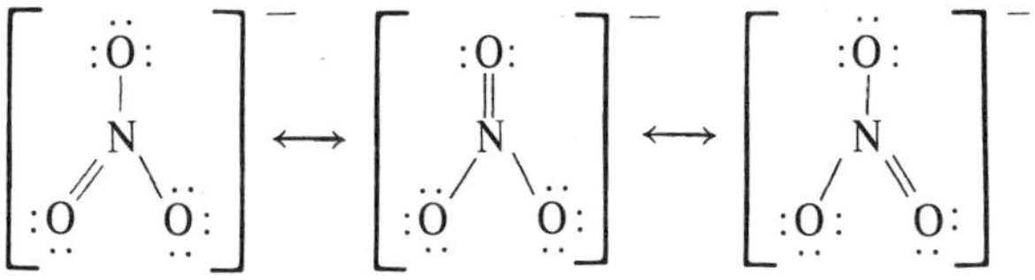

4.4 (a) bromobenzene   (b) o-chloroiodobenzene, 1-chloro-2-iodobenzene   (c) 1,2,4-tribromobenzene
(d) p-iodotoluene, 4-iodotoluene, 1-iodo-4-methylbenzene
(e) m-nitrophenol, 3-nitrophenol, 1-hydroxy-3-nitrobenzene
(f) diphenylmethane, benzylbenzene
(g) m-chloroethylbenzene, 1-chloro-3-ethylbenzene, 3-chloro-1-ethylbenzene
(h) 4-isopropyl-1,2-dimethylbenzene, 4-isopropyl-2-methyltoluene
(i) 1-bromo-3-phenylpropane, 3-bromo-1-phenylpropane
(j) cyclohexylbenzene, phenylcyclohexane
(k) 1-chloro-5-ethyl-2-hydroxy-4-methylbenzene, 2-chloro-4-ethyl-5-methylphenol,
     4-chloro-2-ethyl-5-hydroxytoluene
(l) p-bromobenzyl bromide, 4-bromobenzyl bromide

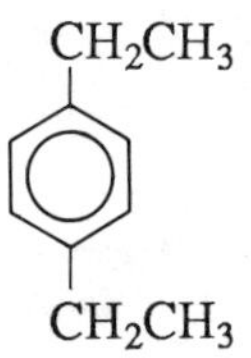

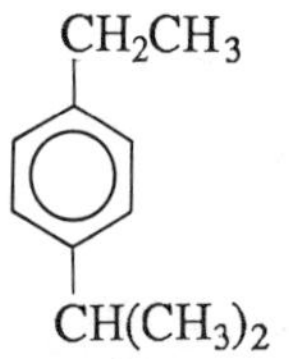

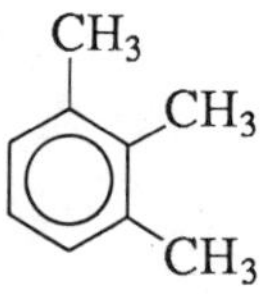

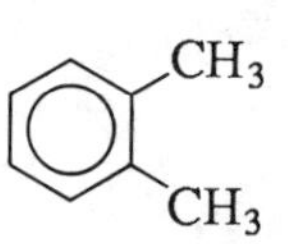

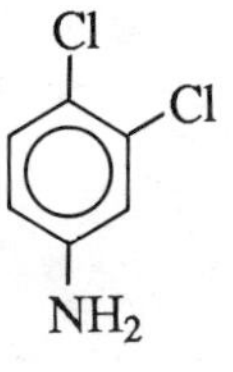

4.5

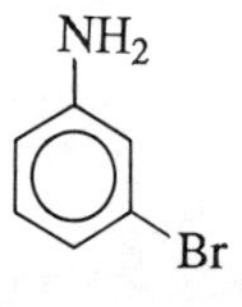

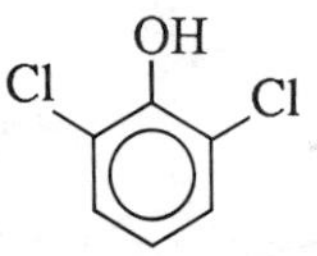

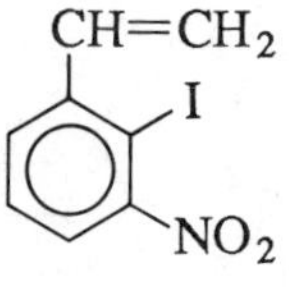

(j) 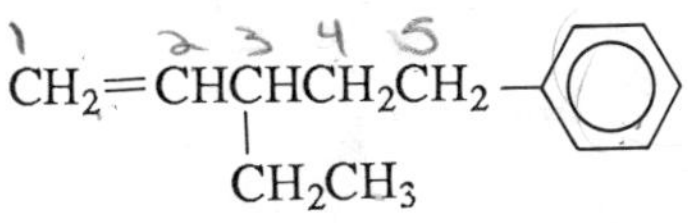 $\overset{1}{CH_2}=\overset{2}{CH}\overset{3}{CH}\overset{4}{CH_2}\overset{5}{CH_2}-$
$\qquad\qquad\;\;\; |$
$\qquad\qquad CH_2CH_3$

(k) 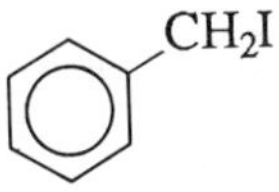 CH_2I

(l) 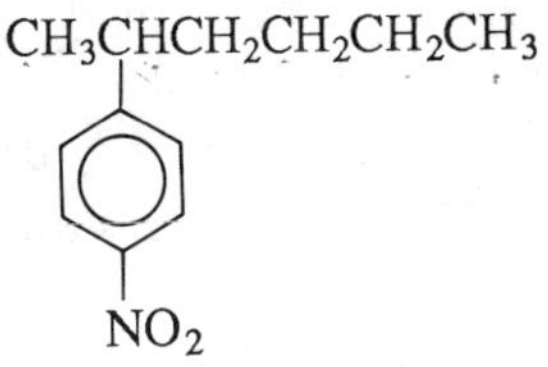 $CH_3CHCH_2CH_2CH_2CH_3$

4.6 Alkyl group: an aliphatic hydrocarbon minus one hydrogen, e.g., CH_3-. Aryl group: an aromatic

hydrocarbon minus one hydrogen, e.g., 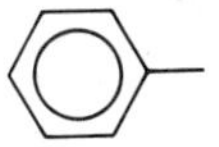 .

4.7

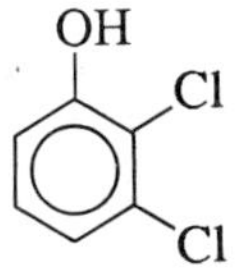

2,3-dichlorophenol 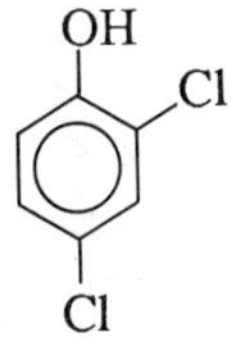   2,4-dichlorophenol 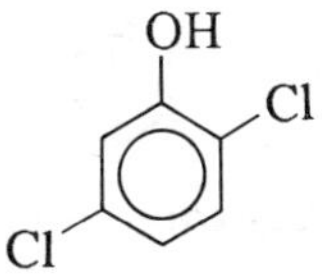   2,5-dichlorophenol 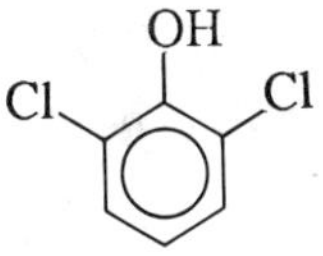   2,6-dichlorophenol

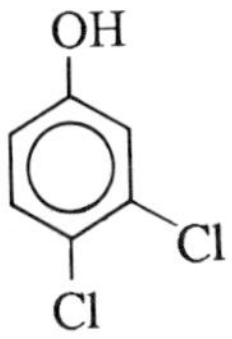

3,4-dichlorophenol 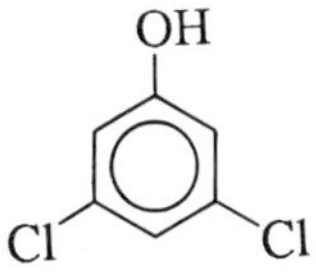   3,5-dichlorophenol

4.8

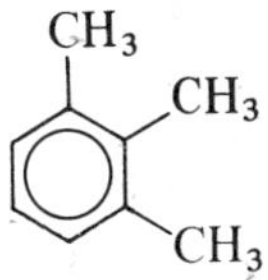

2,3-dimethyltoluene 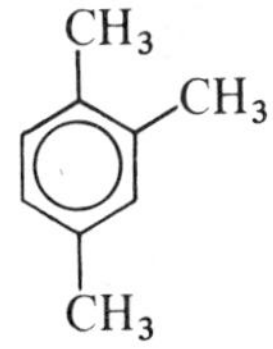   2,4-dimethyltoluene 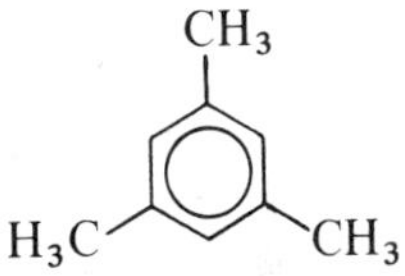   3,5-dimethyltoluene 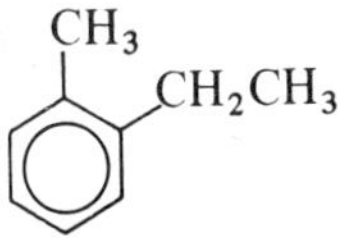   o-ethyltoluene

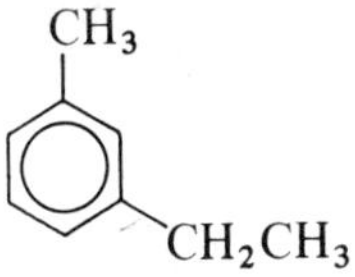

m-ethyltoluene    p-ethyltoluene 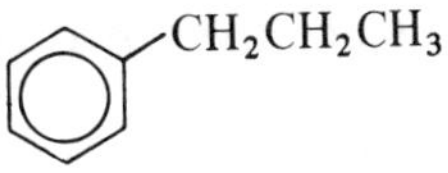   propylbenzene 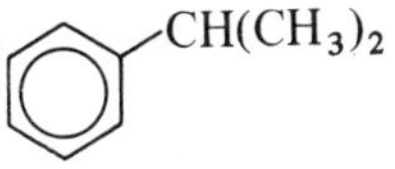   isopropylbenzene

4.9

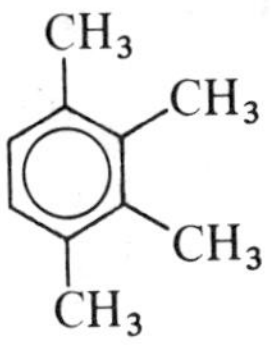

2,3,4-trimethyltoluene 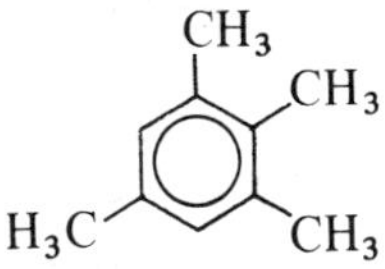   2,3,5-trimethyltoluene 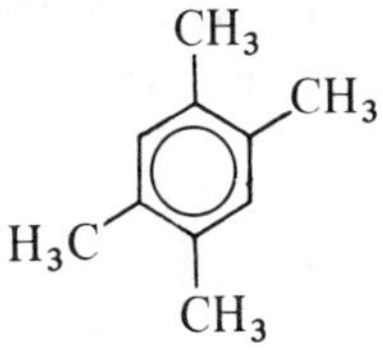   2,4,5-trimethyltoluene

4.10 (a)

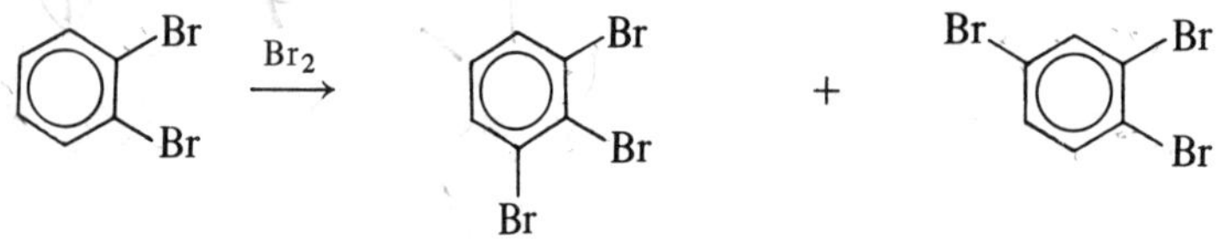

1,2,3-tribromobenzene   1,2,4-tribromobenzene

(b)

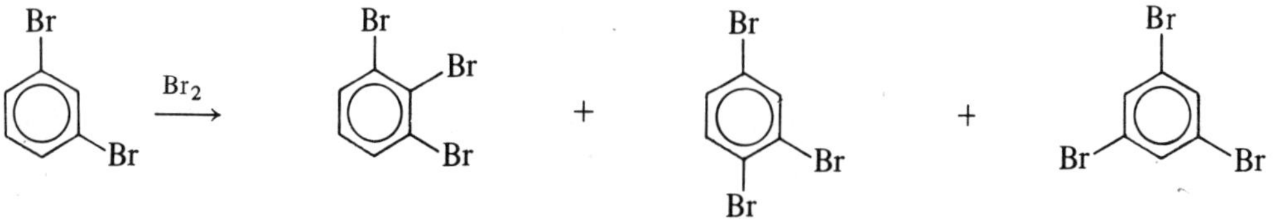

1,2,3-tribromobenzene   1,2,4-tribromobenzene   1,3,5-tribromobenzene

(c)

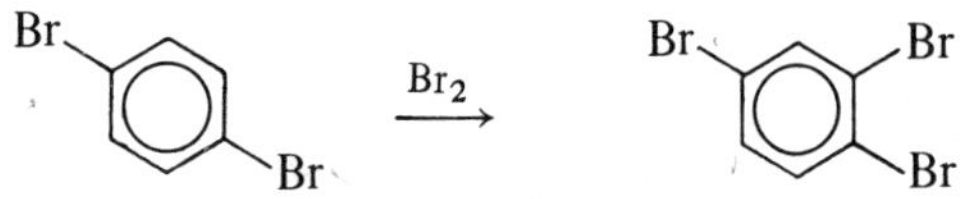

1,2,4-tribromobenzene

4.11 (a) 3   **(b)** 1   **(c)** 1   **(d)** 3

4.12,

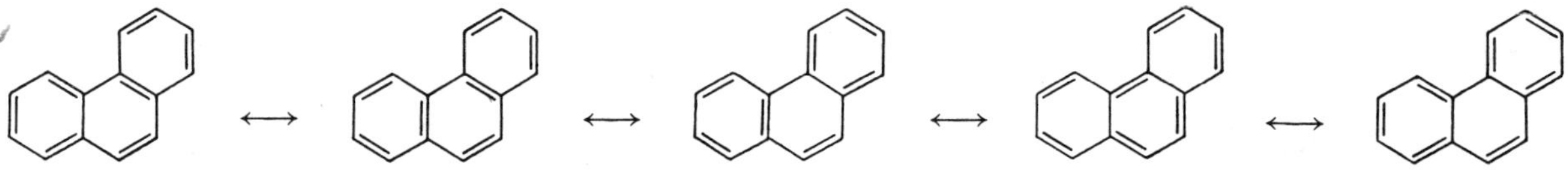

4.13 (a) two:

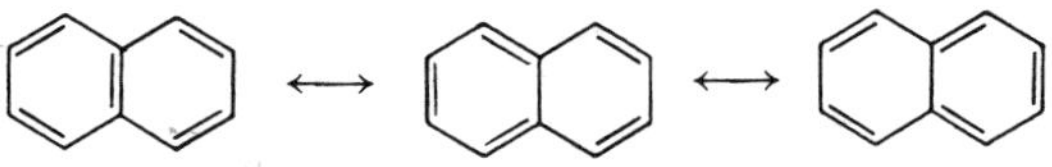

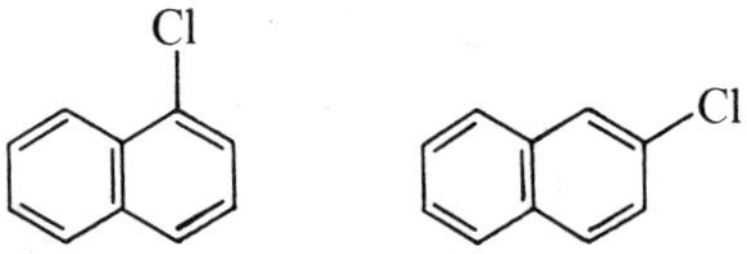

(b) ten

4.14 Both reactions are substitution rather than addition. Bromination of methane proceeds by a free radical mechanism and requires light or heat to initiate the reaction. Bromination of benzene proceeds by electrophilic aromatic substitution and requires a Lewis acid catalyst.

4.15 Ignite a few drops in an evaporating dish. Aliphatic compounds generally burn with a clean flame, whereas aromatic compounds burn with a sooty flame.

4.16 Benzene is relatively inert. When it does react, the reactions are substitution and require catalysts and often elevated temperatures. 1,3-Cyclohexadiene is a reactive compound that readily adds a wide variety of reagents across its double bonds.

4.17 Add a dilute solution of $KMnO_4$. Only cyclohexene will react as observed by the color change from purple to brown. To distinguish cyclohexane from benzene, add Br_2 and expose the solutions to a UV lamp. Cyclohexane will react as evidenced by a color change of the solution. There will be no sign of reaction with benzene.

4.18 (a) A carcinogen is any substance that causes cancer.   **(b)** See Figure 4.3.

CHAPTER 5

Halogen Derivatives of Hydrocarbons

5.1 $CH_3CH_2CH_2CH_2CH_2Br$ $CH_3CH_2CH_2CHBrCH_3$ $CH_3CH_2CHBrCH_2CH_3$
       1-bromopentane   2-bromopentane   3-bromopentane

       $CH_3CH_2CH(CH_3)CH_2Br$ $CH_3CH_2CBr(CH_3)_3$ $(CH_3)_2CHCH_2CH_2Br$
       1-bromo-2-methylbutane   2-bromo-2-methylbutane   1-bromo-3-methylbutane

          $(CH_3)_3CCH_2Br$
       1-bromo-2,2-dimethylpropane

5.2 $CH_3CH_2CH_2CHCl_2$ $CH_3CH_2CHClCH_2Cl$ $CH_3CHClCH_2CH_2Cl$
       1,1-dichlorobutane   1,2-dichlorobutane   1,3-dichlorobutane

       $CH_2ClCH_2CH_2CH_2Cl$ $CH_3CH_2CCl_2CH_3$ $CH_3CHClCHClCH_3$
       1,4-dichlorobutane   2,2-dichlorobutane   2,3-dichlorobutane

         $(CH_3)_2CHCHCl_2$ $(CH_3)_2CClCH_2Cl$ $CH_2ClCH(CH_3)CH_2Cl$
    1,1-dichloro-2-methylpropane   1,2-dichloro-2-methylpropane   1,3-dichloro-2-methylpropane

5.3 (**a**) methylene bromide   (**b**) iodoform   (**c**) carbon tetrafluoride

5.4 (**a**) 3-iodopentane   (**b**) 1,3-dichloro-3-methylbutane   (**c**) 3,3-dibromo-2,2,5,5-tetramethylhexane
    (**d**) 4-fluoro-1-iodo-2,2-dimethylbutane   (**e**) benzyl iodide, iodophenylmethane
    (**f**) 4-bromo-3-iodotoluene, 1-bromo-2-iodo-4-methylbenzene   (**g**) 4-iodo-2-methyl-1-pentene
    (**h**) 3-fluoropentane   (**i**) 2-iodo-2-butene   (**j**) hexafluoroethane, ethylene hexafluoride
    (**k**) 1-chloro-3-ethylcyclopentane   (**l**) 2-bromo-2-phenylbutane

5.5   (a)   (b) $CH_3CH-CCH_2CH_3$ with CH_3 above and CH_3 below, Br above   (c) $CH_3CH=CCHCH_2CH_2CH_3$ with I above and CH_3 below   (d)

(e)   (f)   (g) $CH_2CH=C-I$ with I and I above   (h)

(i) $CH_2=CHBr$ (j) $Br-\bigcirc-CH_3$ (k)   (l)

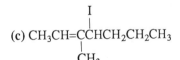

23

5.6

$$\begin{array}{ccc} \overset{\displaystyle Br}{|} & \overset{\displaystyle Br}{|} & \overset{\displaystyle Cl}{|} \\ H-C- & C- & C-H \\ \overset{|}{H} & \overset{|}{H} & \overset{|}{H} \end{array}$$

5.7   Many of them are decomposed by light.

5.8   For the same reason that n-alkanes have higher boiling points than branched-chained alkanes. That is, the molecules are closer together, and thus the van der Waals forces of attraction are greater.

5.9   **(a)** $CHCl_3$ **(b)** ⬡—Br   **(c)** $CH_3CH_2CH_2CH_2Br$ **(d)** $CH_3CH_2CH_2CH_2Cl$

5.10 **(a)** $CH_3CH=CH_2$ **(b)** $(CH_3)_2CHCHBrCHBrCH_3$ **(c)** $CH_3CH_2CHBrCH_3$

(d) $CH_3CH_2CH_2CH(CN)CH_3$ **(e)** CH_3CH_2OH **(f)** ⬡—Br   **(g)** $(CH_3)_2CClCH_2CH_3$

(h) $CH_3CHBrCH_2CH_3$ **(i)** CH_3CHICH_3 **(j)** $(CH_3)_3CBr$ **(k)** CH_3Cl + other products

(l) ⬡—CH_2Br + other products

5.11

$$CH_3CH_2\overset{\displaystyle \overset{Br}{|}}{C}HCH_3 \xrightarrow[\text{KOH}]{\text{conc.}} CH_3CH=CHCH_3$$

$$CH_3CH_2\overset{\displaystyle \overset{Br}{|}}{C}HCH_3 \xrightarrow[\text{KOH}]{\text{dilute}} CH_3CH_2\overset{\displaystyle \overset{OH}{|}}{C}HCH_3$$

5.12 **(a)** A compound in which a halogen atom is bound to an alkyl group, e.g., CH_3Cl.

(b) A compound in which a halogen atom is bound to an aryl group, e.g., ⬡—Br.

(c) Various chlorofluoromethane compounds that are used as refrigerants, e.g., CF_2Cl_2, $CFCl_3$.
(d) Polychlorinated biphenyls are chemicals that had been used as fire-resistant coolants in electric

transformers, e.g.,

(e) A herbicide used as a defoliant in Vietnam and also used to destroy illegal marijuana crops, e.g.,

(f) A reaction in which a small molecule is lost from an organic compound with the concomitant formation of a double bond, e.g.,

$$CH_3CHBrCH_3 \xrightarrow{KOH} CH_3CH=CH_2$$

(g) Species that can donate a pair of electrons in a chemical reaction, e.g., $-\ddot{N}H_2$, $-\ddot{O}H$.

(h) Chemicals that prevent the growth of weeds, e.g., 2,4,-D and 2,4,5-T.

(i) The concentration of some substance increases as it proceeds through the food chain, e.g., the biomagnification of DDT.

(j) Substances that can dissolve oxygen and transport oxygen within the blood vessels, e.g., the perfluoropropylfurans.

5.13 (a) $(CH_3)_2CHCH_2CH_2Br \xrightarrow[\text{alcohol}]{KOH} (CH_3)_2CHCH=CH_2 \xrightarrow{HBr} (CH_3)_2CHCHBrCH_3$

(b) $CH_3CH_2CH_2Br \xrightarrow[\text{alcohol}]{KOH} CH_3CH=CH_2 \xrightarrow[Pt]{H_2} CH_3CH_2CH_3$

(c) $CH_2=CH_2 \xrightarrow{HBr} CH_3CH_2Br \xrightarrow{NaCN} CH_3CH_2CN$

(d)

5.14 Antiseptic: any substance that inhibits the activity of microorganisms, e.g., iodoform. Anesthetic: any substance that produces a loss of sensation, e.g., ether. If given in excessive amounts, chloroform is toxic and can cause permanent liver damage.

5.15 (a) 12   (b) 8   (c) 6   (d) 9   (e) 4   (f) 10
      (g) 11   (h) 7   (i) 5   (j) 1   (k) 3   (l) 2

CHAPTER 6

Alcohols, Phenols, Ethers, and Thiols

6.1 $CH_3CH_2CH_2CH_2CH_2OH$ $CH_3CH_2CH_2CHOHCH_3$ $CH_3CH_2CHOHCH_2CH_3$
1-pentanol (1°)   2-pentanol (2°)   3-pentanol (2°)

$CH_3CH_2CH(CH_3)CH_2OH$ $CH_3CH_2COH(CH_3)_2$ $CH_3CHOHCH(CH_3)_2$
2-methyl-1-butanol (1°)   2-methyl-2-butanol (3°)   3-methyl-2-butanol (2°)

$(CH_3)_2CHCH_2CH_2OH$ $(CH_3)_3CCH_2OH$
3-methyl-1-butanol (1°)   2,2-dimethyl-1-propanol (1°)

6.2 $CH_3OCH_2CH_2CH_2CH_3$ $CH_3OCH(CH_3)CH_2CH_3$ $CH_3OC(CH_3)_3$
methyl butyl ether   methyl sec-butyl ether   methyl t-butyl ether

$CH_3CH_2OCH_2CH_2CH_3$ $CH_3CH_2OCH(CH_3)_2$ $CH_3OCH_2CH(CH_3)_2$
ethyl propyl ether   ethyl isopropyl ether   methyl isobutyl ether

6.3 $CH_3CH_2CH_2CH_2OH$ $CH_3OCH_2CH_2CH_3$ $CH_3CH_2CHOHCH_3$ $CH_3CH_2OCH_2CH_3$
1-butanol   methyl propyl ether   2-butanol   diethyl ether

$CH_3\overset{\underset{|}{CH_3}}{C}HCH_2OH$ $CH_3\overset{\underset{|}{CH_3}}{O}CHCH_3$ $CH_3\overset{\underset{|}{CH_3}}{\underset{\underset{CH_3}{|}}{C}}-OH$
2-methyl-1-propanol   methyl isopropyl ether   2-methyl-2-propanol

6.4 $CH_3CH_2CH_2CH_2SH$ $CH_3SCH_2CH_2CH_3$ $CH_3CH_2CHSHCH_3$ $CH_3CH_2SCH_2CH_3$
1-butanethiol   1-methylthiopropane   2-butanethiol   ethylthioethane

$(CH_3)_2CHCH_2SH$ $(CH_3)_3CSH$ $CH_3SCH(CH_3)_2$
2-methyl-1-propanethiol   2-methyl-2-propanethiol   2-methylthiopropane

6.5 **(a)** 2-methyl-3-hexanol (2°)   **(b)** 4,4-dichloro-2-butanol (2°)
(c) 3,3-dibromo-2-methyl-2-butanol (3°)   **(d)** 5-iodo-3-penten-1-ol (1°)
(e) 1-methylcyclohexanol (3°)   **(f)** cyclopentane-1,2-diol or 1,2-cyclopentanediol (2°),(2°)
(g) 1,3,3-tribromo-2-butanol (2°)   **(h)** 1,3-propanediol (1°), (1°)   **(i)** 3-ethyl-3-hexanol (3°)
(j) 4-chloro-7-methyl-2-octanol (2°)   **(k)** p-iodobenzyl alcohol or p-iodophenylmethanol (1°)
(l) 4-ethylcyclohex-2-en-1-ol (2°)

6.6 (a) 2,6-dinitrophenol   (b) *m*-bromophenol, 3-bromophenol   (c) ethyl methyl ether, methoxyethane
(d) isopropyl propyl ether, 1-isopropoxypropane   (e) methyl phenyl ether, methoxybenzene
(f) propoxycyclohexane, cyclohexyl propyl ether   (g) 5-methoxy-2-methylphenol
(h) *o*-ethylphenol, 2-ethylphenol   (i) 2,4-dimethoxy-3-methylpentane   (j) 2-ethoxy-3-hexene
(k) 2-propanethiol   (l) 4,4-dibromo-2-butanethiol   (m) methyl *t*-butyl sulfide
(n) dipropyl sulfide, 1-propylthiopropane

6.7 (a) 2-isopropyl-5-methylcyclohexanol
(b) 2-isopropyl-5-methylphenol, 3-hydroxy-4-isopropyltoluene

6.8 (a) $CH_3CH_2\overset{\overset{\displaystyle OH}{|}}{\underset{\underset{\displaystyle CH_3}{|}}{C}}CH_2CH_2CH_2CH_3$ (b) $CH_3CH_2\overset{\overset{\displaystyle OH}{|}}{CH}\underset{\underset{\displaystyle CH_3}{|}}{CH}CH(CH_3)_2$ (c) $CH_3\overset{\overset{\displaystyle OH}{|}}{C}(CH_3)CH_2CH_2CH_3$

(d) $\overset{\overset{\displaystyle OH}{|}}{CH_2}CH_2CH=CH_2$ (e)   (f) $\overset{\overset{\displaystyle OH}{|}}{CH}CH(CH_3)CH_2CH_3$ (g) $\overset{\overset{\displaystyle OH}{|}}{CH_2}\overset{\overset{\displaystyle Br}{|}}{\underset{\underset{\displaystyle I}{|}}{C}}CH_2CH_3$

(h) $\overset{\overset{\displaystyle OH}{|}}{CH_2}CH_2\overset{\overset{\displaystyle Cl}{|}}{CH}CH_2\overset{\overset{\displaystyle OCH_3}{|}}{CH}CH_3$ (i) $HOCH_2CH_2CHOHCH_2CH_3$ (j) $CH_3\overset{\overset{\displaystyle OH}{|}}{CH}\underset{\underset{\displaystyle CH_2CH_2CH_3}{|}}{CH}CH_2CH_2CH_3$

(k)   (l)

6.9 (a)   (b)   (c)   (d)   (e) $(CH_3)_3COCH_2CH_2CH_2CH_3$

(f)   (g)   (h)   (i)   (j)

(k)   (l)   (m)   (n) $CH_3SCH_2CH_2CH_2CH_2CH_3$

6.10

coniferyl alcohol   sinapyl alcohol

6.11 $H-\overset{\overset{\displaystyle Cl}{|}}{\underset{\underset{\displaystyle Cl}{|}}{C}}-\overset{\overset{\displaystyle F}{|}}{\underset{\underset{\displaystyle F}{|}}{C}}-O-CH_3$ $F-\overset{\overset{\displaystyle F}{|}}{\underset{\underset{\displaystyle F}{|}}{C}}-\overset{\overset{\displaystyle Cl}{|}}{CH}-O-\overset{\overset{\displaystyle F}{|}}{\underset{\underset{\displaystyle F}{|}}{CH}}$

methoxyflurane   isoflurane

6.12 (a)

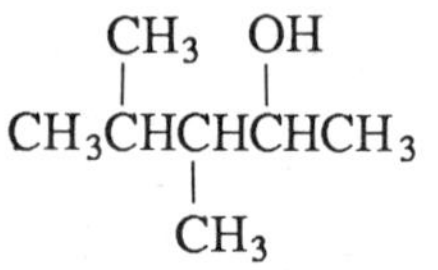

$$CH_3CHCHCHCH_3$$ (with CH₃ and OH on adjacent carbons, CH₃ below)

3,4-dimethyl-2-pentanol

(b)

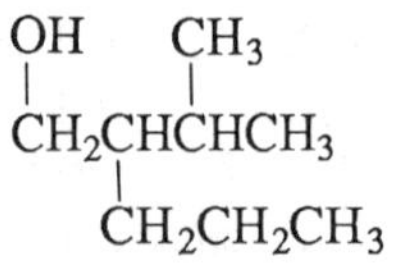

$$CH_2CHCHCH_3$$ (with OH, CH₃, CH₂CH₂CH₃)

2-isopropyl-1-pentanol

(c) 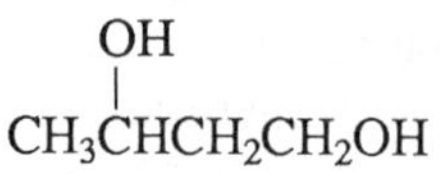 $CH_3CHCH_2CH_2OH$ (OH on second carbon)

1,3-butanediol

(d) 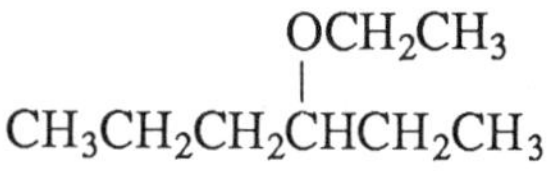 $CH_3CH_2CH_2CHCH_2CH_3$ (with OCH_2CH_3)

3-ethoxyhexane

(e)

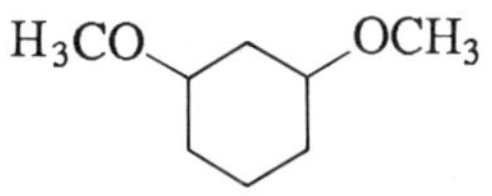

1,3-dimethoxycyclohexane (H_3CO and OCH_3 substituents)

(f) 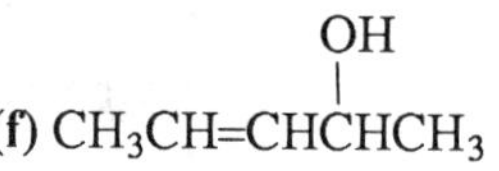 $CH_3CH=CHCHCH_3$ (with OH)

pent-3-en-2-ol

(g)

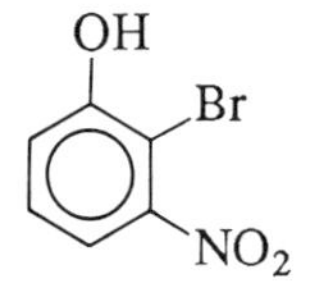

2-bromo-3-nitrophenol (OH, Br, NO_2)

(h)

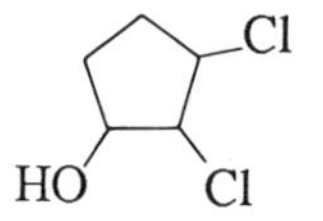

2,3-dichlorocyclopentanol (Cl, HO, Cl)

(i) $CH_3CH_2CH_2SCH_2CH_3$

1-ethylthiopropane

(j)

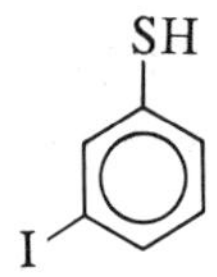

3-iodothiophenol (SH, I)

6.13

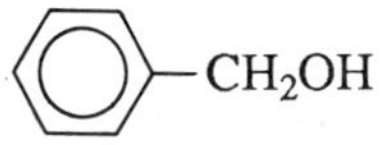

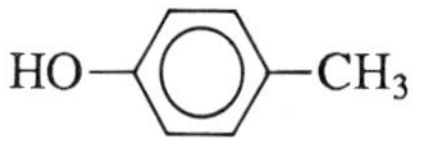

benzyl alcohol ($\bigcirc$—CH_2OH) p-cresol (HO—$\bigcirc$—CH_3)

(a) Both of them are only slightly soluble in water; however benzyl alcohol is more soluble in water than p-cresol.

(b) p-Cresol is more soluble in aqueous NaOH than benzyl alcohol. The former reacts with NaOH to

form a salt, H_3C—$\bigcirc$—O^- Na^+.

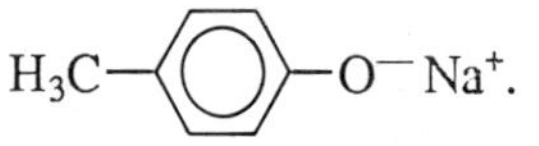

6.14 (a) $CH_3OH > n\text{-}C_4H_9OH > n\text{-}C_4H_9Br$

(b) $CH_2OHCHOHCH_2OH > CH_2OHCH_2CH_2CH_2OH > n\text{-}C_4H_9OH$

(c) $(CH_3)_3CCH_2OH > (CH_3CH_2)_2CHOH > CH_3(CH_2)_4OH$

(d) $(CH_3)_2O > \bigcirc\!\!-OH > CH_3(CH_2)_5OH$

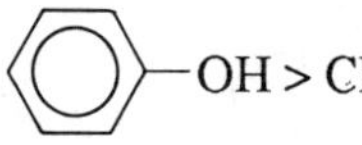

(e) $(CH_3)_3COH > (CH_3CH_2)_2O > \bigcirc\!\!-OH$

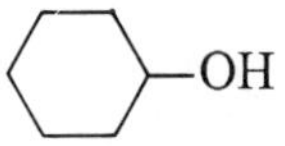

(f) $CH_3SH > CH_3CH_2SH > (CH_3)_2S$

6.15 (a) $C_3H_7OH > CH_3OH > C_4H_{10}$

(b) 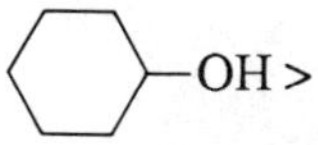 $\bigcirc\!\!-OH > n\text{-}C_4H_9OH > (CH_3)_3COH$

(c) $CH_2OHCHOHCH_2OH > CH_2OHCH_2CH_2OH > C_3H_7OH$

(d) $C_3H_7OH > (CH_3CH_2)_2O > CH_3OCH_2CH_3$

(e) HO—⟨○⟩—OH > ⟨○⟩—OH > ⟨○⟩—O–CH₃

(f) $CH_3CH_2OH > CH_3CH_2SH > CH_3SCH_3$

6.16 (a) $CH_3CH_2OH > n\text{-}C_4H_9OH > CH_3OCH_3$

(b) ⟨○⟩—OH > HOH > ⟨○⟩—OH

(c) H_3C—⟨○⟩—OH > ⟨○⟩—CH_2OH > ⟨○⟩—OCH_3

(d) $CH_3CH_2SH > CH_3CH_2OH > CH_3SCH_3$

6.17 Methyl alcohol is a polar molecule, and it has the capability of hydrogen bonding with water molecules as well as intermolecular hydrogen bonding between alcohol molecules. Therefore it is soluble in water and has a relatively high boiling point. Methyl chloride cannot participate in hydrogen bonding. It is insoluble in water and has a low boiling point.

6.18 (a) diethyl ether   **(b)** 1-butanol

6.19 Place the mixture in a separatory funnel and add aqueous NaOH. Phenol reacts with the base to form a salt that is soluble in water. Add ether in order to extract the 1-octanol. Separate the two layers. Acidify the aqueous layer with HCl to regenerate the phenol.

⟨○⟩OH $\xrightarrow{\text{NaOH}}$ ⟨○⟩O^-Na^+ $\xrightarrow{\text{HCl}}$ ⟨○⟩OH

soluble
in water

6.20 They have relatively high vapor pressures. Thus, they evaporate readily, and the evaporation removes heat from the skin and serves to lower body temperature.

6.21 (a) $CH_3CH_2CH_2CH_2Br \xrightarrow[\text{HOH}]{\text{NaOH}} CH_3CH_2CH_2CH_2OH$

(b) $CH_3CH=CHCH_3 \xrightarrow[\text{H}_2\text{SO}_4]{\text{HOH}} CH_3CHOHCH_2CH_3$

(c) ⟨○⟩—OH $\xrightarrow{\text{KOH}}$ ⟨○⟩—O^-K^+

(d) $CH_3CH_2CH_2OH \xrightarrow[\text{H}_2\text{SO}_4]{\Delta} CH_3CH=CH_2 \xrightarrow[\text{H}_2\text{SO}_4]{\text{HOH}} CH_3CHOHCH_3$

(e) $CH_3CHOHCH_2CH_2CH_3 \xrightarrow[\text{H}_2\text{SO}_4]{\Delta} CH_3CH=CHCH_2CH_3 \xrightarrow[\text{Ni}]{\text{H}_2} CH_3CH_2CH_2CH_2CH_3$

6.22 At temperatures below 130°C, the reaction is so slow that unreacted alcohol distills out of the reaction mixture. Above 150°C, the ethyl sulfate intermediate decomposes to ethylene and H_2SO_4.

6.23

$$CH_3CH_2OH \;+\; H_2SO_4 \xrightarrow{\hspace{1cm}}$$

- $\xrightarrow[\text{temp.}]{\text{room}}$ $CH_3CH_2OSO_3H$
- $\xrightarrow{140°C}$ $CH_3CH_2OCH_2CH_3$
- $\xrightarrow{180°C}$ $CH_2{=}CH_2$

6.24 (a) —OH   (b) $CH_3\overset{\displaystyle O}{\overset{\|}{C}}CH_3$ (c) no reaction   (d) $CH_3CH_2CH_2O^-\,Na^+$

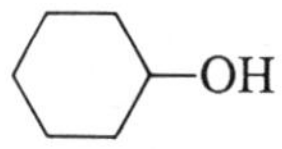

(e) $CH_3CH_2CH_2CH_2I$ (f) —O–S–OH   (g) —CHOHCH$_3$ (h) —CH$_2$Cl

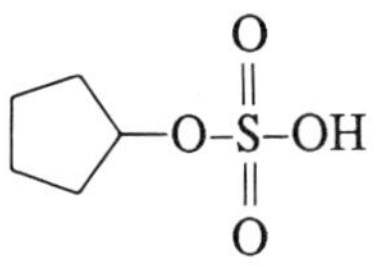

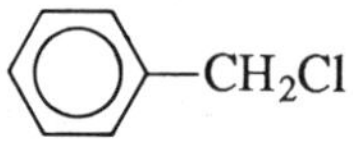

(i) $CH_3CHBrCH(CH_3)_2$ (j) no reaction   (k) Cl——$O^-\,Na^+$ (l) $O{=}$${=}O$

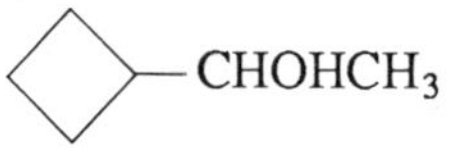

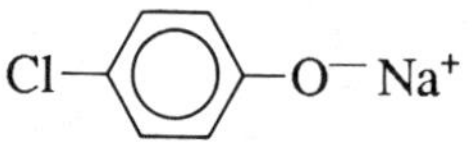

(m) ${=}O$ (n) no reaction   (o) no reaction   (p) $CH_3CH_2I \;+\; CH_3CH_2OH$

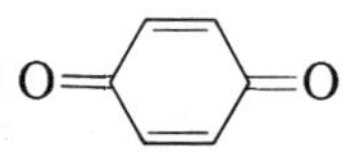

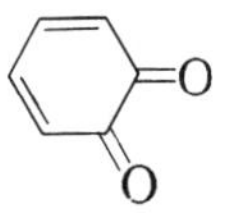

(q) $CH_3CH_2CH_2S^-\,Na^+$ (r) $CH_3{-}S{-}S{-}CH_3$

6.25 (a) $CH_3CH_2C(CH_3)_2OH > CH_3CH_2CH_2CHOHCH_3 > CH_3CH_2CH_2CH_2CH_2OH$

(b) $>$ —OH $>$ —CH$_2$OH

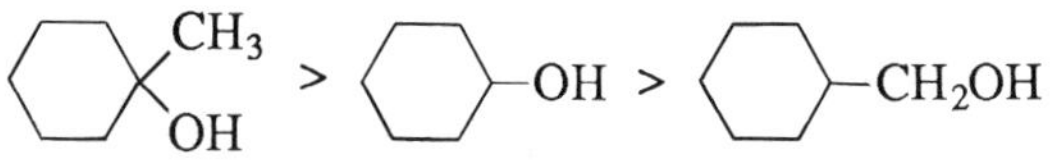

6.26 $\xrightarrow[\text{ether}]{Ag_2O}$

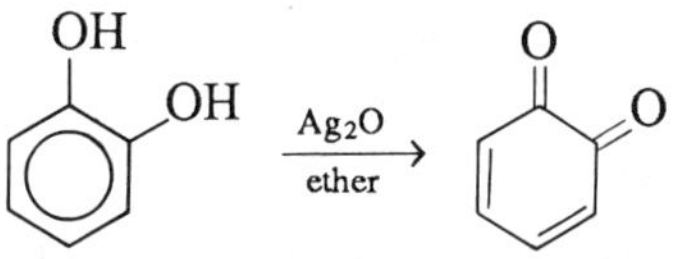

6.27 (a) $CH_3CH{=}CHCH_2CH_3$; $CH_2{=}CHCH_2CH_2CH_3$ (b) $CH_3CH_2CH{=}CHCH_3$ (only product)

(c) $(CH_3)_2C{=}CHCH_3$; $CH_2{=}C(CH_3)CH_2CH_3$ (d) —CH$_2$CH$_3$; —CH$_2$CH$_3$

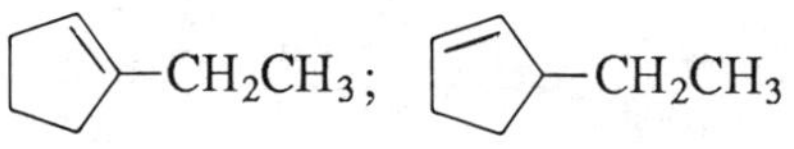

(e) —CH=CHCH$_3$; —CH$_2$CH=CH$_2$ (f) $(CH_3)_2C{=}CHCH_3$; $(CH_3)_2CHCH{=}CH_2$

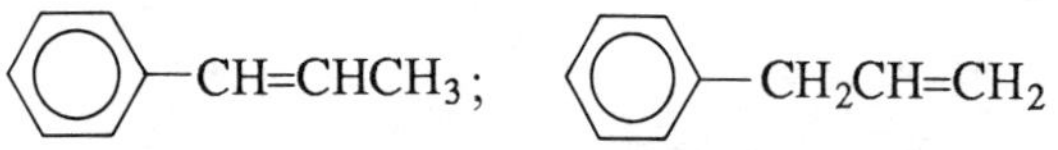

(g) $CH_3CBr_2C(CH_3){=}CH_2$ (only product)   (h) $(CH_3)_3CCH{=}CH_2$ (only product)

(i) —CH$_3$; =CH$_2$ (j) H_3C——CH$_2$CH$_3$; H_3C——CH$_2$CH$_3$

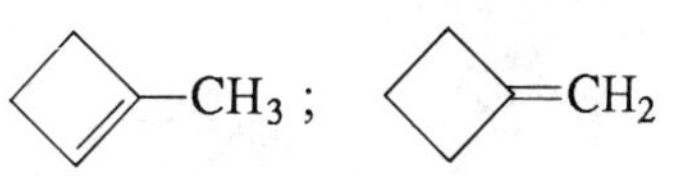

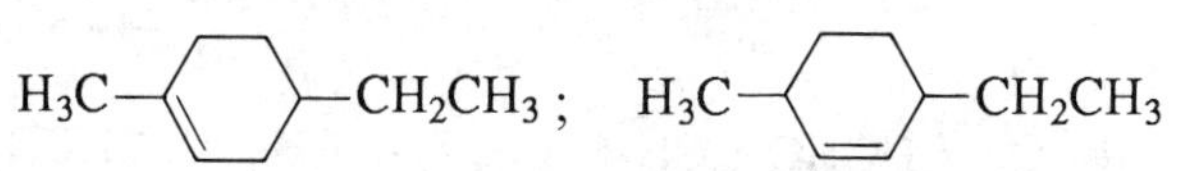

6.28 (a) solvent   (b) alcoholic beverage   (c) rubbing alcohol   (d) radiator antifreeze
(e) lubricant   (f) germicide   (g) wood preservative   (h) photographic film developer
(i) extraction solvent   (j) odorant added to natural gas

6.29 (a) The attraction between a hydrogen atom (that is covalently bonded to fluorine, oxygen, or nitrogen) and a fluorine, oxygen, or nitrogen atom that is in proximity.

(b) 100% alcohol.

(c) Alcohol that has been treated with certain additives, making it unfit to drink.

(d) 43% alcohol by volume.

(e) A measure of the effectiveness of a given germicide as compared to phenol.

(f) A solute dissolved in alcohol (i.e., an alcoholic solution).

(g) Any substance that causes unconsciousness or insensitivity to pain.

(h) Any substance that alleviates pain.

(i) The anion that results from the dissociation of a hydrogen from an alcohol.

(j) The cation that results when an alcohol or an ether binds to a proton.

(k) A constant boiling mixture of two components.

(l) A mixture containing about 10% alcohol and 90% gasoline.

(m) Any substance that dilates blood vessels.

(n) Any substance that kills bacteria and other microorganisms.

(o) An older designation for thiols.

(p) An organic solvent that has been proclaimed to have analgesic properties.

6.30 $CH_3CH_2CHOHCH_3$

6.31 $ROH + K \longrightarrow RO^-K^+ + \frac{1}{2}H_2$

$$560 \text{ mL of } H_2 = \frac{0.56}{22.4} = 0.025 \text{ mole of } H_2$$

therefore moles of alcohol $= 0.050$

$$\text{molecular weight of alcohol} = \frac{1.6}{0.05} = 32$$

6.32 $C_2H_5OH + H_2SO_4 \longrightarrow CH_2{=}CH_2$

$$4.6 \text{ g of ethanol} = \frac{4.6}{46} = 0.10 \text{ mole}$$

Therefore 0.10 mole of ethene is produced $= 2.24$ L

6.33 $X = CH_3\overset{\overset{\displaystyle CH_3}{|}}{C}{=}CH_2 \qquad Y = CH_3\overset{\overset{\displaystyle Br}{|}}{C}(CH_3)_2$

6.34 $CH_3O\overset{\overset{\displaystyle CH_3}{|}}{C}HCH_2CH_3$

CHAPTER 7

Aldehydes and Ketones

7.1

$$CH_3CH_2CH_2CH_2\overset{O}{\underset{H}{\overset{\|}{C}}}$$ $$CH_3CH_2CH_2\overset{O}{\overset{\|}{C}}CH_3$$ $$CH_3CH_2CH(CH_3)\overset{O}{\underset{H}{\overset{\|}{C}}}$$ $$CH_3CH_2\overset{O}{\overset{\|}{C}}CH_2CH_3$$

pentanal   2-pentanone   2-methylbutanal   3-pentanone

$$(CH_3)_2CHCH_2\overset{O}{\underset{H}{\overset{\|}{C}}}$$ $$(CH_3)_2CH\overset{O}{\overset{\|}{C}}CH_3$$ $$(CH_3)_3C\overset{O}{\underset{H}{\overset{\|}{C}}}$$

3-methylbutanal   3-methyl-2-butanone   2,2-dimethylpropanal

7.2 (a) 4,4,4-trichlorobutanal   (b) 2-pentanone   (c) 2-butenal
(d) 4-bromo-2,2-dimethyl-3-hexanone   (e) 2,4-dimethyl-3-pentanone
(f) 2-ethylbutanal   (g) 4-phenyl-2-pentanone   (h) 2,2-dimethyl-4-phenylbutanal
(i) 4,4,5-trichlorohexanal   (j) 5-iodo-2-pentanone   (k) 2,4-hexanedione
(l) 3-hydroxypropanal   (m) 5-bromo-3-hexenal   (n) 3-penten-2-one
(o) 2-methylcyclopentanone   (p) 4-cyclohexylbutanal

7.3 (a) $CH_3CH_2CH_2\overset{O}{\underset{H}{\overset{\|}{C}}}$ (b) $(CH_3)_2CHCH_2\overset{O}{\underset{H}{\overset{\|}{C}}}$ (c) $CH_3CH_2CH=CHCH_2\overset{O}{\underset{H}{\overset{\|}{C}}}$

(d) $CH_3\overset{O}{\overset{\|}{C}}CH_2CH_2CH_2CH_3$ (e) $CH_3CHCl\overset{O}{\underset{H}{\overset{\|}{C}}}$ (f) $CH_3\overset{O}{\overset{\|}{C}}CHBrCH_2CH_2CH_2CH_3$

(g) [cyclohexanone ring with CH₃ and CH₃]   (h) [phenyl]—$CH_2\overset{O}{\overset{\|}{C}}CH_2CH_3$ (i) $CH_3\overset{O}{\overset{\|}{C}}CH_2CH=CHCH_3$

(j) $CH_3\overset{I}{\underset{CH_3}{\overset{|}{C}}}CH_2\overset{O}{\overset{\|}{C}}CH_2CH_2CH_2CH_3$ (k) $CH_3\overset{CH_3}{\overset{|}{C}}=CH\overset{O}{\overset{\|}{C}}CH_3$ (l) $CH_3CH_2CH\underset{CH_2CH_3}{\overset{|}{C}}H_2CH_2CH_2\overset{O}{\underset{H}{\overset{\|}{C}}}$

(m) $CH_3\overset{OH}{\overset{|}{C}}H\overset{CH_3}{\overset{|}{C}}\underset{O}{\overset{\|}{C}}HCH_3$ (n) $CH_3\overset{CH_3}{\overset{|}{C}}HCH_2CH=\overset{CH_3}{\overset{|}{C}}-\overset{O}{\underset{H}{\overset{\|}{C}}}$ (o) $CH_3CH_2\overset{O}{\overset{\|}{C}}$[cyclopentyl]   (p) H_3CO—[phenyl]—$\overset{O}{\underset{H}{\overset{\|}{C}}}$

32

7.4 O_2N—⟨benzene⟩—CH=CH–CH=CH–C(=O)H

7.5 $\overset{OH}{\underset{}{|}}\overset{OH}{\underset{}{|}}$ $CH_2CHC(=O)H$ $HOCH_2CCH_2OH$ (with O)

7.6 $O=\overset{}{C}CH_2CH_2CH_2C=O$ (both with H)

7.7   (a) 3 $CH_3CH_2CH_2OH$ + 2 CrO_3 + 6 HCl $\xrightarrow[\text{CH}_2\text{Cl}_2]{\text{pyridine}}$ 3 $CH_3CH_2C(=O)H$ + 2 $CrCl_3$ + 6 H_2O

(b) 3 [cyclohexane with OH and CH_3] + 2 $KMnO_4$ ⟶ 3 [cyclohexanone with CH_3] + 2 MnO_2 + 2 KOH + 2 H_2O

7.8   The aldehydes that are formed by the oxidation of primary alcohols are themselves easily oxidized further to carboxylic acids, and therefore some aldehyde is lost through this process. Ketones, formed from the oxidation of secondary alcohols, cannot be further oxidized.

7.9   (a) CH_3—⟨cyclohexane⟩—OH   (b) $CH_3-\overset{CH_3}{\underset{CH_3}{\overset{|}{\underset{|}{C}}}}-CH_2OH$ (c) $CH_3CH_2CHBrCH_2CH_2OH$

(d) $CH_3CHOHCH_2CH_2CH_3$ (e) ⟨Ph⟩–$\overset{OH}{\underset{H}{\overset{|}{\underset{|}{C}}}}$–⟨Ph⟩   (f) ⟨benzene with CH_2OH and CH_3⟩

7.10 (a) $CH_3CH_2CH_2CH_2OH$ > $CH_3CH_2CH_2C(=O)H$ > $CH_3CH_2OCH_2CH_3$

(b) $CH_3CHOHCH_3$ > $CH_3\overset{O}{\overset{||}{C}}CH_3$ > $CH_3OCH_2CH_3$

(c) ⟨cyclohexane⟩=O > $CH_3CH_2\overset{O}{\overset{||}{C}}CH_3$ > $CH_3C(=O)H$

7.11 (a) $CH_3C(=O)H$ > $CH_3CH_2CH_2CH_2OH$ > $CH_3CH_2\overset{O}{\overset{||}{C}}CH_2CH_3$

(b) $H-C(=O)H$ > $CH_3CH_2\overset{O}{\overset{||}{C}}CH_3$ > ⟨benzene⟩–C(=O)H

(c) CH_3OCH_3 > $CH_3CHOHCH_2CH_3$ > $CH_3\overset{O}{\overset{||}{C}}CH_2CH_2CH_3$

7.12 (a) *o*-hydroxybenzaldehyde 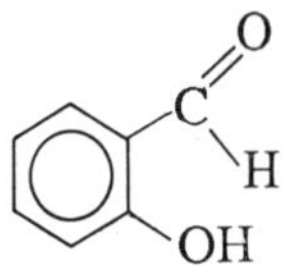

(b) *p*-Hydroxybenzaldehyde has the higher boiling point because it only forms intermolecular hydrogen bonds. The intramolecular hydrogen bonding that occurs with the *o*-hydroxybenzaldehyde decreases the intermolecular hydrogen bonding because it ties up both the hydroxyl group and the carbonyl group.

7.13 They have lower boiling points because there are only weak polar interactions (no intermolecular hydrogen bonding). They have comparable solubilities to alcohols because water molecules can form hydrogen bonds with the carbonyl oxygens.

7.14 (a) carbonyl (b) electrophilic reagent (c) nucleophilic reagent (d) the carbon atom

7.15 (a) 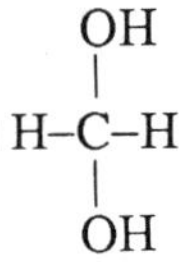 (b) 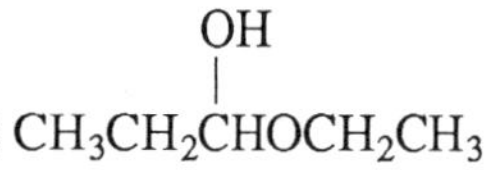 (c) 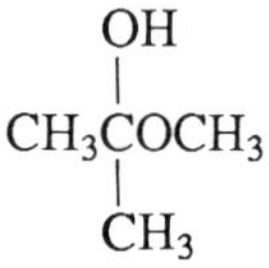 (d) 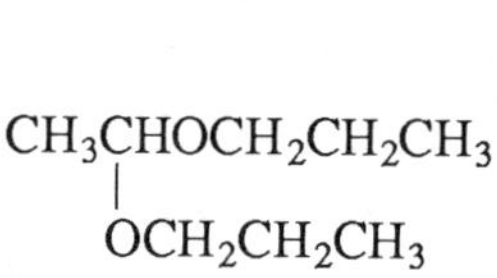

(e) 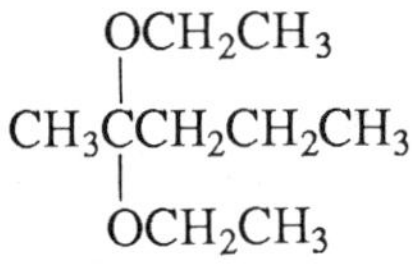 (f) 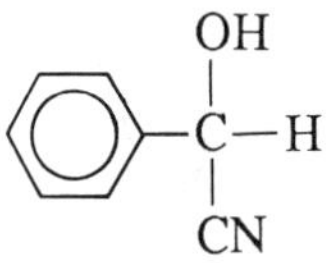

(g) 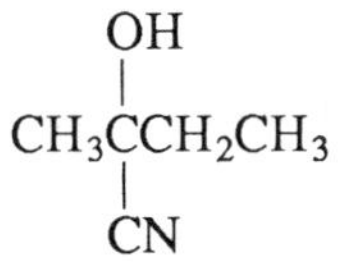 (h) 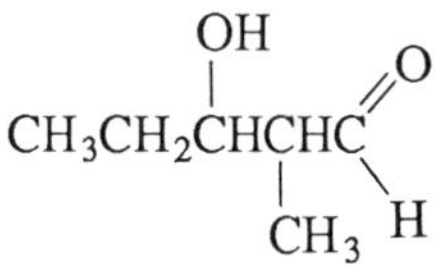 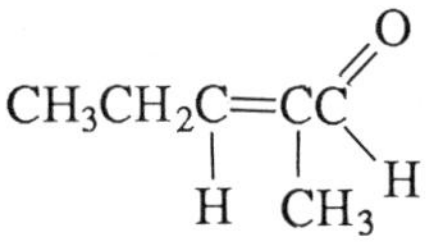

(i) 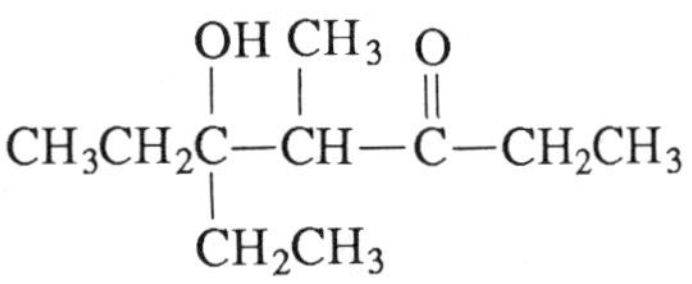 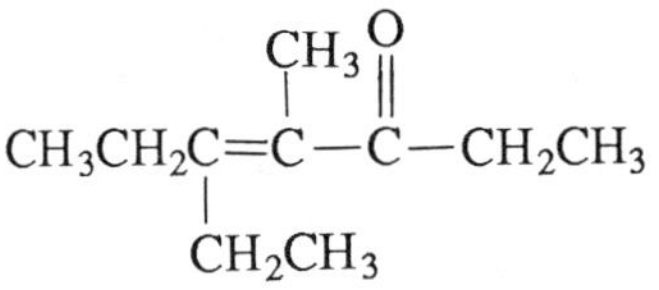

(j) 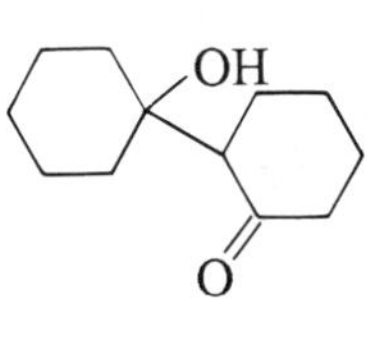 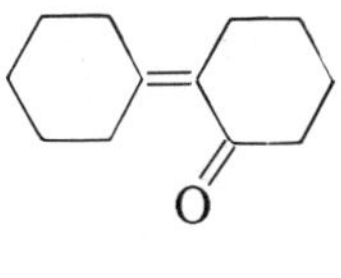 (k) 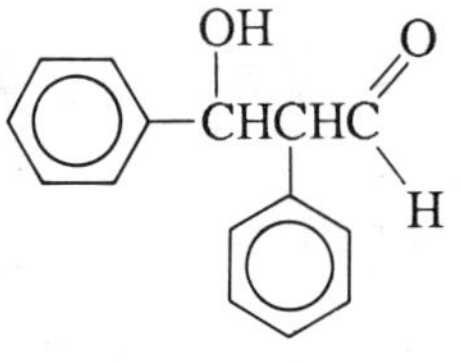

(l) 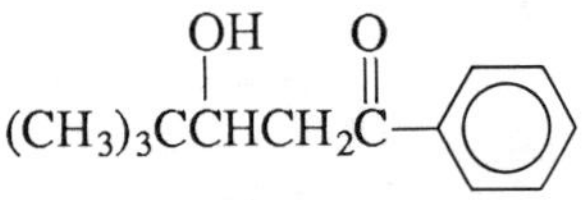 (m) 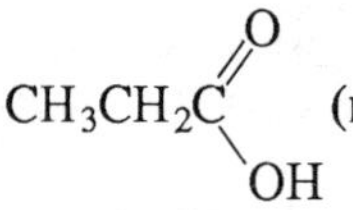 (n) no reaction (o) $CH_3CH_2CH_2CH_2OH$

(p) 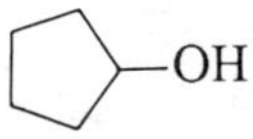 (q) 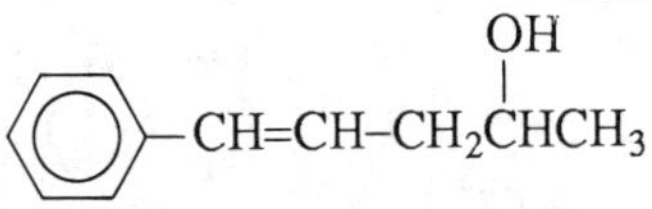 (r) 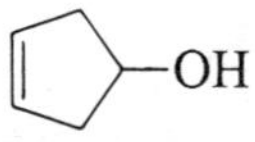

7.16 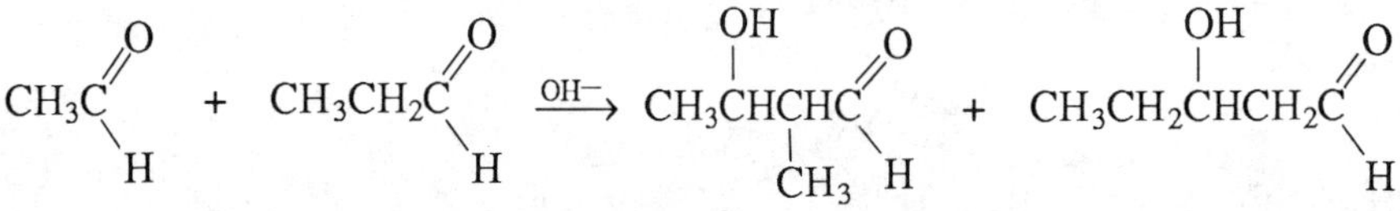

7.17 These compounds do not contain α-hydrogens.

7.18 (a) 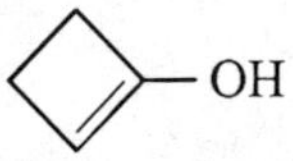 (b) 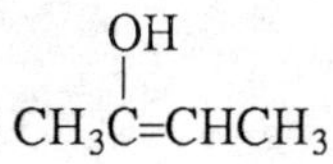 (c) 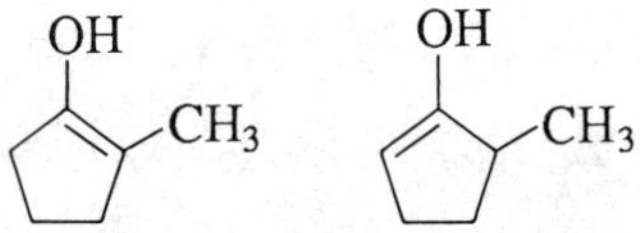

$$\overset{\displaystyle OH}{\underset{}{}}$$

(d) $CH_2=\overset{OH}{\underset{|}{C}}CH_2CH_2CH_3 \qquad CH_3\overset{OH}{\underset{|}{C}}=CHCH_2CH_3$

7.19 (a) $CH_3\overset{-}{C}-\overset{\displaystyle O}{\overset{\|}{C}}\underset{H}{\overset{}{}}\,\rightleftharpoons\, CH_3\ C=\overset{\displaystyle O^-}{\overset{}{C}}\underset{H}{}$

(b) $CH_3CH_2\overset{\displaystyle O}{\overset{\|}{C}}-\overset{-}{C}CH_3 \,\rightleftharpoons\, CH_3CH_2\overset{\displaystyle O^-}{\overset{}{C}}=CCH_3$
$\underset{H}{}\underset{H}{}$

7.20 (a) CH_3C ... (b) CH_3C ... CH_3 ... (c) ...

(d) $\qquad$ no intramolecular H-bond

7.21 $\ 2\ \underset{}{\bigcirc}\!-\!\overset{\displaystyle O}{\overset{\|}{C}}\underset{H}{} + \ 3\ OH^- \longrightarrow \underset{}{\bigcirc}\!-\!\overset{\displaystyle O}{\overset{\|}{C}}\underset{O^-}{} + \ 2\ H_2O \ + \ 2\ e^-$

$Ag(NH_3)_2^+ \ + \ e^- \longrightarrow Ag \ + \ 2\ NH_3$

7.22 (a) $CH_2=CHC\overset{\displaystyle O}{\underset{H}{}} \xrightarrow[\text{ether}]{LiAlH_4} \xrightarrow{\text{water}} CH_2=CHCH_2OH$

(b) $CH_3C\overset{\displaystyle O}{\underset{H}{}} + Na_2Cr_2O_7 \xrightarrow{H^+} CH_3C\overset{\displaystyle O}{\underset{OH}{}}$

(c) $CH_3CHOHCH_2CH_2CH_2CH_3 + KMnO_4 \xrightarrow{H^+} CH_3\overset{\displaystyle O}{\overset{\|}{C}}CH_2CH_2CH_2CH_3$

(d) $\bigcirc\!-\!C\overset{\displaystyle O}{\underset{H}{}} + H_2 \xrightarrow{Pt} \bigcirc\!-\!CH_2OH$

(e) $CH_3-\overset{\displaystyle O}{\overset{\|}{C}}-CH_3 + 2\ CH_3CH_2OH \xrightarrow{H^+} (CH_3)_2C(OCH_2CH_3)_2$

(f) $CH_3CH_2C\overset{\displaystyle O}{\underset{H}{}} + CH_3CH_2C\overset{\displaystyle O}{\underset{H}{}} \xrightarrow{OH^-} CH_3CH_2\overset{OH}{\underset{H}{C}}-\overset{}{\underset{CH_3}{CH}}-C\overset{\displaystyle O}{\underset{H}{}} \xrightarrow{\Delta}$

$CH_3CH_2\underset{H}{\overset{}{C}}=\underset{CH_3}{\overset{}{C}}-C\overset{\displaystyle O}{\underset{H}{}} \xrightarrow[\text{ether}]{LiAlH_4} \xrightarrow{H_2O} CH_3CH_2\underset{H}{\overset{}{C}}=\underset{CH_3}{\overset{}{C}}-CH_2OH$

7.23 (a) The carbon–oxygen double bond, e.g., $CH_3\text{-}\overset{\displaystyle O}{\underset{\displaystyle \|}{C}}\text{-}H$

(b) The product of the reaction between one mole of an aldehyde and one mole of an alcohol, e.g.,

$$CH_3\overset{\displaystyle OH}{\underset{\displaystyle OCH_2CH_3}{\overset{|}{\underset{|}{C}}}}\text{-}H$$

(c) The product of the reaction between one mole of a ketone and one mole of an alcohol, e.g.,

$$CH_3\overset{\displaystyle OH}{\underset{\displaystyle OCH_2CH_3}{\overset{|}{\underset{|}{C}}}}CH_3$$

(d) The product of the reaction between one mole of an aldehyde and two moles of an alcohol, e.g.,

$$CH_3\overset{\displaystyle OCH_2CH_3}{\underset{\displaystyle OCH_2CH_3}{\overset{|}{\underset{|}{C}}}}\text{-}H$$

(e) The product of the reaction between one mole of a ketone and two moles and an alcohol, e.g.,

$$CH_3\overset{\displaystyle OCH_2CH_3}{\underset{\displaystyle OCH_2CH_3}{\overset{|}{\underset{|}{C}}}}CH_3$$

(f) The product of the reaction between HCN and an aldehyde or a ketone, e.g.,

$$CH_3\text{-}\overset{\displaystyle OH}{\underset{\displaystyle CN}{\overset{|}{\underset{|}{C}}}}\text{-}CH_3$$

(g) The combination of two carbonyl compounds in the presence of an aqueous base, e.g.,

$$2\ CH_3\overset{O}{\underset{H}{C{<}}}\ \xrightarrow{OH^-}\ CH_3\overset{OH}{\underset{H}{\overset{|}{\underset{|}{C}}}}\text{-}CH_2\text{-}\overset{O}{\underset{H}{C{<}}}$$

(h) Hydrogens that are bound to the alpha carbon atom, e.g.,

$$CH_3\text{-}\overset{\boxed{H}}{\underset{\boxed{H}}{\overset{|}{\underset{|}{C}}}}\text{-}\overset{O}{\underset{H}{C{<}}}$$

(i) The process that converts the keto form of a carbonyl compound to the enol form, e.g.,

$$CH_3\text{-}\overset{O}{\underset{H}{C{<}}}\ \underset{or\ H^+}{\overset{OH^-}{\rightleftharpoons}}\ H_2C{=}\overset{}{\underset{H}{\overset{}{\underset{|}{C}}}}\text{-}OH$$

(j) The unsaturated alcohol that arises from an enolization reaction, e.g.,

$$\underset{\text{(enol form of ethanal)}}{CH_2=\overset{\displaystyle OH}{\overset{|}{C}}-H}$$

(k) This is the carbonyl form of an aldehyde or ketone, e.g., $CH_3\overset{\displaystyle O}{\overset{||}{C}}CH_3$ (keto form of acetone)

(l) The resonance-stabilized ion that is an intermediate in the enolization process, e.g.,

$$H_2\bar{C}-\overset{\displaystyle O}{\underset{\displaystyle H}{C}} \rightleftharpoons H_2C=\overset{\displaystyle O^-}{\overset{|}{C}}-H$$

(m) The interconversion of isomers that differ only in the position of a hydrogen and the corresponding location of the double bond, e.g.,

$$CH_3\overset{\displaystyle O}{\overset{||}{C}}CH_2\overset{\displaystyle O}{\overset{||}{C}}CH_3 \rightleftharpoons CH_3\overset{\displaystyle O}{\overset{||}{C}}CH=\overset{\displaystyle OH}{\overset{|}{C}}CH_3$$

(n) The isomers that are interconvertible via tautomerism, e.g., the keto and enol forms of a carbonyl compound.

7.24 (a) Cu(II) **(b)** Cu(II) **(c)** Ag(I)

7.25 (a) $CH_3CHOHCH_2CH_3 \xrightarrow[\text{reflux}]{K_2Cr_2O_7} CH_3\overset{\displaystyle O}{\overset{||}{C}}CH_2CH_3 \xrightarrow{\text{blue litmus}}$ no effect on litmus

color change from orange to green

$(CH_3)_2CHCH_2OH \xrightarrow[\text{reflux}]{K_2Cr_2O_7} (CH_3)_2CH\overset{\displaystyle O}{\overset{/\!/}{C}}\!\!\diagdown_{OH} \xrightarrow{\text{blue litmus}}$ litmus turns red

color change from orange to green

$(CH_3)_3COH \xrightarrow[\text{reflux}]{K_2Cr_2O_7}$ no color change

(b) Only 1-butanol will yield bubbles of H_2 gas upon addition of Na. Butanal can be distinguished from butanone either by oxidation, followed by testing with litmus, or by use of Tollens' reagent.

(c) $CH_3CH=CH_2 \xrightarrow{\text{dil KMnO}_4} CH_3CHOHCH_2OH$
no effect on litmus

$CH_3CH_2\overset{\displaystyle O}{\overset{/\!/}{C}}\!\!\diagdown_{H} \xrightarrow{\text{dil KMnO}_4} CH_3CH_2COOH$
turns blue litmus red

(d) Benzaldehyde will react with Tollens' reagent to yield a silver precipitate. Cyclohexanone will not react with Tollens' reagent.

7.26 (a) CH_3OH $\xrightarrow[>300°]{Cu\ or\ Ag}$ H–C$\diagup^{O}_{\diagdown H}$

Used as a reagent for the preparation of other organic compounds.

(b) CH_3CH_2OH $\xrightarrow{Ag}$ $CH_3C\diagup^{O}_{\diagdown H}$

Used as a reagent for the preparation of other organic compounds.

(c) $CH_3CHOHCH_3$ $\xrightarrow{Ag}$ $CH_3\overset{\overset{\textstyle O}{\|}}{C}CH_3$

Used as a solvent.

(d) $Cl_3CC\diagup^{O}_{\diagdown H}$ + HOH $\longrightarrow$ $Cl_3C\overset{\overset{\textstyle OH}{|}}{C}H$ over OH

Used as a sedative.

7.27 A = $CH_2{=}\overset{\overset{\textstyle CH_3}{|}}{C}CH_2C\diagup^{O}_{\diagdown H}$ or $CH_3\overset{\overset{\textstyle CH_3}{|}}{C}{=}CHC\diagup^{O}_{\diagdown H}$

B = $CH_2{=}\overset{\overset{\textstyle CH_3}{|}}{C}CH_2C\diagup^{O}_{\diagdown OH}$ or $CH_3\overset{\overset{\textstyle CH_3}{|}}{C}{=}CHC\diagup^{O}_{\diagdown OH}$

CHAPTER 8

Acids and Esters

8.1

$\underset{\text{butanoic acid}}{CH_3CH_2CH_2\overset{\displaystyle O}{\underset{\displaystyle OH}{C}}}$ $\underset{\text{2-methylpropanoic acid}}{CH_3CH(CH_3)\overset{\displaystyle O}{\underset{\displaystyle OH}{C}}}$ $\underset{\text{methyl propanoate}}{CH_3CH_2\overset{\displaystyle O}{\underset{\displaystyle OCH_3}{C}}}$ $\underset{\text{ethyl acetate (ethanoate)}}{CH_3\overset{\displaystyle O}{\underset{\displaystyle OCH_2CH_3}{C}}}$

$\underset{\text{propyl formate (methanoate)}}{H-\overset{\displaystyle O}{\underset{\displaystyle OCH_2CH_2CH_3}{C}}}$ $\underset{\text{isopropyl formate (methanoate)}}{H-\overset{\displaystyle O}{\underset{\displaystyle OCH(CH_3)_2}{C}}}$

8.2 **(a)** 3-methylbutanoic acid   **(b)** methyl acetate (ethanoate)   **(c)** 3,4,4-trimethylpentanoic acid
   (d) propyl pentanoate   **(e)** 4-hydroxybutanoic acid   **(f)** sodium propanoate   **(g)** 4-hexenoic acid
   (h) 4,4-dichloro-5-methylhexanoic acid   **(i)** cyclopentane carboxylic acid   **(j)** *p*-bromobenzoic acid
   (k) trimethylphosphate   **(l)** potassium hexanoate   **(m)** 4-hydroxy-2-iodopentanoic acid
   (n) succinic acid (butanedioic acid)   **(o)** isopropyl formate (methanoate)   **(p)** acetyl chloride
   (q) ethyl 2-ethylbutanoate   **(r)** 3-ethyl-4-methoxypentanoic acid
   (s) phenyl propanoate   **(t)** ethyl benzoate

8.3 **(a)** $CH_3\underset{\displaystyle CH_3}{\overset{\displaystyle CH_3}{C}}-\overset{\displaystyle O}{\underset{\displaystyle OH}{C}}$ **(b)** $CH_3\underset{\displaystyle Cl}{C}HCH_2\underset{\displaystyle CH_3}{C}H\overset{\displaystyle O}{\underset{\displaystyle OH}{C}}$ **(c)** $\underset{\displaystyle HO}{\overset{\displaystyle O}{C}}-\underset{\displaystyle OH}{\overset{\displaystyle O}{C}}$ **(d)** $CH_3\overset{\displaystyle O}{\underset{\displaystyle OCH_2CH_3}{C}}$

   (e) $CH_3\overset{\displaystyle O}{\overset{\displaystyle \|}{C}}-O-\overset{\displaystyle O}{\overset{\displaystyle \|}{C}}CH_3$ **(f)** $CH_3CH_2\underset{\displaystyle OH}{C}HI\overset{\displaystyle O}{C}$ **(g)** $\left(CH_3\overset{\displaystyle O}{\underset{\displaystyle O^-}{C}}\right)_2 Ca^{2+}$ **(h)** Br—(benzene ring)—$\overset{\displaystyle O}{\underset{\displaystyle OH}{C}}$

   (i) (benzene ring)—$\overset{\displaystyle O}{\underset{\displaystyle O-CH(CH_3)_2}{C}}$ **(j)** $CH_3\underset{\displaystyle CH_3}{C}HCH_2CH_2CH_2\overset{\displaystyle O}{\underset{\displaystyle O-\text{(phenyl)}}{C}}$ **(k)** $CH_3CH_2\overset{\displaystyle O}{\underset{\displaystyle O^- K^+}{C}}$

   (l) $CH_3\underset{\displaystyle CH_3}{\overset{\displaystyle CH_3}{C}}CH_2\overset{\displaystyle O}{\underset{\displaystyle OCH_3}{C}}$ **(m)** $HO-\overset{\displaystyle O}{\underset{\displaystyle OH}{P}}-OCH_2CH_3$ **(n)** $CH_3CH_2CH_2\overset{\displaystyle O}{\underset{\displaystyle OCH(CH_3)_2}{C}}$

   (o) $CH_3\underset{\displaystyle Br}{C}=CHCH_2\overset{\displaystyle O}{\underset{\displaystyle OH}{C}}$ **(p)** $CH_3CHOH\overset{\displaystyle O}{\underset{\displaystyle OCH_3}{C}}$ **(q)** O_2N-(benzene ring)$-\overset{\displaystyle O}{\underset{\displaystyle OH}{C}}$

(r) 2-hydroxybenzoic acid structure (salicylic acid with COOH and OH)

(s) $C_6H_5-CH_2CH_2C(=O)OH$

(t) $CH_3CH_2OCH_2CH_2C(=O)OH$

8.4 (a) $CH_3CH_2CH_2CH_2OH \xrightarrow[H^+]{K_2Cr_2O_7} CH_3CH_2CH_2C(=O)OH$

(b) $CH_3CH_2CH_2C(=O)H \xrightarrow[H^+]{K_2Cr_2O_7} CH_3CH_2CH_2C(=O)OH$

(c) $CH_3CH_2CH_2Br \xrightarrow{KCN} CH_3CH_2CH_2CN \xrightarrow[H^+]{HOH} CH_3CH_2CH_2C(=O)OH$

(d) $CH_3CH_2CH_2C(=O)OCH_2CH_3 \xrightarrow[H^+]{HOH} CH_3CH_2CH_2C(=O)OH + CH_3CH_2OH$

(e) $CH_3CH_2CH_2C(=O)O^-Na^+ \xrightarrow{HCl} CH_3CH_2CH_2C(=O)OH + NaCl$

8.5 (a) Acetic acid is more acidic than ethyl alcohol because the carboxylate anion that is formed upon dissociation is stabilized by resonance. The alkoxide ion formed upon dissociation or ethanol is not stabilized by any factors.

(b) The inductive effects of the three chlorine atoms aid in the release of the hydrogen ion, and they more effectively stabilize the carboxylate anion.

(c) Methyl acetate does not have a hydrogen atom directly bonded to an oxygen atom and therefore cannot participate in intermolecular hydrogen bonding. There is considerable intermolecular hydrogen bonding in methyl alcohol and in formic acid.

(d) Because formic acid is a smaller molecule, there is much more extensive intermolecular hydrogen bonding.

(e) Water molecules can hydrogen bond to the hydroxyl hydrogen atom of butyric acid (as well as to both oxygens). Esters lack a hydroxyl hydrogen and, therefore, there is less hydrogen bonding between water and ethyl acetate.

(f) Such a solution is a buffer solution, and it resists changes in pH because the components of the buffer will react with hydrogen ions or with hydroxide ions.

(g) Salts are ionic compounds, and water molecules readily surround the individual cations and anions. Hydrogen bonding between the carboxyl group of benzoic acid and water is not sufficient to separate benzoic acid molecules.

(h) The carboxyl group contains a carbon-oxygen double bond (carbonyl group) which is shorter than the carbon-oxygen single bond. The carboxylate anion is a resonance hybrid. Both carbon-oxygen bonds are thus of equal length.

(i) The alkaline hydrolysis of an ester yields an alcohol and a salt. These products cannot react to reform the ester. The acid hydrolysis of esters yields alcohols and acids, which can react to reform the ester.

(j) Salicylic acid is a bifunctional molecule. In the preparation of aspirin, the reaction occurs at the hydroxyl group.In the preparation of oil of wintergreen, the reaction occurs at the carboxyl group.

(k) Succinylcholine is an inhibitor of acetylcholine, whereas neostigmine increases the effects of acetylcholine by inhibiting the enzyme acetylcholinesterase.

8.6 **(a)** CH$_3$CH$_2$C(=O)OH > CH$_3$CH$_2$CH$_2$CH$_2$C(=O)OH > C$_6$H$_5$C(=O)OH

(b) CH$_3$CH$_2$C(=O)OH > CH$_3$CH$_2$CHOHCH$_3$ > CH$_3$C(=O)OCH$_2$CH$_2$CH$_3$

(c) C$_6$H$_5$C(=O)O$^-$Na$^+$ > CH$_3$CH$_2$CH$_2$CH$_2$C(=O)OH > C$_6$H$_{11}$C(=O)OH

8.7 **(a)** C$_6$H$_5$C(=O)OH > CH$_3$CH$_2$CH$_2$CH$_2$C(=O)OH > CH$_3$CH$_2$C(=O)OH

(b) CH$_3$CH$_2$C(=O)OH > CH$_3$CH$_2$CHOHCH$_3$ > CH$_3$C(=O)OCH$_2$CH$_2$CH$_3$

(c) C$_6$H$_5$C(=O)O$^-$Na$^+$ > C$_6$H$_{11}$C(=O)OH > CH$_3$CH$_2$CH$_2$CH$_2$C(=O)OH

8.8

$$R-C(=O)OH + HOH \rightleftharpoons R-C(=O)O^- + H_3O^+$$

The strength of an acid depends upon the extent to which the acid dissociates. The greater the percent dissociation, the stronger the acid.

8.9 $RCOOH \rightleftharpoons RCOO^- + H^+$

$[RCOO^-] = [H^+] = (0.034)(0.10\,M) = 3.4 \times 10^{-3}\,M$

$$K_a = \frac{(3.4 \times 10^{-3})(3.4 \times 10^{-3})}{0.10 - (3.4 \times 10^{-3})} = 1.2 \times 10^{-4}$$

8.10 $CH_3COOH \rightleftharpoons CH_3COO^- + H^+$

$$K_a = \frac{[CH_3COO^-][H^+]}{[CH_3COOH]}$$

Let $x = [CH_3COO^-] = [H^+]$; then $1.0\,M - x = [CH_3COOH]$.

$$1.8 \times 10^{-5} = \frac{(x)(x)}{1.0 - x}$$

$$x = 4.2 \times 10^{-3} = [H^+]$$

$$[OH^-] = \frac{1.0 \times 10^{-14}}{4.2 \times 10^{-3}} = 2.4 \times 10^{-12}$$

$$pH = 3 - \log 4.2 = 3 - 0.62 = 2.4$$

8.11 $ClCH_2COOH \rightleftharpoons ClCH_2COO^- + H^+$

$$K_a = 1.4 \times 10^{-3} = \frac{[ClCH_2COO^-][H^+]}{[ClCH_2COOH]}$$

$$1.4 \times 10^{-3} = \frac{(x)(x)}{0.01 - x}$$

Solving the quadratic equation

$$x^2 + (1.4 \times 10^{-3})x - 1.4 \times 10^{-5} = 0$$

gives

$$x = [H^+] = 7.0 \times 10^{-4} \, M$$

$$pH = 4 - \log 7.0 = 3.15$$

8.12

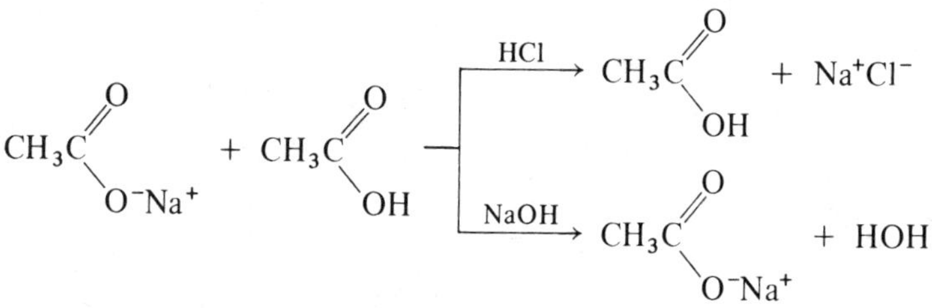

8.13 The hydrogen ion concentration of a 0.1 M solution of acetic acid is 1.3×10^{-3} (you can easily calculate this). Hence the pH of the solution is 2.9. The pH of a solution containing 0.1 M acetic acid and 0.1 M sodium acetate can be calculated from the Henderson–Hasselbalch equation.

$$pH = pK_a + \log \frac{[\text{anion}]}{[\text{acid}]}$$

$$= 4.7 + \log \frac{0.1}{0.1} = 4.7$$

In a solution of acetic acid, there is only one reaction with the solvent water molecules and that reaction tends to increase the hydrogen ion concentration.

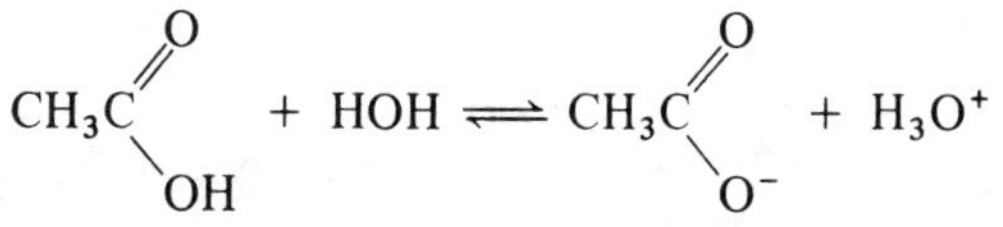

In a solution of acetic acid and sodium acetate, there are two reactions that tend to balance each other. The net effect is only a slight increase in the hydrogen ion concentration.

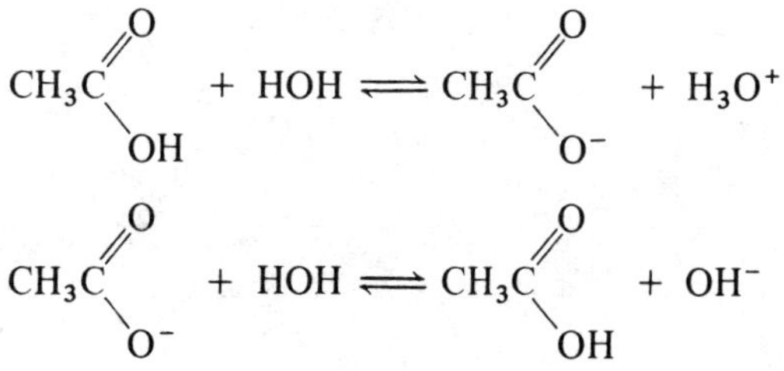

8.14 $H_2PO_4^- \longrightarrow HPO_4^{2-} + H^+$

$$pH = pK_a + \log \frac{[HPO_4{}^{2-}]}{[H_2PO_4{}^-]}$$

$$7.4 = 7.2 + \log \frac{[HPO_4{}^{2-}]}{[H_2PO_4{}^-]}$$

$$\log \frac{[HPO_4{}^{2-}]}{[H_2PO_4{}^-]} = 0.2$$

$$\frac{[HPO_4{}^{2-}]}{[H_2PO_4{}^-]} = \frac{1.6}{1}$$

8.15 (a) $CH_3CH_2CH_2COOH$ < $CH_2ClCH_2CH_2COOH$ < $CH_3CHClCH_2COOH$
       < $CH_3CH_2CHClCOOH$

(b) $(CH_3)_3CCOOH$ < $CH_3CH_2CH_2CH_2COOH$ < $CH_3CH_2CH_2CHClCOOH$
       < $CH_3CH_2CH_2CHFCOOH$

(c) CH_3CH_2COOH < CH_3COOH < $CH_3CHClCOOH$ < CH_3CCl_2COOH

(d) CH_3COOH < $CH_2ClCH_2CH_2COOH$ < $CH_2BrCOOH$ < $CHCl_2COOH$

(e) H_2O < $CH_3CH_2CH_2COOH$ < $CH_3CH_2CHClCOOH$ < CCl_3COOH

(f) CH_3OH < CH_3COOH < $HCOOH$ < CH_2FCOOH

(g) $CH_3CH_2C(CH_3)_2COOH$ < $CH_3CH_2CH(CH_3)COOH$ < $CH_3CH_2CH_2COOH$
       < $CH_3CH_2CHBrCOOH$

(h) [cyclohexyl]—OH < [phenyl]—OH < [cyclohexyl]—COOH < [phenyl]—COOH

(i) [phenyl]—COOH < $CH_3CHOHCOOH$ < $CH_3\overset{O}{\overset{\|}{C}}CH_2COOH$ < $CH_3\overset{O}{\overset{\|}{C}}COOH$

(j) CH_3COOH < [phenyl]—CH_2COOH < [phenyl]—COOH < $HCOOH$

8.16 (a) CH_3CH_2COOH (b) [phenyl]—COOH   (c) [phenyl]—COOH   (d) $HOOC$—[phenyl]—COOH

(e) no reaction   (f) CH_3CH_2CN; CH_3CH_2COOH (g) no reaction   (h) $CH_3C(CH_3)_2CH_2\overset{O}{\overset{\|}{C}}\underset{O^-Na^+}{}$

(i) CH_3—[phenyl]—$\overset{O}{\overset{\|}{C}}\underset{OH}{}$ (j) $CH_3CH=CH\overset{O}{\overset{\|}{C}}\underset{O^-Na^+}{}$ (k) [phenyl]—$CH_2CH_2CH_2\overset{O}{\overset{\|}{C}}\underset{Cl}{}$

(l) $2\ CH_3COOH$ (m) [phenyl]—$\overset{O}{\overset{\|}{C}}\underset{OCH_3}{}$ (n) $(CH_3)_2CH\overset{O}{\overset{\|}{C}}\underset{O^-K^+}{}$ + CH_3—[phenyl]—OH

(o) $CH_3CHBr\overset{O}{\overset{\|}{C}}\underset{OH}{}$ + $(CH_3)_2CHOH$ (p) $2\ CH_3CH_2OH$ (q) [phenyl with CH_2OH and Br]

(r) [phenyl with $\overset{O}{\overset{\|}{C}}$—$OCH_3$ and OH]   (s) [phenyl with $\overset{O}{\overset{\|}{C}}$—OH and O–C–CH_3 (O)]   (t) $CH_3\overset{O}{\overset{\|}{C}}\underset{OH}{}$ + $(CH_3)_3\overset{+}{N}CH_2CH_2OH$

8.17 **(a)** methyl salicylate   **(b)** acetylsalicylic acid   **(c)** formic acid   **(d)** acetic acid
(e) octyl acetate   **(f)** amyl acetate   **(g)** butyric acid   **(h)** oxalic acid   **(i)** citric acid
(j) lactic acid   **(k)** tartaric acid   **(l)** salicylic acid   **(m)** terephthalic acid
(n) tricresyl phosphate   **(o)** adenosine triphosphate   **(p)** acetylcholine

8.18 $RCOOH + ROH \rightleftharpoons RCOOR + HOH$

$$K = 3.5 = \frac{[RCOOR][HOH]}{[RCOOH][ROH]}$$

(a) shift to the right;   **(b)** no change;   **(c)** shift to the right;   **(d)** shift to the left

8.19 **(a)** $+ HOH$ **(b)** $+ H^{18}OH$

(c) $+ HOH$ **(d)** $+ {}^{14}CH_3OH$

(e)   **(f)** $-CH_2{}^{18}OH + CH_3OH$

(g) $(CH_3)_2CHCH_2{}^{18}OH +$ **(h)** $CH_3CH_2OH + $

8.20 **(a)** and **(c)** no reaction; **(b)** will react with water

8.21 **(a)** Add a solution of sodium bicarbonate to each. Evolution of bubbles of CO_2 identifies the acid.
(b) Treat both with an oxidizing agent such as $KMnO_4$. Only the aldehyde will be oxidized, as evidenced by a color change from purple to pink.
(c) Treat both with an oxidizing agent such as $K_2Cr_2O_7$ and heat. Only the alcohol will be oxidized, as evidenced by a color change from orange to green.
(d) Add a dilute $KMnO_4$ solution to each. The alkene will react to form a glycol; the acid will not react. Reaction is observed by decoloration of the permanganate solution.
(e) Add a solution of sodium bicarbonate to each. Evolution of bubbles of carbon dioxide identifies benzoic acid.

8.22 **(a)** 100 mL of 0.1 M CH_3COOH and 100 mL of 0.1 M $CH_3COO^-Na^+$
(b) $CH_3COOH + NaOH \longrightarrow CH_3COO^-Na^+ + HOH$

(c) $CH_3(CH_2)_{16}COO^-Na^+$ **(d)** $CH_3-\overset{O}{\overset{\|}{C}}-O-\overset{O}{\overset{\|}{C}}-CH_3$ **(e)** CH_3CN **(f)** CH_3NCO

(g)   **(h)**   **(i)**

(j) 100% acetic acid (CH_3COOH)   (k) $K^{+\ -}OOCCHOHCHOHCOOH$

(l) $CH_3C\overset{O}{\underset{OH}{<}}$ + CH_3OH $\overset{H^+}{\rightleftharpoons}$ $CH_3C\overset{O}{\underset{OCH_3}{<}}$ + HOH

(m) $CH_3C\overset{O}{\underset{OCH_3}{<}}$ + HOH $\overset{H^+}{\rightleftharpoons}$ $CH_3C\overset{O}{\underset{OH}{<}}$ + CH_3OH

(n) $CH_3C\overset{O}{\underset{OCH_3}{<}}$ + NaOH $\longrightarrow$ $CH_3C\overset{O}{\underset{O^-Na^+}{<}}$ + CH_3OH

(o) $CH_3C\overset{O}{\underset{OCH_3}{<}}$ + NH_3 $\longrightarrow$ $CH_3C\overset{O}{\underset{NH_2}{<}}$ + CH_3OH

(p) The space (gap) that separates two nerve cells (or nerve cells and muscle, or gland).

(q) $CH_3C\overset{O}{\underset{OCH_2CH_2\overset{+}{N}(CH_3)_3}{<}}$ (r) The enzyme that catalyzes the hydrolysis of acetylcholine.

(s) $(CH_3CH_2O)_2-\overset{S}{\underset{}{\overset{||}{P}}}-O-\bigcirc-NO_2$ (t) $CH_3\overset{H}{\underset{CH_3}{\overset{|}{C}}}-O-\overset{O}{\underset{F}{\overset{||}{P}}}-CH_3$

8.23 $CH_3CH_2CH_2CH_2C\overset{O}{\underset{OCH_2CH_3}{<}}$ $CH_3CH_2CH(CH_3)C\overset{O}{\underset{OCH_2CH_3}{<}}$

$(CH_3)_2CHCH_2C\overset{O}{\underset{OCH_2CH_3}{<}}$ $(CH_3)_3CC\overset{O}{\underset{OCH_2CH_3}{<}}$

8.24 $CH_3CH_2C\overset{O}{\underset{OCH_2CH_2CH_3}{<}}$

8.25 $RCOOH + NaOH \longrightarrow RCOO^-Na^+ + HOH$

moles of NaOH = (0.4)(0.125) = 0.0500 = moles of acid

molecular weight of acid = $\dfrac{5.1}{0.050}$ = 102

Empirical formula of acid, $C_5H_{10}O_2$; possible structural formulas are

$CH_3CH_2CH_2CH_2COOH$ $CH_3CH_2CH(CH_3)COOH$ $(CH_3)_2CHCH_2COOH$ $(CH_3)_3CCOOH$

8.26
$$CH_3C\overset{O}{\underset{OH}{<}} + CH_3OH \rightleftharpoons CH_3C\overset{O}{\underset{OCH_3}{<}} + HOH$$

moles of acetic acid $= \dfrac{3.0}{60} = 0.050$

moles of methyl acetate formed $= 0.050$
grams of methyl acetate $= (0.050)$(molecular weight of methyl acetate)
$= (0.050)(74) = 3.7$ g

8.27
$$2\ \underset{Cl}{\overset{Cl}{H\!C\!C}}\overset{O}{\underset{OH}{<}} + Ba(OH)_2 \longrightarrow \left(\underset{Cl}{\overset{Cl}{H\!C\!C}}\overset{O}{\underset{O^-}{<}}\right)_2 Ba^{2+} + 2\ HOH$$

moles of dichloroacetic acid $= \dfrac{0.5}{129} = 4 \times 10^{-3}$

moles of $Ba(OH)_2$ required to neutralize the acid $= 2 \times 10^{-3}$

volume of $Ba(OH)_2$ required $= \dfrac{2 \times 10^{-3}}{0.10} = 0.02$ L $= 20$ mL

CHAPTER 9

Amides and Amines

9.1

$$CH_3CH_2CH_2C(=O)NH_2$$
butanamide

$$CH_3CH(CH_3)C(=O)NH_2$$
2-methylpropanamide

$$CH_3CH_2C(=O)N(H)CH_3$$
N-methylpropanamide

$$CH_3C(=O)N(H)CH_2CH_3$$
N-ethylethanamide

$$H-C(=O)N(H)CH(CH_3)_2$$
N-isopropylmethanamide

$$CH_3C(=O)N(CH_3)CH_3$$
N,*N*-dimethylethanamide

$$H-C(=O)N(H)CH_2CH_2CH_3$$
N-propylmethanamide

$$H-C(=O)N(CH_3)CH_2CH_3$$
N-ethyl-*N*-methylmethanamide

9.2

$$CH_3CH_2CH_2CH_2NH_2$$
butylamine
(1-aminobutane)

$$CH_3CH_2CHNH_2CH_3$$
sec-butylamine
(2-aminobutane)

$$(CH_3)_3CNH_2$$
t-butylamine
(2-amino-2-methylpropane)

$$CH_3CH_2CH_2NHCH_3$$
methylpropylamine

$$(CH_3)_2CHNHCH_3$$
methylisopropylamine

$$(CH_3)_2CHCH_2NH_2$$
1-amino-2-methylpropane

$$CH_3CH_2NHCH_2CH_3$$
diethylamine

$$CH_3CH_2N(CH_3)_2$$
dimethylethylamine

9.3 (a) 2-methylbutanamide (b) methylamine (c) 3-methylbutanamide
(d) diethylamine (e) *N*-ethylethanamide (*N*-ethylacetamide) (f) 2-aminopentane
(g) *N*-ethyl-*N*-methylpropanamide (h) 1,3-diaminopropane (i) 3-hydroxybutanamide
(j) tetraethylammonium iodide (k) pyridine (l) *p*-bromobenzamide (4-bromobenzamide)
(m) cyclohexylamine (aminocyclohexane) (n) *N*,*N*-dimethylbenzamide
(o) *p*-ethylaniline (4-ethylaniline) (p) *p*-aminobenzoic acid (4-aminobenzoic acid)
(q) methylphenylamine (*N*-methylaniline) (r) pyrrole

9.4 (a) $CH_3C(=O)NH_2$ (b) $C_6H_5C(=O)NH_2$ (c) $CH_3CHOHCH_2C(=O)NH_2$ (d) $CH_3C(=O)N(CH_2CH_3)CH_2CH_3$

(e) $CH_3CH=CHCH_2CH(CH_3)C(=O)NH_2$ (f) $C_6H_5C(=O)N(CH_2CH_3)H$ (g) $C_6H_5CH_2CH_2C(=O)N(H)CH_3$

(h) $H_2N-CH_2C(=O)N(H)-C_6H_5$ (i) $(CH_3)_3N$ (j) $CH_3CH_2CHCH(NH_2)CH_3$ with OH   (k) $H_2N-CH_2CH_2CH_2CH_2-NH_2$

47

(l) ⟨benzene⟩—CH$_2$—$\overset{+}{\underset{CH_3}{\overset{CH_3}{N}}}$—CH$_3$ Br$^-$ (m) ◁—N$\underset{H}{}$—CH$_3$ (n) ⟨benzene⟩—N$\overset{CH_2CH_3}{\underset{CH_2CH_3}{}}$

(o) $\underset{Br\qquad\quad Br}{\overset{NH_2}{⟨benzene⟩}}$ (p) ⟨benzene⟩—N$\underset{H}{}$—⟨benzene⟩   (q) ⟨imidazole⟩   (r) ⟨indole⟩

9.5 A primary amine has only one organic group bonded to the nitrogen atom, a secondary amine contains two organic groups bonded to nitrogen, and in a tertiary amine all three hydrogens have been replaced by organic groups.

$$R—\underset{H}{N}—H \qquad\qquad R—\underset{H}{N}—R \qquad\qquad R—\underset{R}{N}—R$$

primary amine   secondary amine   tertiary amine

9.6 $CH_3\overset{O}{\underset{Cl}{C}}$ + NH_3 ⟶ $CH_3\overset{O}{\underset{NH_2}{C}}$

$CH_3\overset{O}{\overset{\|}{C}}—O—\overset{O}{\overset{\|}{C}}CH_3$ + NH_3 ⟶ $CH_3\overset{O}{\underset{NH_2}{C}}$

9.7 $CH_3—C\equiv N$ + $LiAlH_4 \xrightarrow{ether} CH_3CH_2NH_2$

$CH_3—\overset{O}{\underset{NH_2}{C}}$ + $LiAlH_4 \xrightarrow{ether} CH_3CH_2NH_2$

9.8 Amides, with the exception of formamide, are all solids. Most are colorless and odorless. The lower members are soluble in both water and alcohol, water solubility decreasing as the carbon chain increases. Amides are neutral compounds with abnormally high boiling points and melting points.

Amines exist as gases, liquids, or solids. All have pronounced odors and are basic. The lower members are soluble in water; water solubility decreases with increasing length of carbon chain.

9.9 Intermolecular hydrogen bonding is much more extensive between ethyl alcohol molecules than between ethylamine molecules (a result of the greater electronegativity of oxygen). More energy is required to separate the individual ethyl alcohol molecules, hence the higher boiling point.

9.10 Trimethylamine is a tertiary amine and thus it lacks a hydrogen bonded to nitrogen. There is no intermolecular hydrogen bonding between trimethylamine molecules. Hydrogen bonding does occur between propylamine molecules.

9.11 (a) $CH_3CH_2CH_2\overset{O}{\underset{NH_2}{C}}$ > $CH_3CH_2CH_2\overset{O}{\underset{\underset{H}{N}}{C}}CH_3$ > $CH_3CH_2CH_2\overset{O}{\underset{\underset{CH_3}{N}}{C}}CH_3$

(b) $CH_3\overset{\displaystyle O}{\overset{\|}{C}}$ $>$ $CH_3CH_2CH_2CH_2\overset{\displaystyle O}{\overset{\|}{C}}$ $>$ (benzene ring)$\overset{\displaystyle O}{\overset{\|}{C}}$
$\quad\quad\quad\;\;NH_2$ $\quad\quad\quad\quad\quad\quad\quad\;\;NH_2$ $\quad\quad\quad\quad\quad\quad\;\;NH_2$

(c) $CH_3CH_2NH_2$ $>$ $(CH_3CH_2)_2NH$ $>$ (benzene ring)$-NH_2$

(d) $(CH_3CH_2)_4N^+Cl^-$ $>$ $CH_3CH_2CH_2CH_2\overset{\displaystyle O}{\overset{\|}{C}}$ $>$ $[(CH_3)_2CH]_3N$
$\quad\quad\quad\quad\quad\quad\quad\quad\quad\quad\quad\quad\quad\quad\quad\;\;NH_2$

9.12 (a) $CH_3CH_2CH_2\overset{\displaystyle O}{\overset{\|}{C}}$ $>$ $CH_3\overset{\displaystyle O}{\overset{\|}{C}}$ $>$ $CH_3\overset{\displaystyle O}{\overset{\|}{C}}$
$\quad\quad\quad\quad\quad\quad\;\;NH_2$ $\quad\quad\quad\;\;NH_2$ $\quad\quad\;N(CH_3)_2$

(b) $CH_3\overset{\displaystyle O}{\overset{\|}{C}}$ $>$ $H_2NCH_2CH_2NH_2$ $>$ $CH_3CH_2CH_2NH_2$
$\quad\quad\quad\;NHCH_3$

(c) $CH_3CH_2CH_2NH_2$ $>$ $CH_3CH_2NHCH_3$ $>$ $(CH_3)_3N$

(d) (benzene ring)$\overset{\displaystyle O}{\overset{\|}{C}}$ $>$ (benzene ring)$-NH_2$ $>$ $CH_3(CH_2)_3NH_2$
$\quad\quad\quad\quad\;NH_2$

9.13 The oxygen of the amide can form hydrogen bonds with the hydrogens of water molecules. *N,N*-Dimethylformamide is a fairly small, compact molecule that can readily be surrounded by water molecules.

9.14 $CH_3\overset{..}{N}HCH_3 + HOH \rightleftharpoons CH_3\overset{+}{N}H_2CH_3 + OH^-$

9.15 The presence of the nitrogen atom with its unshared pair of electrons makes pyridine a polar molecule. The unshared electron pair can hydrogen bond to the hydrogens of water molecules, and thus water molecules are attracted to, and can surround pyridine molecules. Benzene is a nonpolar hydrocarbon. There is very little attraction between benzene and water molecules.

9.16 (a) $CH_3CH_2CH_2NH_2$ $>$ $(CH_3)_3N$ $>$ $CH_3CH_2\overset{\displaystyle O}{\overset{\|}{C}}$
$\quad\quad\quad\quad\quad\quad\quad\quad\quad\quad\quad\quad\quad\quad\quad\quad\;\;NH_2$

(b) NH_3 $>$ (benzene ring)$-NH_2$ $>$ $CH_3\overset{\displaystyle O}{\overset{\|}{C}}$
$\quad\quad\quad\quad\quad\quad\quad\quad\quad\quad\quad\quad\quad\quad\;NH_2$

(c) $(CH_3)_2NH$ $>$ $(CH_3CH_2)_2NH$ $>$ $(CH_3CH_2)_3N$

(d) (pyridine ring) $>$ (benzene ring)$-NH_2$ $>$ (benzene ring)$-\overset{H}{\overset{|}{N}}-$(cyclopentane ring)

(e) (imidazole ring) $>$ (pyridine ring) $>$ (pyrrole ring)

(f) $CH_3CH_2NH_2$ > $CH_3CHClNH_2$ > $CH_3CH_2NH_3^+Cl^-$

(g) ⟨◯⟩—CH_2NH_2 > ⟨◯⟩—NH_2 > ⟨◯⟩—C(=O)—NH_2

9.17 Dissociation of the hydrogen is facilitated by resonance stabilization of the anion.

9.18 (a) ⟨◯⟩—C(=O)—NH_2 (b) H—C(=O)—OH + $CH_3NH_3^+$ Cl^- (c) $CH_3(CH_2)_5NH_3^+$ Br^-

(d) CH_3C(=O)—O^-Na^+ + $(CH_3CH_2)_2NH$ (e) $[H_3N^+CH_2CH_2CH_2CH_2NH_3^+]SO_4^{2-}$

(f) $CH_3N^+(CH_2CH_3)_3$ Br^- (g) ⟨◯⟩—$CH_2CH_2NH_2$ (h) $(CH_3)_2CH-N(CH_3)-N=O$

(i) H_2N—⟨◯⟩—NH_2 (j) CH_3C(=O)—$N(CH_2CH_3)(CH_2CH_3)$ (k) ⟨◯⟩—CH_2NH_2 (l) [pyridinium] N^+—CH_3 I^-

9.19 (a) CH_3—C(=O)—$N(CH_3)(H)$

(b) The bond between the carbonyl carbon and the nitrogen.
(c) A polymer formed from the condensation polymerization of adipic acid and hexamethylenediamine.

(d) $H_2N-C(=O)-NH_2$ (e) The $-NH_2$ group   (f) $H_2N-(CH_2)_5-NH_2$
(g) $(CH_3)_4 N^+ Cl^-$ (h) The reaction of a compound with nitrous acid.   (i) $(CH_3)_2 N-N=O$
(j) A ring compound containing a nitrogen atom within the ring.
(k) Any chemical substance that affects an individual in such a way as to bring about physiological, emotional, or behavorial change.
(l) Any nitrogen-containing compound that is obtained from plants and that has physiological activity.
(m) A substance that alleviates pain.
(n) A substance that has a soothing, calming effect.
(o) Substances that produce physical addiction.
(p) Term used to signify the body's ability to adapt to a drug.
(q) A feeling of great happiness or well-being.
(r) Pentapeptides that are neurotransmitters.
(s) Polypeptides that are neurotransmitters.
(t) Drugs that have sedative and hypnotic effects.
(u) A substance that induces sleep.
(v) A slang term for marijuana.
(w) Synthetic stimulants.
(x) Certain neurotransmitters, e.g., epinephrine

(y) The altered physiological state produced by repeated taking of a drug, so that, on discontinuing use of the drug, certain withdrawal symptoms occur.

(z) An uncontrollable desire for a drug.

(aa) Drugs that stimulate sensory perceptions that have no basis in physical reality.

(bb) An inhibitory neurotransmitter.

(cc) A substance that causes birth defects.

(dd) Drugs that are taken (1) to modify psychotic behavior without inducing sleep or (2) to reduce anxiety, excitement, and restlessness.

(ee) Valium and Librium are examples.

(ff) Synthetic drugs that are analogs of compounds that have some proven pharmacological activity.

9.20 (a) 15 **(b)** 4 **(c)** 5 **(d)** 9 **(e)** 7 **(f)** 11 **(g)** 8 **(h)** 1 **(i)** 10 **(j)** 3
(k) 6 **(l)** 14 **(m)** 16 **(n)** 17 **(o)** 2 **(p)** 18 **(q)** 12 **(r)** 13

9.21 In meperidine the ester linkage occurs to the right of the carbonyl group (i.e., it is an ethyl ester). In MPPP the ester linkage occurs to the left of the carbonyl group (i.e., it is a complex ester of propanoic acid).

9.22

9.23 $CH_3CH_2CH_2CH_2NH_2 + HCl \longrightarrow CH_3CH_2CH_2CH_2NH_3^+Cl^-$

$$\text{moles of butylamine} = \frac{0.25}{73} = 0.0034 \text{ mole}$$

$$\text{volume of HCl required} = \frac{0.0034}{0.15} = 0.023 \text{ L} = 23 \text{ mL}$$

9.24 $RNH_2 + HOH \rightleftharpoons RNH_3^+ + OH^-$

$$1.0 \times 10^{-6} = \frac{[RNH_3^+][OH^-]}{[RNH_2]}$$

$$1.0 \times 10^{-6} = \frac{(x)(x)}{0.030 - x}$$

Assume x is small compared to 0.030; then

$$x^2 = 3.0 \times 10^{-8}$$
$$x = [OH^-] = 1.7 \times 10^{-4}$$
$$pOH = 3.77$$
$$pH = 10.23$$

9.25 $pK_b = 8.7$. Therefore $K_b = 2.0 \times 10^{-9}$.

$$2.0 \times 10^{-9} = \frac{x^2}{0.1}$$
$$x^2 = 2.0 \times 10^{-10}$$
$$x = [OH^-] = 1.4 \times 10^{-5}$$
$$pOH = 4.85$$
$$pH = 9.15$$

CHAPTER 10

Stereoisomerism

10.1 (a) Isomers that have the same structural formulas, but differ in the arrangement of their atoms.
    (b) Stereoisomers that can be interconverted only by the breaking and reformation of bonds.
    (c) Stereoisomers that can be interconverted by rotation about a bond.
    (d) An instrument that detects and measures the effects of various substances upon plane-polarized light.
    (e) Light vibrating in a single plane.
    (f) Any substance that rotates plane-polarized light.
    (g) The amount of rotation caused by an optically active substance.
    (h) The amount of rotation caused by 1 g of an optically active substance per cubic centimeter in a 1 dm sample tube.
    (i) Rotates plane-polarized light in a counterclockwise direction.
    (j) Rotates plane-polarized light in a clockwise direction.
    (k) Stereoisomers that are nonsuperimposable mirror images.
    (l) Stereoisomers that are not enantiomers.
    (m) A molecule that is not symmetric.
    (n) A mixture containing equal amounts of the two members of a pair of enantiomers.
    (o) Compounds that have different configurations because of the presence of a rigid structure in the molecule.
    (p) A compound that is optically inactive because of internal compensation.
    (q) Chemicals that are used for communication between members of the same species of insects.
    (r) Isomerization that is caused by the energy of light.

10.2 The compound must contain chiral (asymmetrical) molecules.

10.3 3-methylhexane and 2,3-dimethylpentane

10.4 (a)

$$CH_3CH_2-\overset{\overset{\displaystyle H}{|}}{\underset{\underset{\displaystyle Cl}{|}}{C}}-CH_3 \qquad CH_3CH_2-\overset{\overset{\displaystyle Cl}{|}}{\underset{\underset{\displaystyle H}{|}}{C}}-CH_3$$

(b)

$$CH_3CH_2-\overset{\overset{\displaystyle Br}{|}}{\underset{\underset{\displaystyle H}{|}}{C}}-CH_2Br \qquad CH_3CH_2-\overset{\overset{\displaystyle H}{|}}{\underset{\underset{\displaystyle Br}{|}}{C}}-CH_2Br \qquad CH_3-\overset{\overset{\displaystyle H}{|}}{\underset{\underset{\displaystyle Br}{|}}{C}}-\overset{\overset{\displaystyle Br}{|}}{\underset{\underset{\displaystyle H}{|}}{C}}-CH_3$$

$$CH_3-\overset{\overset{\displaystyle Br}{|}}{\underset{\underset{\displaystyle H}{|}}{C}}-CH_2CH_2Br \qquad CH_3-\overset{\overset{\displaystyle H}{|}}{\underset{\underset{\displaystyle Br}{|}}{C}}-CH_2CH_2Br \qquad CH_3-\overset{\overset{\displaystyle Br}{|}}{\underset{\underset{\displaystyle H}{|}}{C}}-\overset{\overset{\displaystyle H}{|}}{\underset{\underset{\displaystyle Br}{|}}{C}}-CH_3$$

10.5 (a) $CH_3CH_2CH_2CH_3$ and $CH_3CH(CH_3)_2$

(b) CH_3CH_2OH and $CH_3–O–CH_3$

(c) and

(d)
$$
\begin{array}{c}
CH_3 \\
| \\
H—C—OH \\
| \\
CH_2OH
\end{array}
\quad \text{and} \quad
\begin{array}{c}
CH_3 \\
| \\
HO—C—H \\
| \\
CH_2OH
\end{array}
$$

(e) same as (d)

(f)
$$
\begin{array}{c}
CH_3 \\
| \\
H—C—OH \\
| \\
H—C—Br \\
| \\
CH_3
\end{array}
\quad \text{and} \quad
\begin{array}{c}
CH_3 \\
| \\
H—C—OH \\
| \\
Br—C—H \\
| \\
CH_3
\end{array}
$$

(g)
$$
\begin{array}{c}
CH_3 \quad\quad CH_3 \\
\diagdown \quad / \\
C=C \\
/ \quad\quad \diagdown \\
H \quad\quad\quad H
\end{array}
\quad \text{and} \quad
\begin{array}{c}
CH_3 \quad\quad H \\
\diagdown \quad / \\
C=C \\
/ \quad\quad \diagdown \\
H \quad\quad\quad CH_3
\end{array}
$$

(h) and

10.6 Enantiomers have the same physical and chemical properties. They rotate plane-polarized light the same number of degrees *but* in opposite directions.

10.7 A racemic mixture is a *mixture* of *equal* amounts of two enantiomers. A meso compound is a single substance. Both a racemic mixture and a meso compound are optically inactive.

10.8 (a) 2   (b) 0   (c) 3   (d) 0   (e) 1   (f) 1

10.9 (a), (c), (d), (e), (g), (i)

10.10 (a)
$$
\begin{array}{c}
H \\
| \\
CH_3—C—CH_2CH_3 \\
| \\
Br
\end{array}
\quad\quad\quad
\begin{array}{c}
Br \\
| \\
CH_3—C—CH_2CH_3 \\
| \\
H
\end{array}
$$

(b)
$$
\begin{array}{c}
Br \\
| \\
CH_3CH_2—C—I \\
| \\
H
\end{array}
\quad\quad\quad
\begin{array}{c}
H \\
| \\
CH_3CH_2—C—I \\
| \\
Br
\end{array}
$$

(e)
$$
\begin{array}{c}
H \\
| \\
CH_3CH_2—C—COOH \\
| \\
NH_2
\end{array}
\quad\quad\quad
\begin{array}{c}
NH_2 \\
| \\
CH_3CH_2—C—COOH \\
| \\
H
\end{array}
$$

(f)

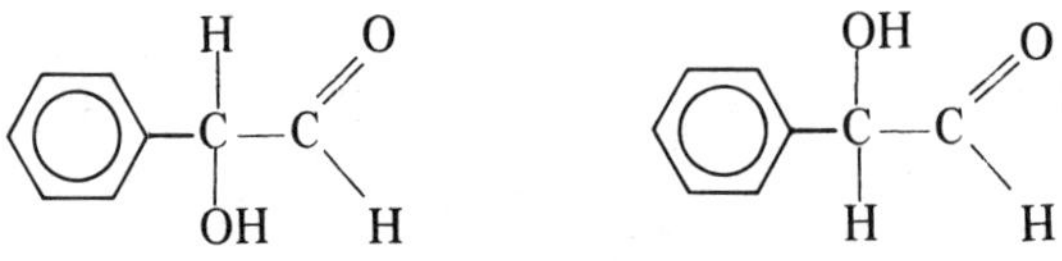

$C_6H_5-\overset{\displaystyle H}{\underset{\displaystyle OH}{C}}-\overset{\displaystyle O}{\overset{\|}{C}}-H$ $C_6H_5-\overset{\displaystyle OH}{\underset{\displaystyle H}{C}}-\overset{\displaystyle O}{\overset{\|}{C}}-H$

(h) $CH_3CH_2\overset{\displaystyle O}{\overset{\|}{C}}-\overset{\displaystyle H}{\underset{\displaystyle Cl}{C}}-CH_3$ $CH_3CH_2\overset{\displaystyle O}{\overset{\|}{C}}-\overset{\displaystyle Cl}{\underset{\displaystyle H}{C}}-CH_3$

(i) (cyclopentene with CH_3 and H substituents)   (cyclopentene with H and CH_3 substituents)

(k) $C_6H_{11}-\overset{\displaystyle H}{\underset{\displaystyle Br}{C}}-CH_3$ $C_6H_{11}-\overset{\displaystyle Br}{\underset{\displaystyle H}{C}}-CH_3$

(l)

CH_3	CH_3	CH_3	CH_3
$H-C-I$	$H-C-I$	$I-C-H$	$I-C-H$
CH_2	CH_2	CH_2	CH_2
$H-C-Br$	$Br-C-H$	$H-C-Br$	$Br-C-H$
CH_3	CH_3	CH_3	CH_3

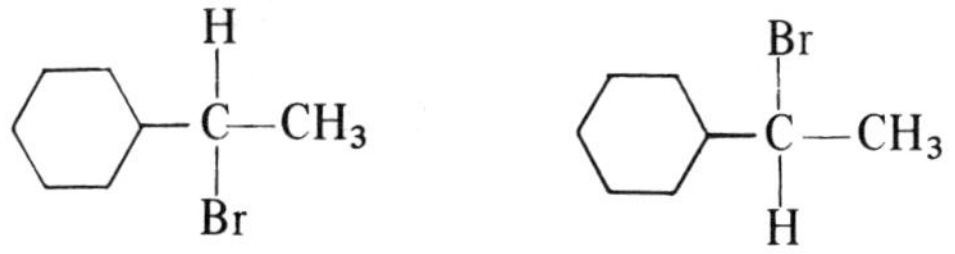

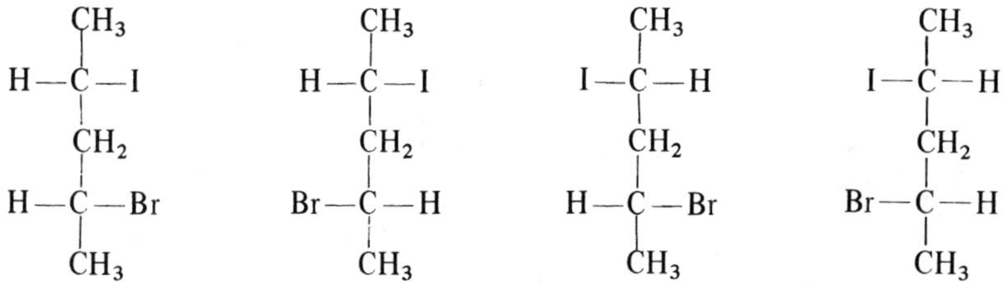

(m)

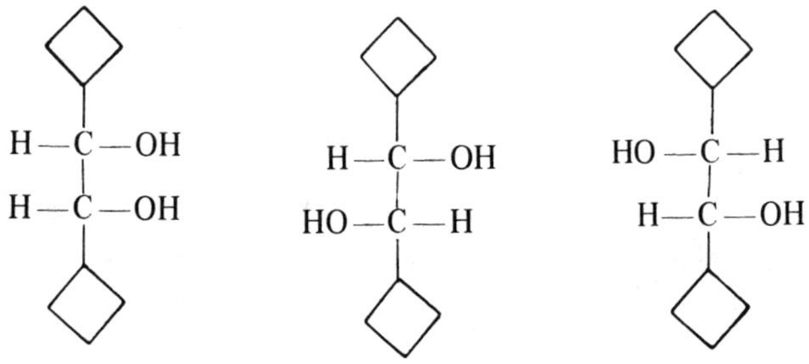

(cyclobutyl)	(cyclobutyl)	(cyclobutyl)
$H-C-OH$	$H-C-OH$	$HO-C-H$
$H-C-OH$	$HO-C-H$	$H-C-OH$
(cyclobutyl)	(cyclobutyl)	(cyclobutyl)

(n)

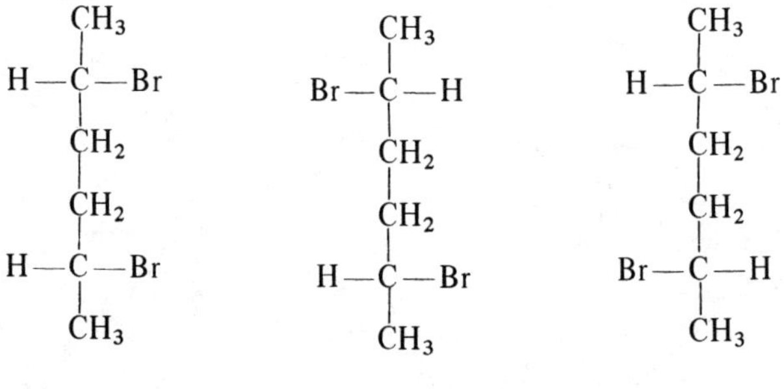

CH_3	CH_3	CH_3
$H-C-Br$	$Br-C-H$	$H-C-Br$
CH_2	CH_2	CH_2
CH_2	CH_2	CH_2
$H-C-Br$	$H-C-Br$	$Br-C-H$
CH_3	CH_3	CH_3

(o)

CH_3	CH_3	CH_3
$H-C-OH$	$H-C-OH$	$HO-C-H$
$H-C-CH_3$	$H-C-CH_3$	$H-C-CH_3$
$H-C-OH$	$HO-C-H$	$H-C-OH$
CH_3	CH_3	CH_3

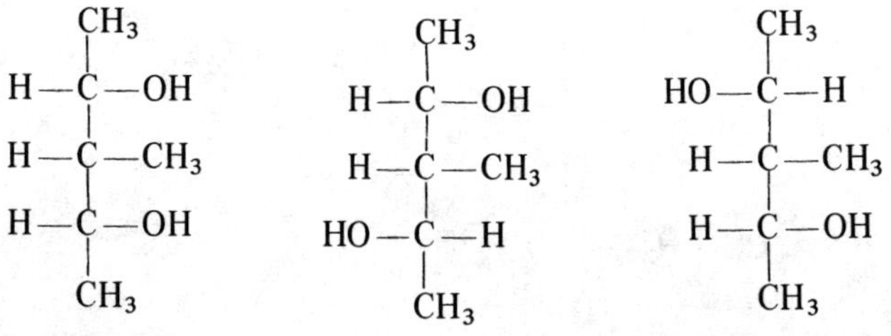

10.11 (a)

$$CH_2\!\!=\!\!CH-\underset{\overset{|}{Br}}{\overset{\overset{H}{|}}{C}}-CH_3 \qquad\qquad CH_2\!\!=\!\!CH-\underset{\overset{|}{H}}{\overset{\overset{Br}{|}}{C}}-CH_3$$

enantiomers, both optically active

(b)

$$CH_3-\underset{\overset{|}{OH}}{\overset{\overset{H}{|}}{C}}-\bigcirc \qquad\qquad CH_3-\underset{\overset{|}{H}}{\overset{\overset{OH}{|}}{C}}-\bigcirc$$

enantiomers, both optically active

(c)

$$CH_3CH_2CH_2-\underset{\overset{|}{I}}{\overset{\overset{H}{|}}{C}}-CH_2I \qquad\qquad CH_3CH_2CH_2-\underset{\overset{|}{H}}{\overset{\overset{I}{|}}{C}}-CH_2I$$

enantiomers, both optically active

(d)

$$CH_3CH_2-\underset{\overset{|}{CH_3}}{\overset{\overset{H}{|}}{C}}-CH_2OH \qquad\qquad CH_3CH_2-\underset{\overset{|}{H}}{\overset{\overset{CH_3}{|}}{C}}-CH_2OH$$

enantiomers, both optically active

(e)

$$(CH_3)_2CH-\underset{\overset{|}{H}}{\overset{\overset{OH}{|}}{C}}-CH_3 \qquad\qquad (CH_3)_2CH-\underset{\overset{|}{OH}}{\overset{\overset{H}{|}}{C}}-CH_3$$

entaniomers, both optically active

(f)

COOH	COOH	COOH
H—C—Cl	H—C—Cl	Cl—C—H
H—C—Cl	Cl—C—H	H—C—Cl
COOH	COOH	COOH
I	II	III

II and III are enantiomers and are optically active. I is a meso compound
and is optically inactive. I and II, I and III are diastereomers.

(g)

CH₂CH₃	CH₂CH₃	CH₂CH₃
HO—C—CH₃	HO—C—CH₃	H₃C—C—OH
HO—C—CH₃	H₃C—C—OH	HO—C—CH₃
CH₂CH₃	CH₂CH₃	CH₂CH₃
I	II	III

Same as responses to (f).

(h)

CH₃	CH₃	CH₃
H—C—Br	H—C—Br	Br—C—H
CH₂	CH₂	CH₂
H—C—Br	Br—C—H	H—C—Br
CH₃	CH₃	CH₃
I	II	III

Same as responses to (f).

(i)

```
     CH3            CH3            CH3            CH3
 H—C—Cl        Cl—C—H         H—C—Cl        Cl—C—H
 H—C—Cl        Cl—C—H         Cl—C—H        H—C—Cl
    COOH           COOH           COOH           COOH
     I              II             III            IV
```

All isomers are optically active; enantiomers: I and II, III and IV; diastereomers: I and III, I and IV, II and III, II and IV.

(j)

```
     CH3            CH3            CH3            CH3
 H—C—Br        H—C—Br         H—C—Br        Br—C—H
 H—C—I         I—C—H          H—C—I         I—C—H
 H—C—Br        H—C—Br         Br—C—H        H—C—Br
    CH3            CH3            CH3            CH3
     I              II             III            IV
```

I and II are both meso forms and each is optically inactive; III and IV are enantiomers and are optically active; I and II, I and III, I and IV, II and III, II and IV are diastereomers.

(k)

```
   H   O         H   O         H   O         H   O
    \ //          \ //          \ //          \ //
     C             C             C             C
    CH2           CH2           CH2           CH2
 H—C—OH       HO—C—H        H—C—OH    HO—C—H
 H—C—OH       HO—C—H        HO—C—H    H—C—OH
    CH3           CH3           CH3           CH3
     I             II            III           IV
```

Same as responses to (i).

(l)

```
    CH3           CH3           CH3           CH3
    C=O           C=O           C=O           C=O
 H—C—OH       HO—C—H        H—C—OH    HO—C—H
 H—C—OH       HO—C—H        HO—C—H    H—C—OH
    CH3           CH3           CH3           CH3
     I             II            III           IV
```

Same as responses to (i).

10.12

```
    CH2OH          CH2OH          CH2OH          CH2OH
 H—C—OH        H—C—OH         H—C—OH        HO—C—H
 H—C—OH        HO—C—H         H—C—OH        HO—C—H
 H—C—OH        H—C—OH         HO—C—H        H—C—OH
    CH2OH          CH2OH          CH2OH          CH2OH
     I              II             III            IV
```

Enantiomers: III and IV; diastereomers: I and II, I and III, I and IV, II and III, II and IV; meso forms: I and II.

10.13

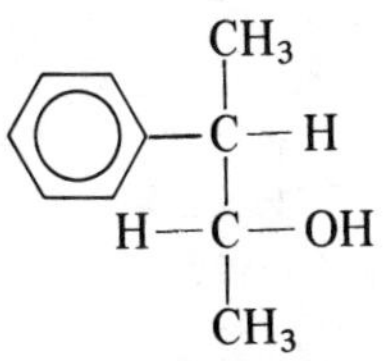

bp(25 mmHg) = 118°C
$[\alpha]_D^{25}$ = -30.9°

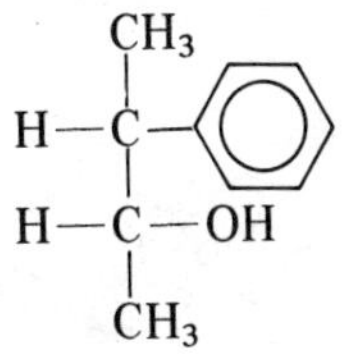

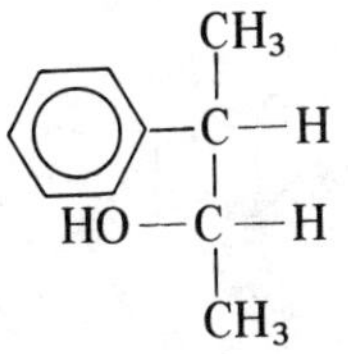

These two isomers are diastereomers of
of the other two. Therefore they will
have different boiling points and different
specific rotations.

10.14 (a)

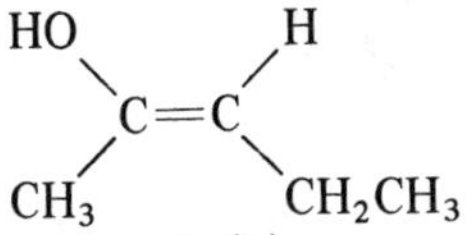

cis (E)

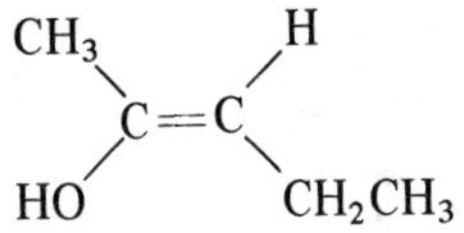

trans (Z)

(b)

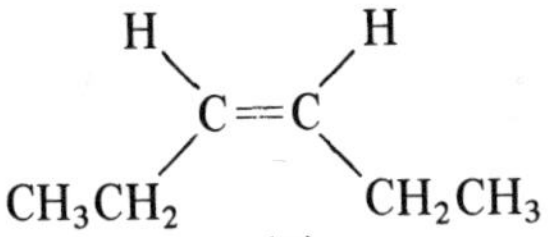

cis (Z)

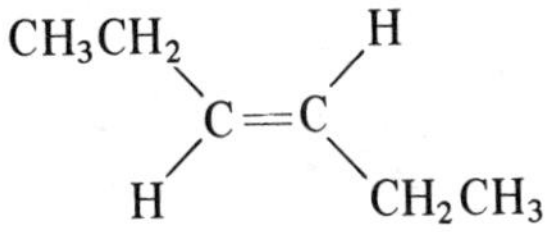

trans (E)

(c) none

(d) none

(e)

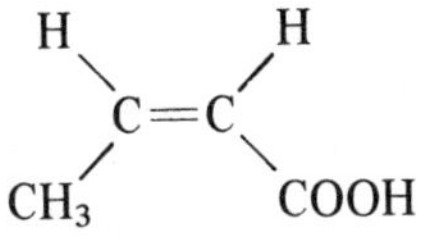

cis (Z)

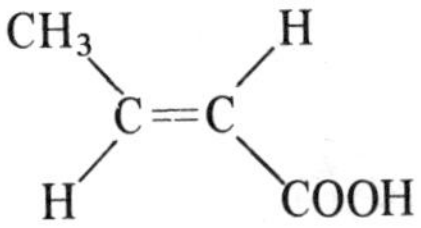

trans (E)

(f)

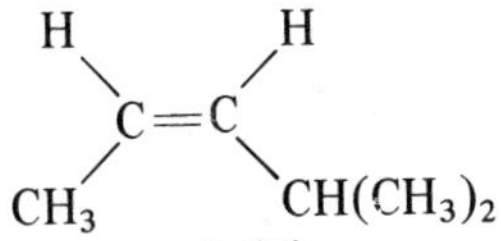

cis (Z)

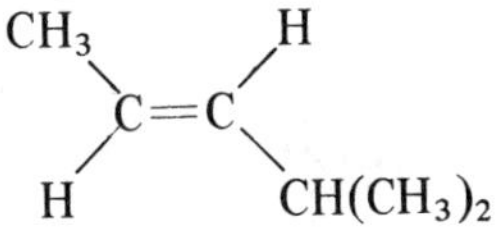

trans (E)

(g) none

(h) none

(i) none

(j)

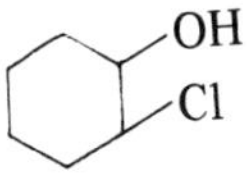

cis (Z)

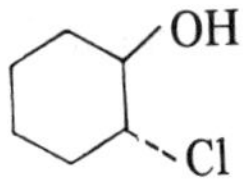

trans (E)

(k)

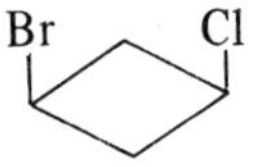

cis (Z)

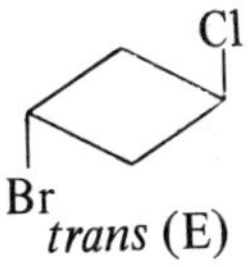

trans (E)

(l)

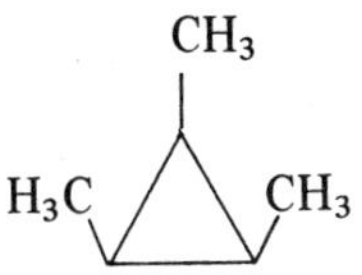

cis (Z)

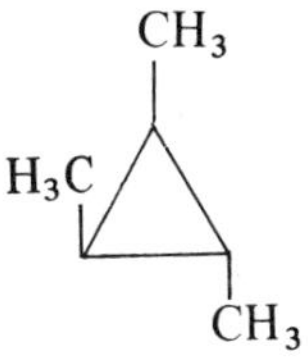

trans (E)

10.15

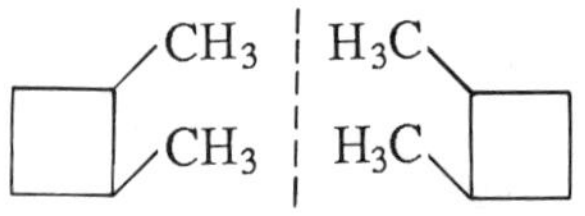

mirror

cis-1,2-Dimethylcyclobutane is achiral.
The mirror images are identical.

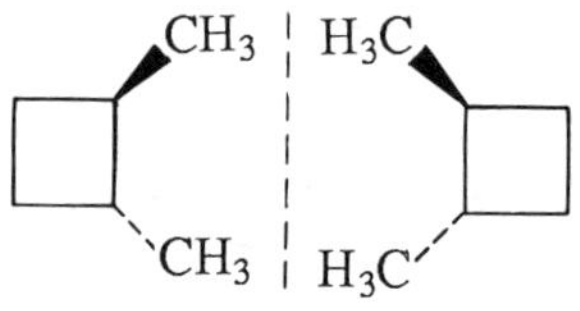

mirror

trans-1,2-Dimethylcyclobutane is chiral.
The mirror images are not identical.

cis- and *trans*-1,2-Dimethylcyclobutanes are diastereomers.

10.16

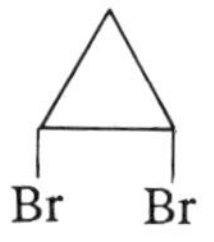

meso

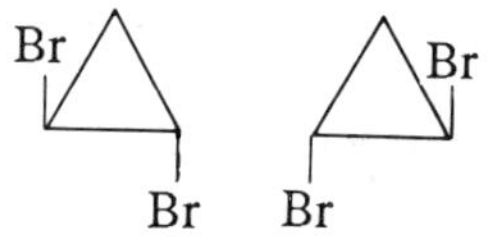

enantiomers

10.17 (a)

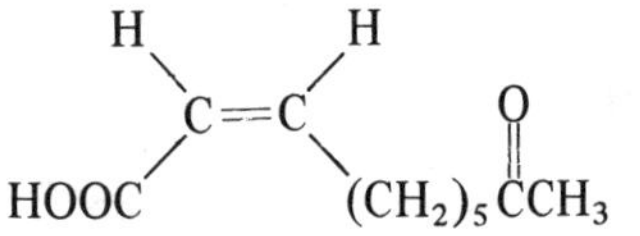

9-oxo-*cis*-2-decenoic acid

(b)

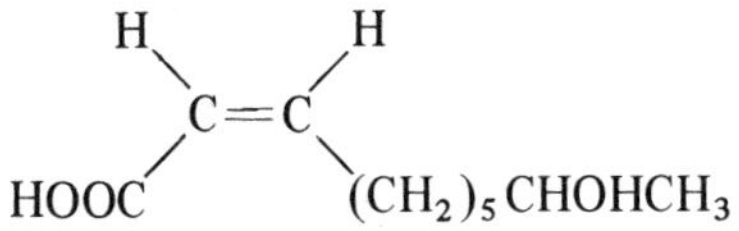

9-hydroxy-*cis*-2-decenoic acid

(c)

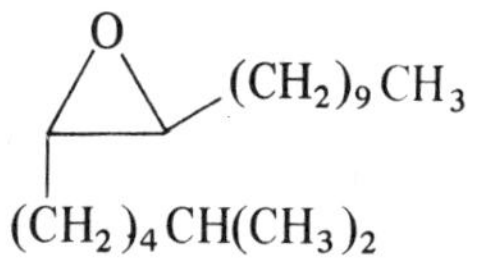

trans-7,8-epoxy-2-methyloctadecane

(d)

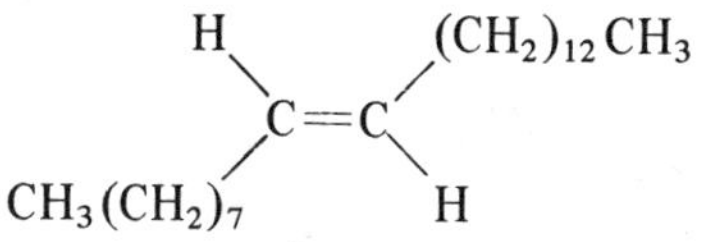

trans-9-tricosene

(e)

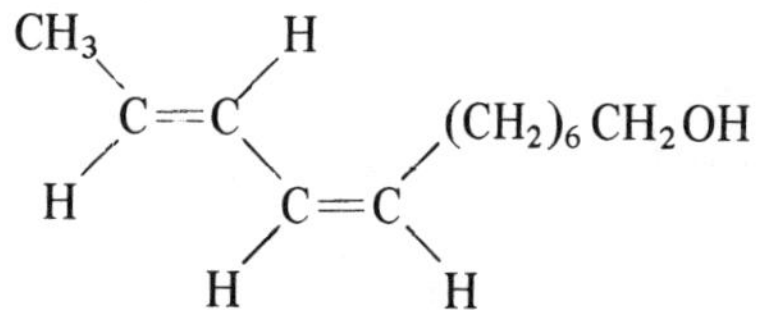

cis-8-*trans*-10-dodecadien-1-ol

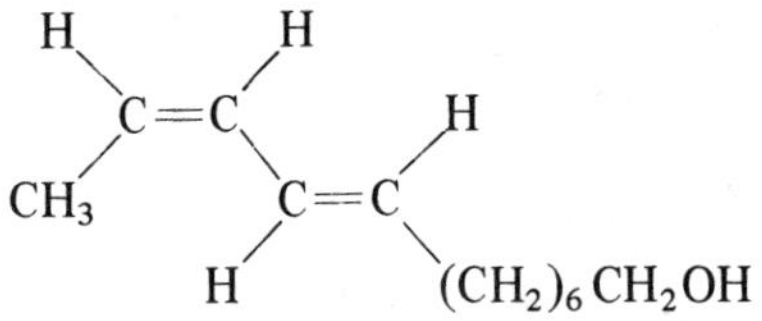

trans-8-*cis*-10-dodecadien-1-ol

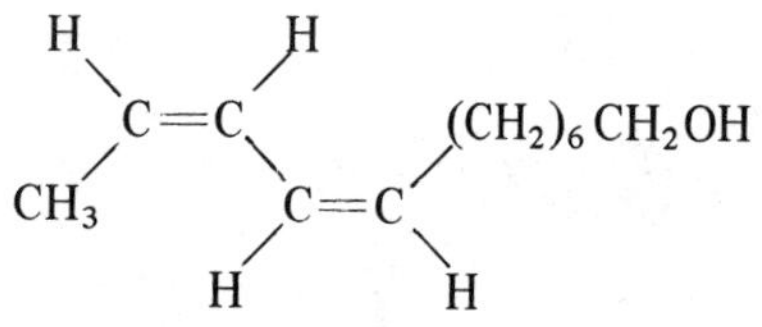

cis-8-*cis*-10-dodecadien-1-ol

(f)

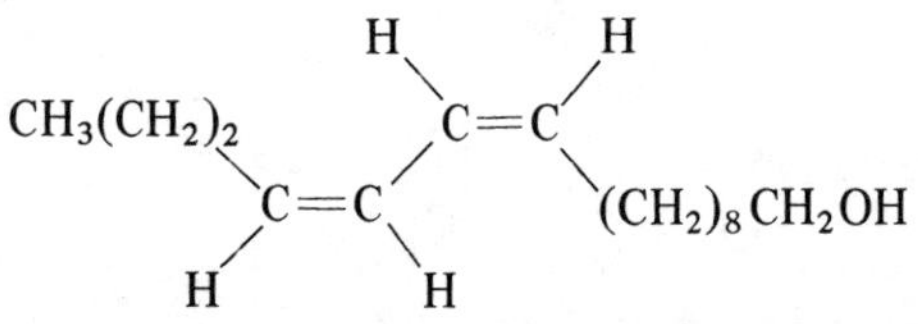

cis-10-*cis*-12-hexadecadien-1-ol

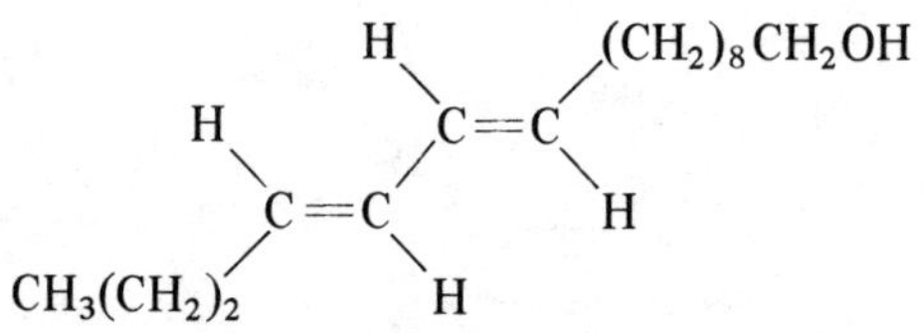

trans-10-*trans*-12-hexadecadien-1-ol

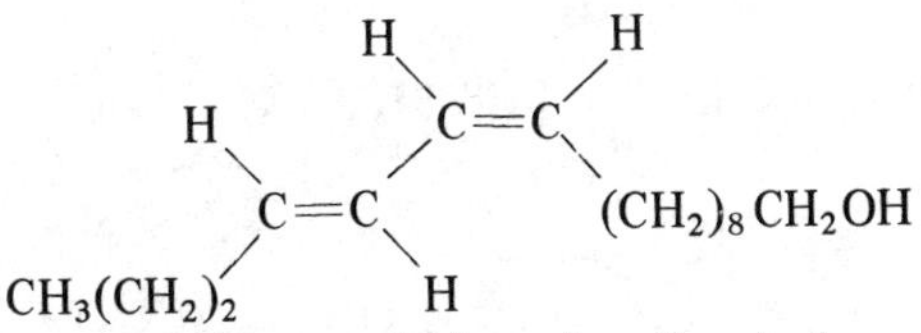

H
|
C=C
/   \
H   H
|
C=C
/   \
CH₃(CH₂)₂   H   (CH₂)₈CH₂OH

cis-10-*trans*-12-hexadecadien-1-ol

10.18 HO   OH
(structure) —(CH₂)₇
C=C
H   H
/
C=C
H
CH₂   CH₂CH₂CH₃

HO   OH
—(CH₂)₇
C=C
H   H
CH₂CH₂CH₃
/
C=C
H
CH₂   H

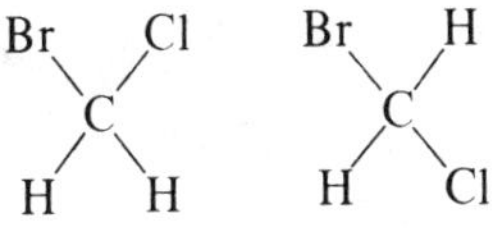

HO   OH
—(CH₂)₇
H   H   H   H
C=C   C=C
CH₂   CH₂CH₂CH₃

HO   OH
—(CH₂)₇
H   H   H   CH₂CH₂CH₃
C=C   C=C
CH₂   H

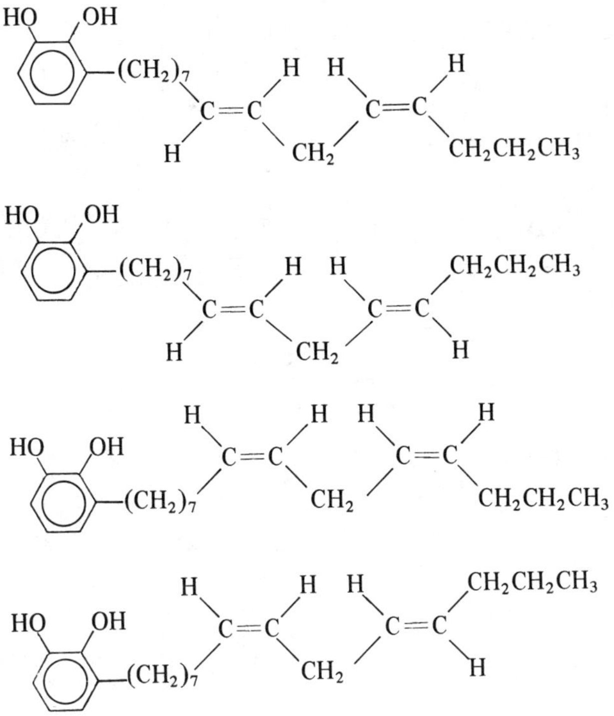

10.19 Two:
Br   Cl   Br   H
\ /   \ /
C   C
/ \   / \
H   H   H   Cl

10.20 CH₃CH₂   CH₃   CH₃CH₂   H
\   /   \   /
C=C   C=C
/   \   /   \
H   H   H   CH₃

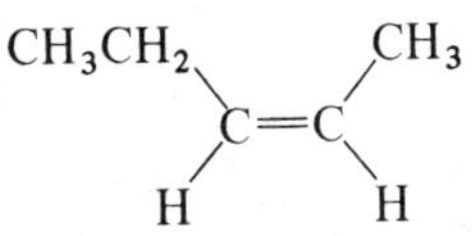

cis-2-pentene 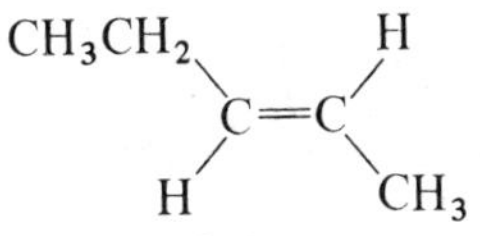   *trans*-2-pentene

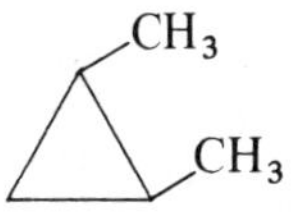

CH₃   CH₃
(cyclopropane)   (cyclopropane)
CH₃   CH₃

cis-1,2-dimethyl-   *trans*-1,2-dimethyl-
cyclopropane   cyclopropane

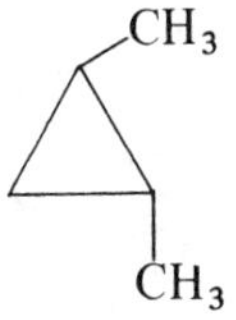

10.21 CH₃CH₂CH(CH₃)CHO

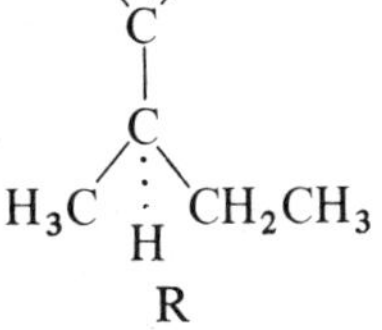

H   O   H   O
\ //   \ //
C   C
| |
C   C
/ ⋮ \   / ⋮ \
H₃C   CH₂CH₃   H₃CH₂C   CH₃
H   H
R 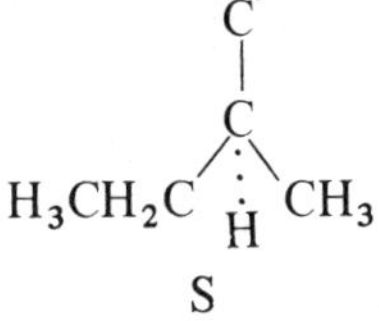   S

10.22

C_7H_{16}: CH₃CH₂CH₂—C(CH₃)(H)—CH₂CH₃ and (CH₃)₂CH—C(CH₃)(H)—CH₂CH₃

(structures with CH₃ on central C and H below)

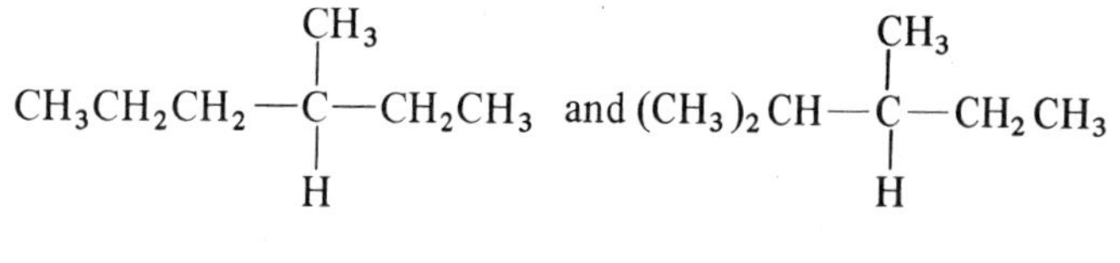

10.23 (a) $CH_3CH_2CH_2CH_2OH$ $CH_3CH_2CHOHCH_3$ $(CH_3)_2CHCH_2OH$ $(CH_3)_3COH$

(b) $CH_3CH_2CHOHCH_3$ (c)

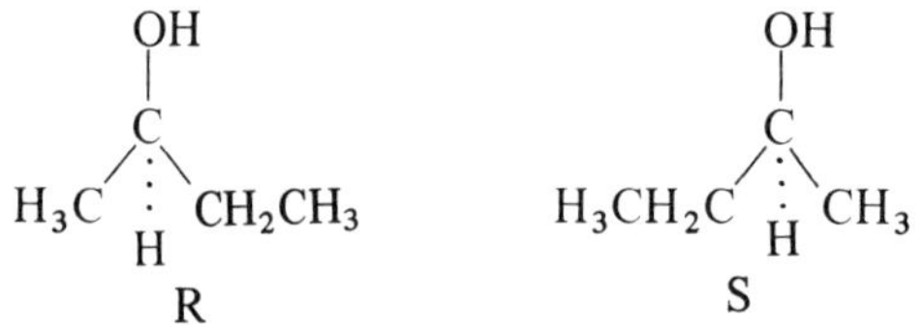

OH
C
H₃C ⋮ CH₂CH₃
    H
    R

OH
C
H₃CH₂C ⋮ CH₃
     H
     S

10.24 (a)

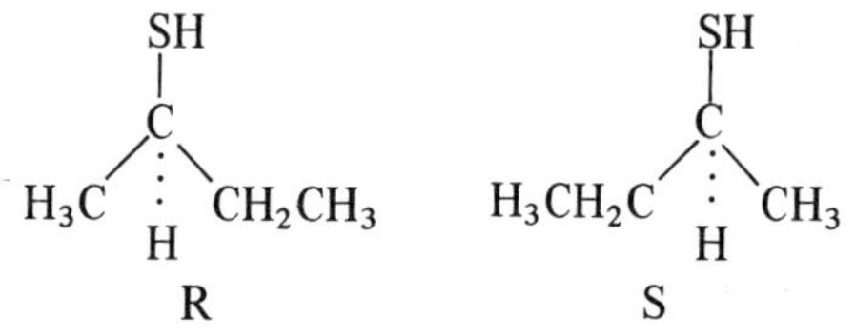

SH
C 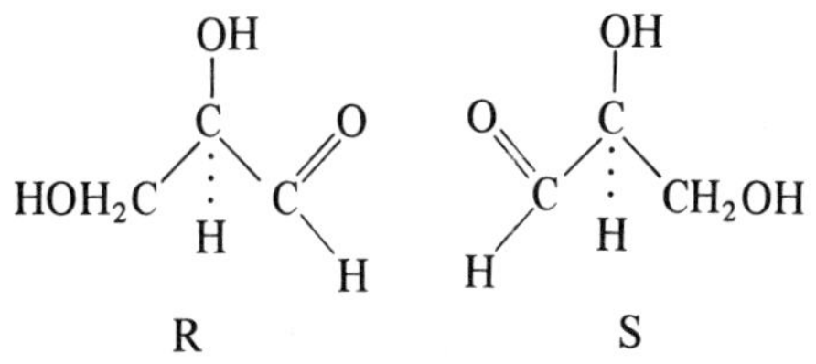
H₃C ⋮ CH₂CH₃
    H
    R

SH
C
H₃CH₂C ⋮ CH₃
     H
     S

(b)

OCH₃
C
H₃C ⋮ COOH
    H
    R

OCH₃
C
HOOC ⋮ CH₃
    H
    S

(c)

OH
C
HOH₂C ⋮ C=O
    H   H
    R

OH
C
O=C ⋮ CH₂OH
  H H
     S

(d)

Br
C
H₃CH₂C ⋮ COOH
     H
     R

Br
C
HOOC ⋮ CH₂CH₃
    H
    S

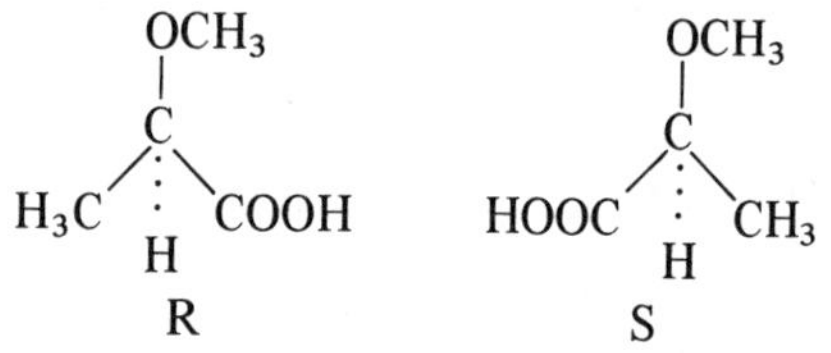

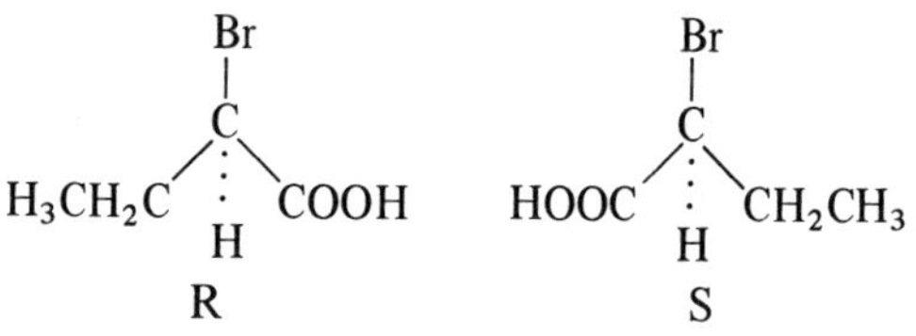

10.25 $[\alpha] = \dfrac{\alpha}{(l)(g/mL)} = \dfrac{+1.4°}{(1)(0.012)} = +1.2°$

10.26 $[\alpha] = \dfrac{-25}{(1)(0.2)} = -12.5°$

10.27

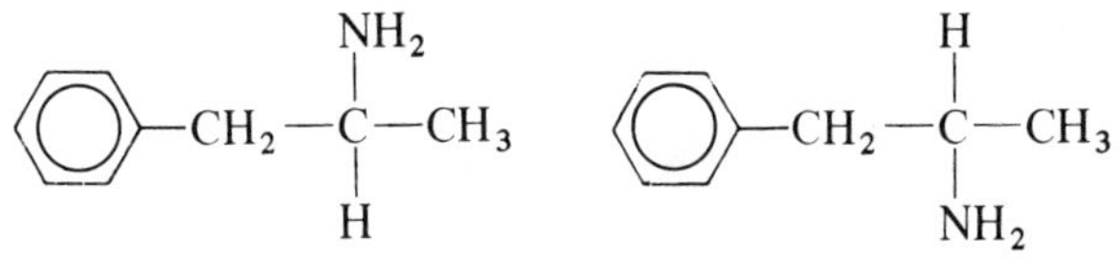

⬡—CH₂—C(NH₂)(CH₃)—H   ⬡—CH₂—C(H)(CH₃)—NH₂

10.28

(four ergoline-type ring structures with H₅C₂—N—C=O substituents and N—CH₃ groups)

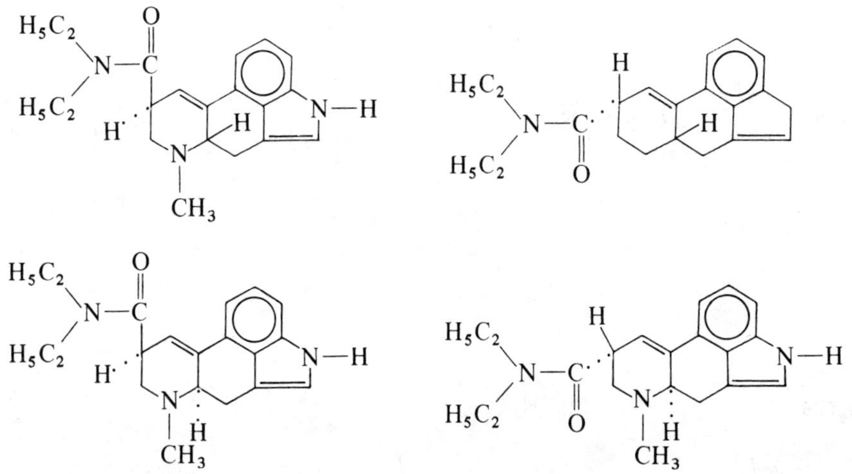

CHAPTER 11

Carbohydrates

11.1 **(a)** A substance that is a polyhydroxy aldehyde, polyhydroxy ketone, or a substance that yields such compounds upon acid hydrolysis.
(b) A simple sugar; one that cannot be broken down into smaller molecules by hydrolysis.
(c) A carbohydrate that yields two monosaccharides upon hydrolysis.
(d) A carbohydrate that yields many monosaccharides upon hydrolysis.
(e) A six-carbon monosaccharide.
(f) A four-carbon monosaccharide containing an aldehyde group.
(g) A five-carbon monosaccharide containing a ketone group.
(h) The gradual change of specific rotation that accompanies the attainment of an equilibrium mixture of the α- and β-forms of a sugar.

(i)

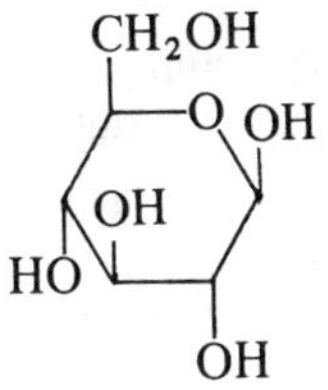

CH_2OH
O OH
OH
HO
OH

(j) The carbon-oxygen-carbon linkage that joins two monosaccharides.

(k)

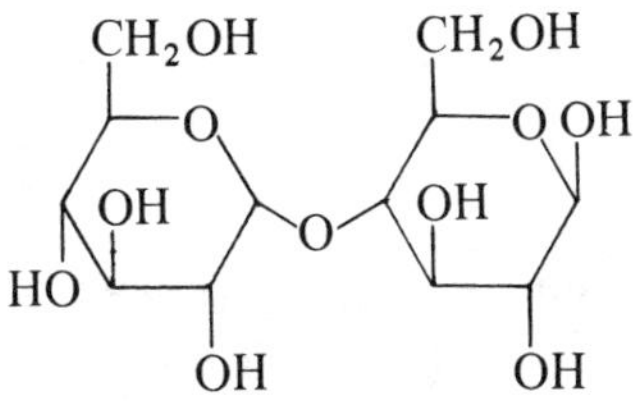

CH_2OH CH_2OH
O O OH
OH OH
HO
OH OH

(l)

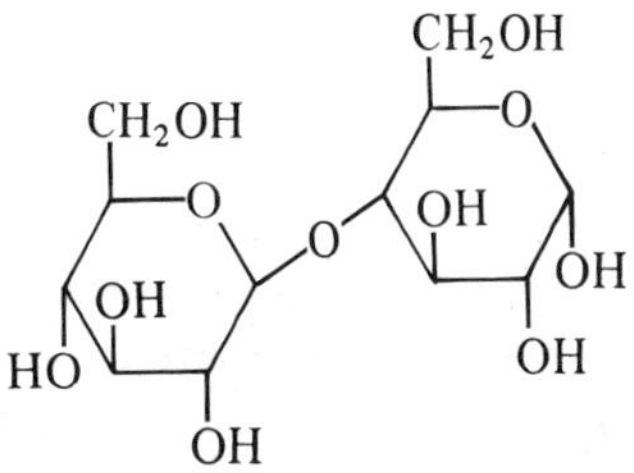

CH_2OH CH_2OH
O O
OH
OH OH
HO
OH

(m)

CH_2OH CH_2OH
HO O O
OH OH
OH
OH

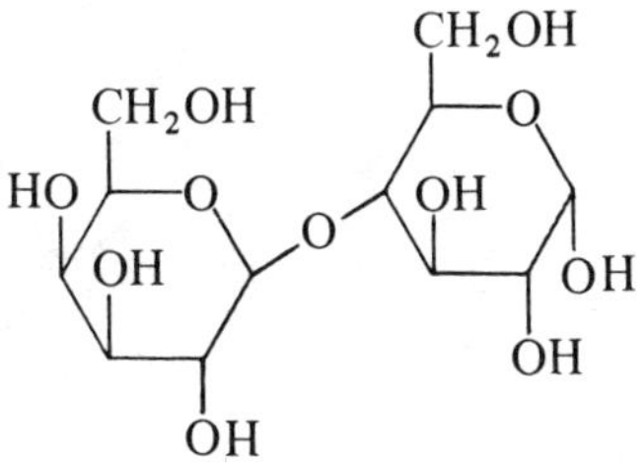

(n) An inability to digest lactose.
(o) Glucose polysaccharides of varying lengths.
(p) A carbohydrate that is capable of acting as a reducing agent. It must have an anomeric carbon that is capable of opening to a free carbonyl form.

11.2 (a) An aldose is a carbohydrate that contains an aldehyde group, whereas a ketose contains a ketone group.

(b) Saccharin was the only synthetic sweetener that was permitted in the United States between 1969 and 1977. For many years the debate raged over the potential cancer-causing properties of saccharin. The American Medical Association has since concluded that there is no significant correlation between the use of saccharin and increased risk of bladder cancer. Aspartame is a low-calorie sweetener that was approved for use in the United States in 1981. Aspartame is comprised of two amino acids: aspartic acid and phenylalanine. Aspartame is better known to the American public as Nutrasweet.

(c) A D-sugar is an aldose or a ketose whose −OH group on the penultimate carbon (next to last) is oriented to the right of its Fischer projection. In an L-sugar, the −OH group on the penultimate carbon is oriented to the left of its Fischer projection.

(d) The penultimate carbon is the chiral carbon that is farthest from the aldehyde or ketone group. The anomeric carbon is a hemiacetal or hemiketal carbon from which alpha and beta cyclic isomers are derived.

(e) Anomer refers to the alpha or beta cyclic isomers of a sugar. An epimer is a member of a diastereomeric pair of sugars that differ in configuration about a single carbon atom.

(f) Epimers are a pair of diastereomers that differ in configuration about a single carbon atom, whereas enantiomers are a pair of stereoisomers that are non-superimposable mirror images.

(g) Furanose sugars are cyclic sugars with five-membered rings, whereas pyranoses are sugars with six-membered rings.

(h) Dextrose is a common name for glucose and refers to the fact that the predominant natural form of the molecule is dextrorotatory. Levulose is a common name for fructose whose optical rotation is strongly levorotatory.

(i) A glucoside is the acetal that results when glucose reacts with any hydroxy compound. Glycoside is a general term used to designate the acetal that is formed when any carbohydrate has reacted with a hydroxy compound.

(j) Sucrose is a disaccharide that is comprised of glucose and fructose linked by an α-1,2-glucosidic linkage. It is a nonreducing sugar. Invert sugar is an equimolar mixture of D-glucose and D-fructose.

11.3

formaldehyde   acetic acid   lactic acid

11.4 Yes, all monosaccharides and disaccharides are relatively soluble in water. These compounds are polyhydroxylated and thus readily form hydrogen bonds with water molecules. In contrast to polysaccharides, mono- and disaccharides are relatively small and easily surrounded (extensively hydrated) by water molecules.

11.5

D-erythrulose   L-erythrulose

11.6

α-D-glucose   D-glucose β-D-glucose

α-D-mannose   D-mannose β-D-mannose

α-D-galactose   D-galactose β-D-galactose

α-D-fructose   D-fructose β-D-fructose

(Mannose is the least significant of the hexoses for humans.)

11.7

```
   CH₂OH          CH₂OH          CH₂OH          CH₂OH
    |              |              |              |
    C=O            C=O            C=O            C=O
    |              |              |              |
 H—C—OH        HO—C—H         HO—C—H          H—C—OH
    |              |              |              |
 H—C—OH        HO—C—H          H—C—OH        HO—C—H
    |              |              |              |
 H—C—OH        HO—C—H         HO—C—H          H—C—OH
    |              |              |              |
   CH₂OH          CH₂OH          CH₂OH          CH₂OH


   CH₂OH          CH₂OH          CH₂OH          CH₂OH
    |              |              |              |
    C=O            C=O            C=O            C=O
    |              |              |              |
 H—C—OH        HO—C—H         HO—C—H          H—C—OH
    |              |              |              |
 H—C—OH        HO—C—H          H—C—OH        HO—C—H
    |              |              |              |
HO—C—H          H—C—OH          H—C—OH        HO—C—H
    |              |              |              |
   CH₂OH          CH₂OH          CH₂OH          CH₂OH
```

11.8 A D-sugar is a sugar in which the –OH group on the penultimate carbon is oriented to the right of its Fischer projection. Although they share this structural relationship, D(+)-glucose and D(–)-fructose rotate plane-polarized light in opposite directions.

11.9

```
       CH₂OH
        |
        C=O
        |
   HO—C—H
        |
    H—C—OH
        |
   HO—C—H
        |
       CH₂OH
```

L-sorbose

11.10 (a) epimers   **(b)** none of these   **(c)** diastereomers   **(d)** enantiomers   **(e)** epimers
(f) diastereomers   **(g)** epimers   **(h)** anomers   **(i)** enantiomers   **(j)** epimers

11.11 Dissolve some pure α-D-glucose in water, place the solution in a polarimeter, and record its specific rotation. Periodically, record the specific rotation of the glucose solution, and observe the gradual change in that value as an equilibrium mixture of α- and β-D-glucose is attained. Yes. They could form furanose ring structures that may exist in both the alpha and beta configurations.

11.12 Ribose has a hydroxyl group at carbon-2, whereas deoxyribose lacks a hydroxyl group at carbon-2.

β-D-ribose   β-D-deoxyribose

11.13

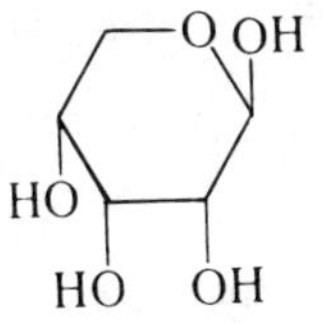

β-D-ribose
(pyranose form)

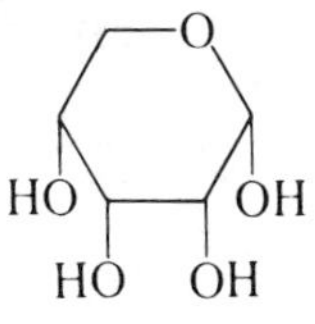

α-D-ribose
(pyranose form)

11.14

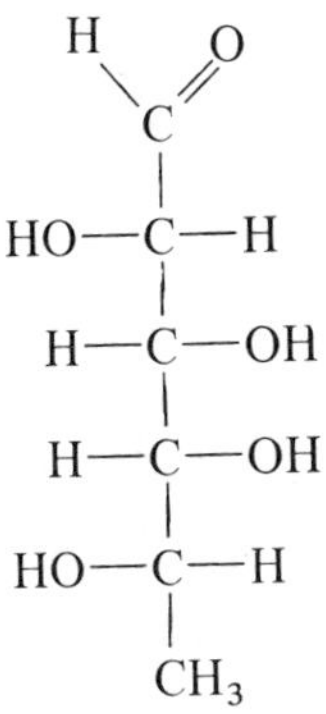

L-fucose

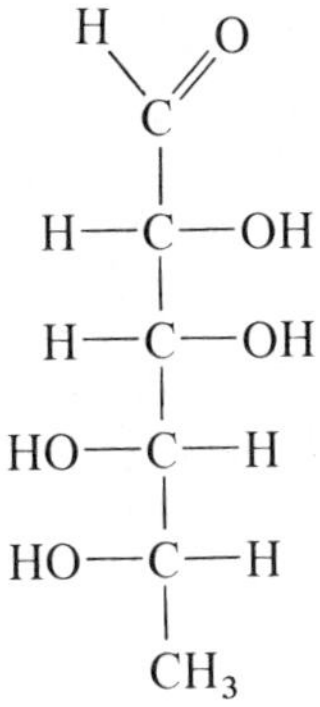

L-rhamnose

11.15 (a)

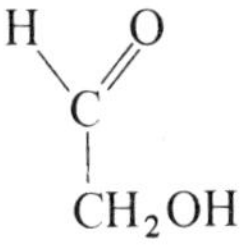

glycolaldehyde

(b)

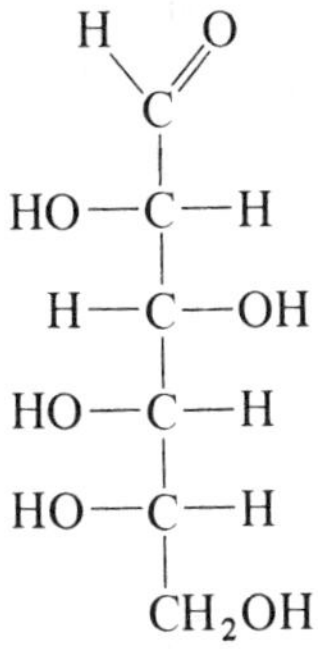

L-(−)-glucose

(c)

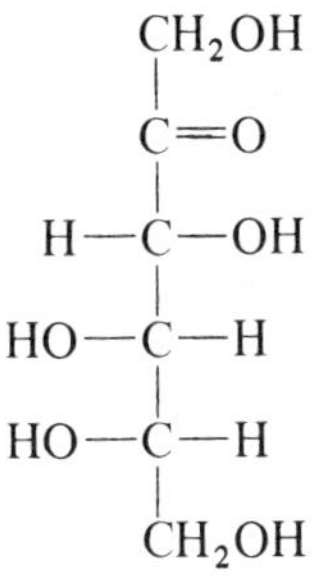

L(+)-fructose

(d)

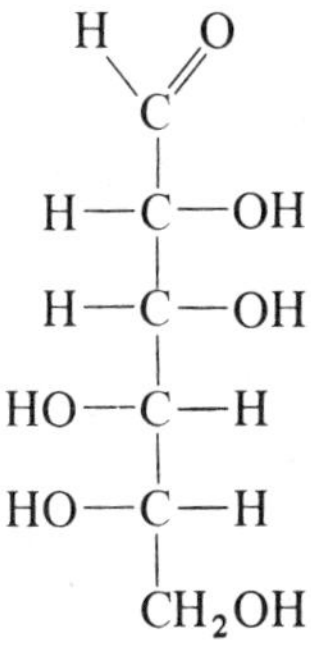

(e)

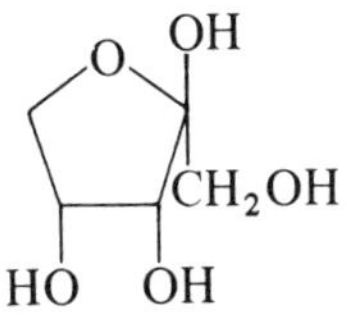

(f)

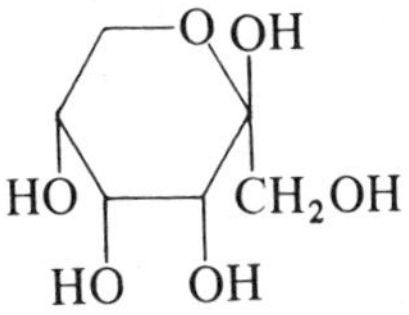

(g)

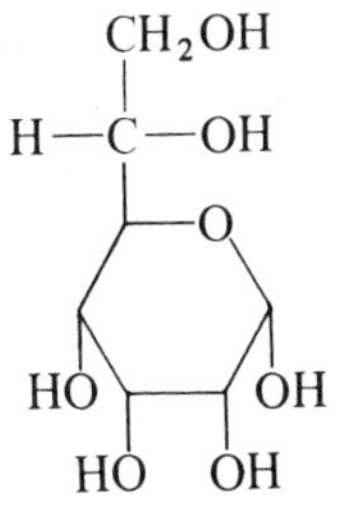

(h)

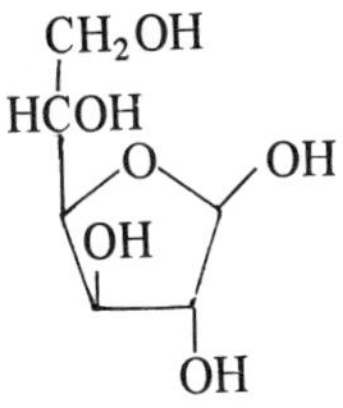

β-D-galactofuranose

(i)

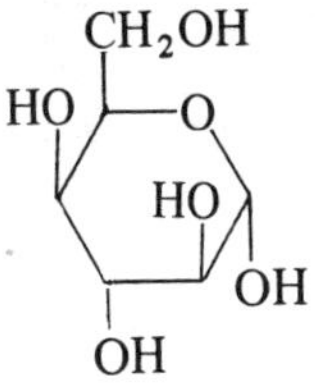

α-D-idose

(j)

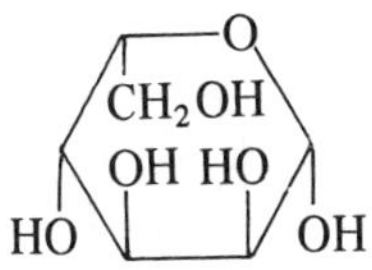

β-L-gulose

(k)

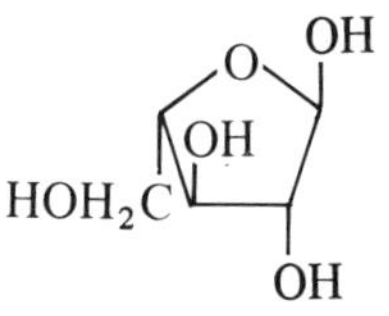

α-L-arabinofuranose

(l)

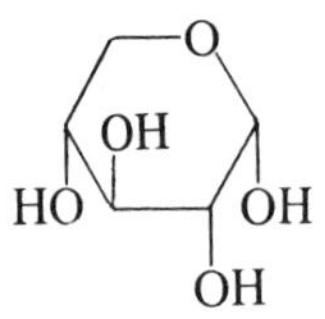

α-D-xylopyranose

(m)

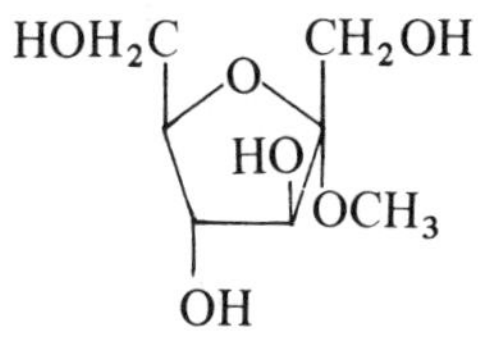

α-methyl-D-fructoside

(n)

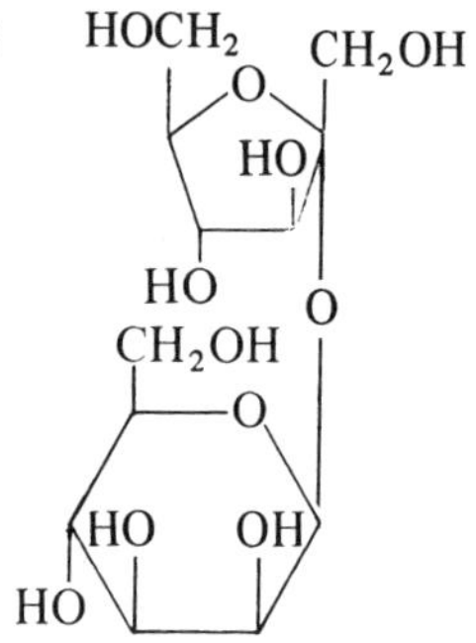

(o)

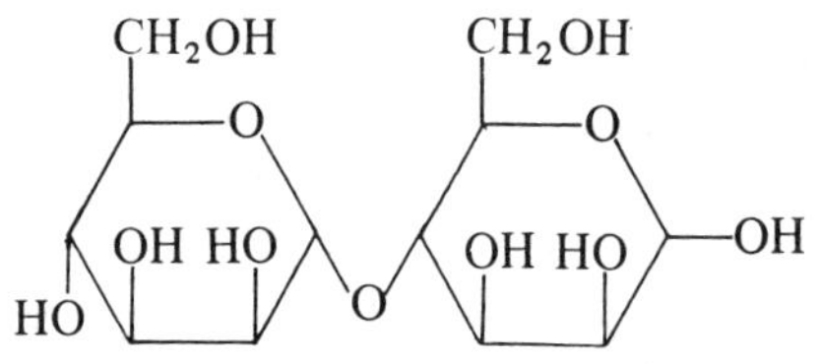

(p)

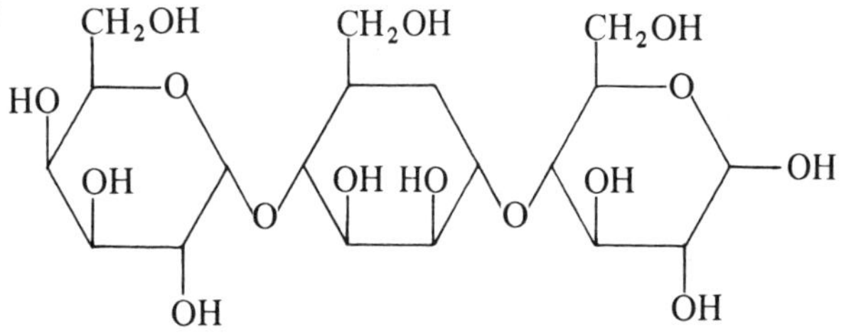

11.16 (a) lactose   **(b)** maltose   **(c)** cellobiose   **(d)** sucrose

11.17

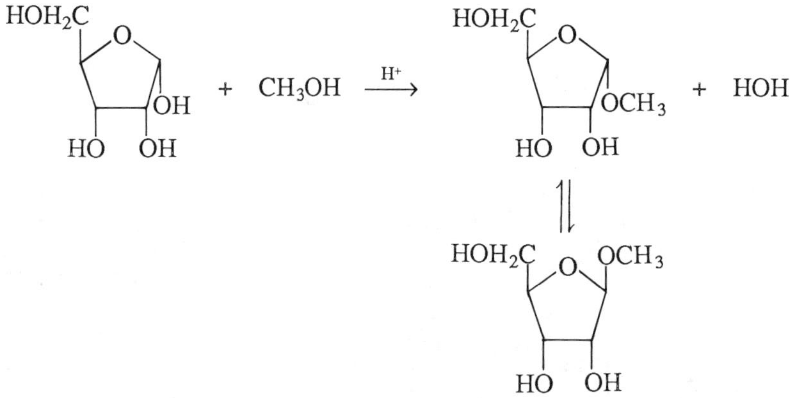

11.18

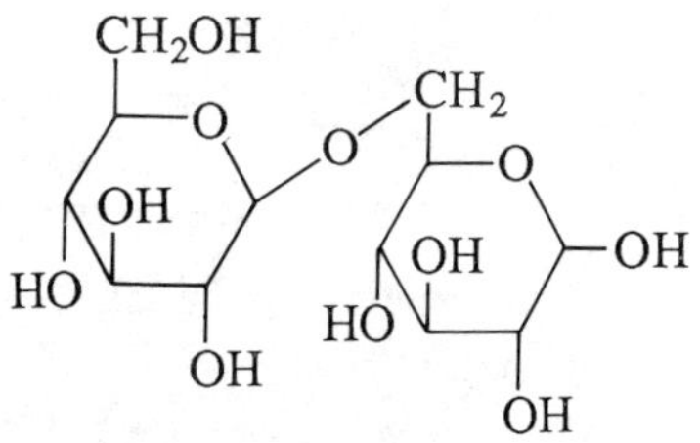

gentiobiose

11.19

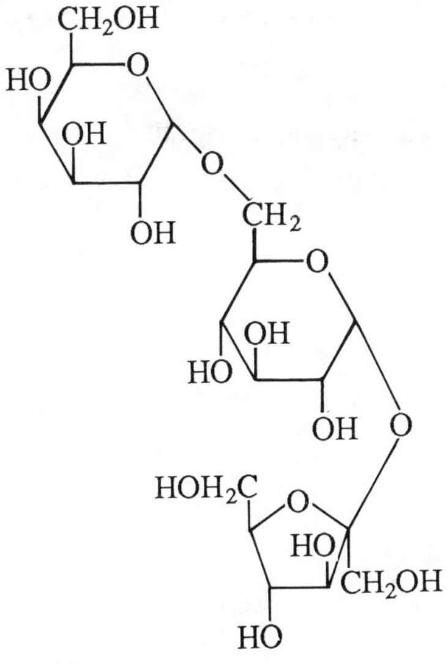

raffinose

11.20 (a) Both are disaccharides containing two glucose units. In maltose, the glucose units are joined by a α-1,4-glucosidic linkage; in cellobiose, the glucose units are joined by a β-1,4-glucosidic linkage.

(b) Both are polysaccharides containing only glucose. In starch, the glucoside linkages are α-1,4; in cellulose, β-1,4. Starch also contains 1,6-glucosidic linkages, and it is more highly branched than cellulose.

(c) Both are polysaccharide components of starch. Amylose is a straight-chain polysaccharide having only α-1,4-glucosidic linkages. Amylopectin is a branched-chain polysaccharide with occasional α-1,6-glucosidic linkages in addition to the α-1,4-linkages.

(d) Both are branched-chain polysaccharides containing glucose units. Glycogen is more highly branched, and its branches are shorter.

11.21 Humans possess the enzymes necessary to hydrolyze the α-1,4 and α-1,6-glucosidic linkages in starch but lack the enzymes that hydrolyze the β-1,4-glucosidic linkages that characterize the cellulose polymer.

11.22

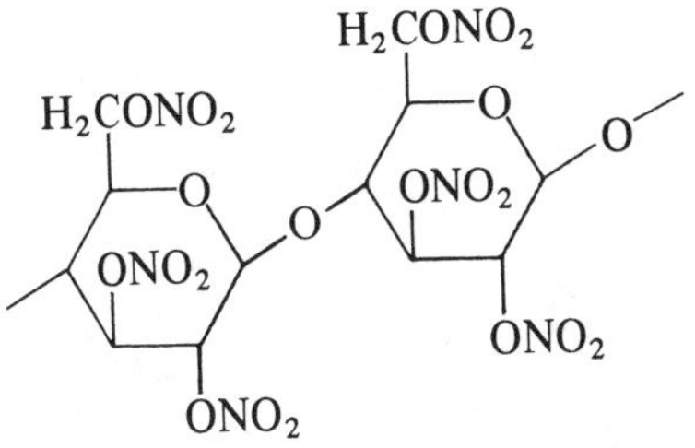

11.23

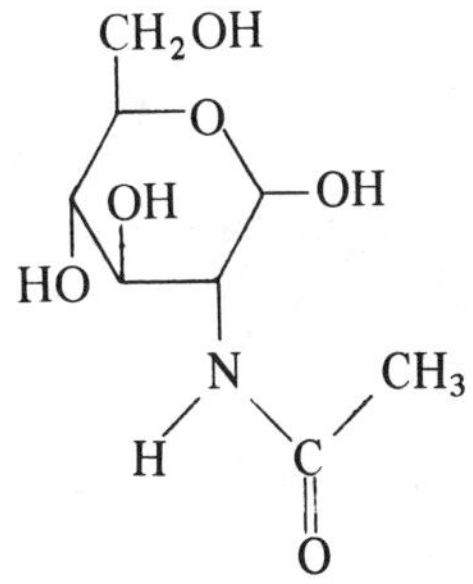

11.24 Starch serves as a storage form of carbohydrates, and cellulose serves as the structural component of the plant's cell wall. Glycogen serves as the storage form of carbohydrates in animals.

11.25 A reducing sugar must contain a free anomeric carbon (potential aldehyde group) that is not involved in a glycosidic linkage.

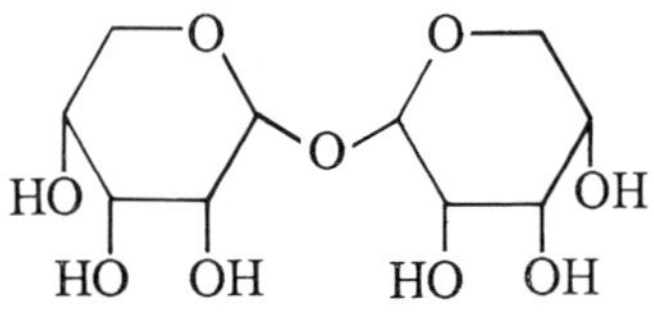

nonreducing disaccharide comprised of 2 aldopentoses

11.26

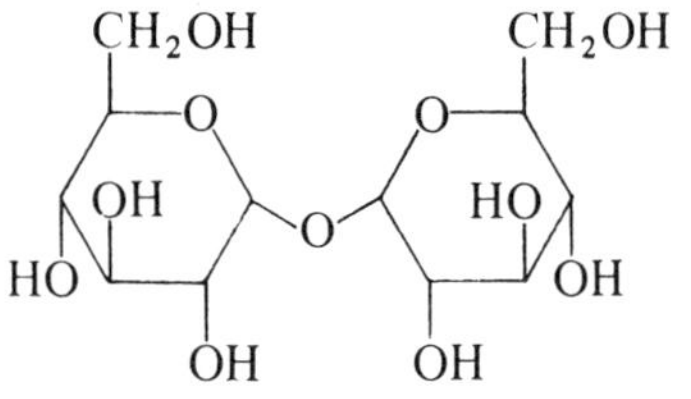

11.27

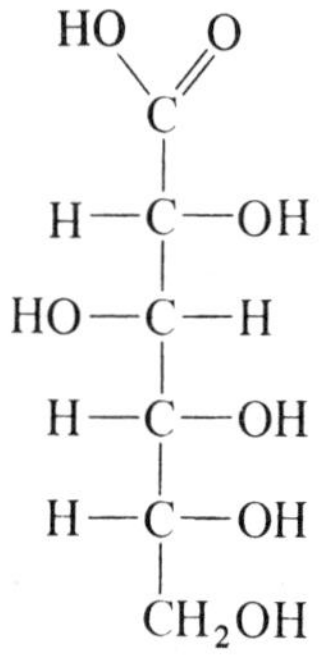

D-gluconic acid 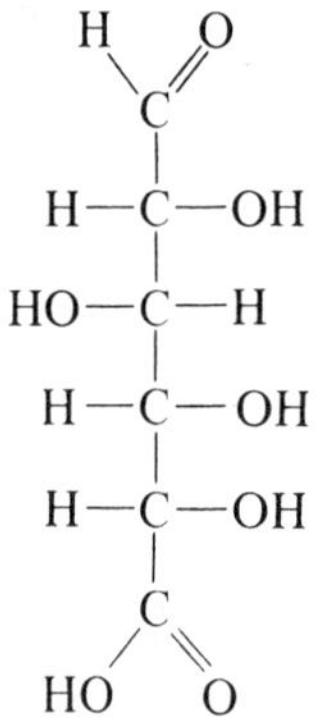   D-glucuronic acid 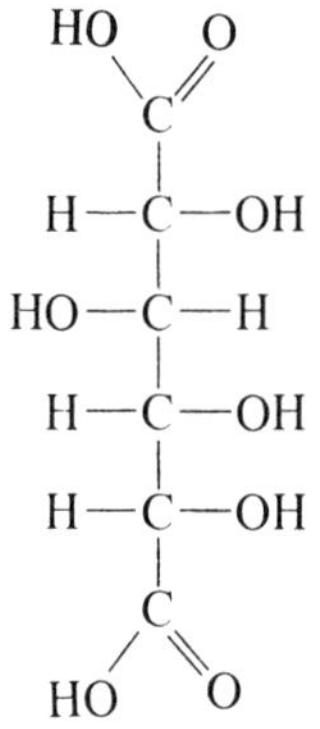   D-glucaric acid

11.28

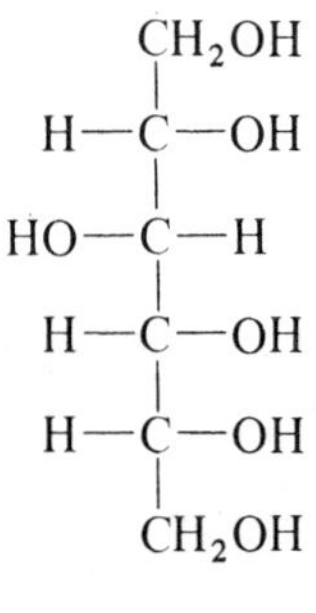

D-sorbitol 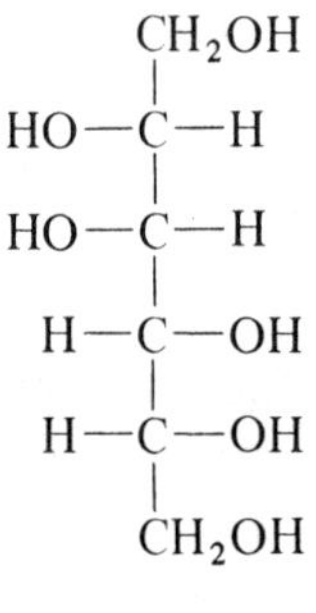   D-mannitol

Note that D-sorbitol and D-mannitol are epimers in that they differ in configuration only about carbon-2. When the carbonyl group at position-2 of D-fructose is reduced, an equilibrium mixture of the two epimers is obtained.

11.29 The alkaline reaction conditions facilitate enolization, and by this process frutose is converted into glucose. Glucose is then oxidized by the cupric ions in the reagent.

11.30

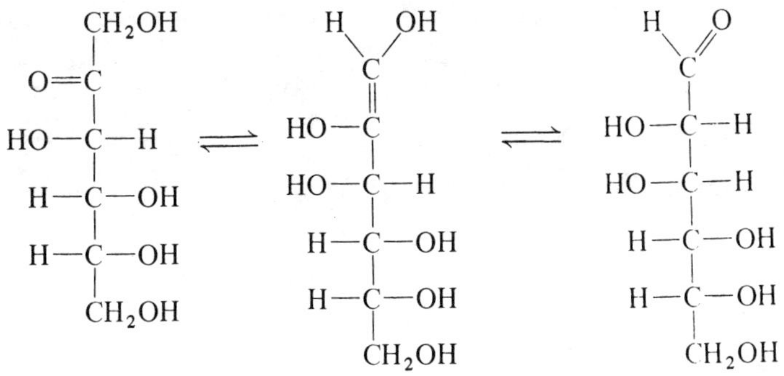

CH$_2$OH
O=C
HO—C—H
H—C—OH
H—C—OH
CH$_2$OH

D-fructose

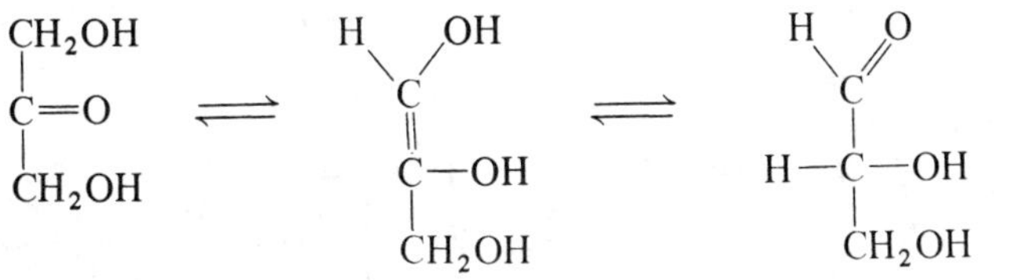

⇌

H   OH
C
HO—C
HO—C—H
H—C—OH
H—C—OH
CH$_2$OH

enediol

⇌

H   O
C
HO—C—H
HO—C—H
H—C—OH
H—C—OH
CH$_2$OH

D-mannose

CH$_2$OH
C=O
CH$_2$OH

dihydroxyacetone

⇌

H   OH
C
C—OH
CH$_2$OH

enediol

⇌

H   O
C
H—C—OH
CH$_2$OH

D-glyceraldehyde

11.31 (a)–(d) and (f)–(h) give positive Benedict's test.

11.32 (d) glucose   **(e)** glucose and fructose   **(g)** glucose and galactose   **(h)** glucose
(i) fructose   **(j)** glucose   **(k)** glucose   **(l)** glucose

11.33 The Molisch test. It gives a positive test (indicated by the appearance of a purple ring) with all carbohydrates larger than tetroses. Concentrated H_2SO_4 will first hydrolyze all di- and polysaccharides to monosaccharides; the monosaccharides then form furfural or hydroxymethylfurfural, which condense with α-naphthol (in the test reagent) to form a purple condensation product.

11.34 (a) Benedict's   **(b)** Seliwanoff's   **(c)** Benedict's   **(d)** Barfoed's   **(e)** Bial's   **(f)** Molisch

11.35 (a)

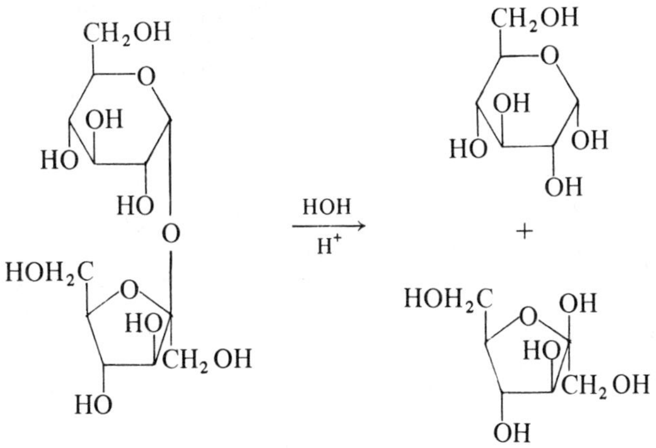

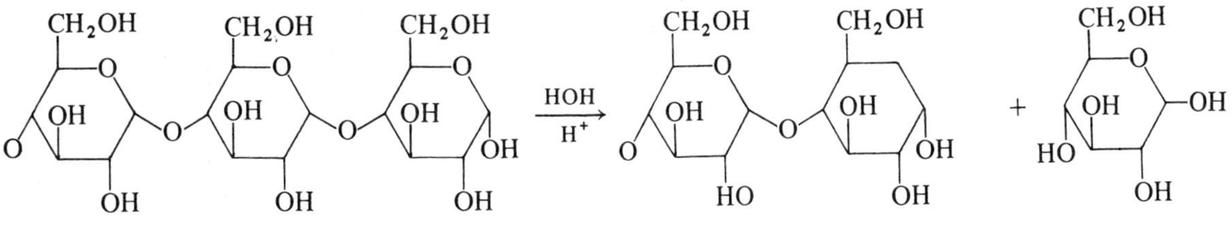

(c)

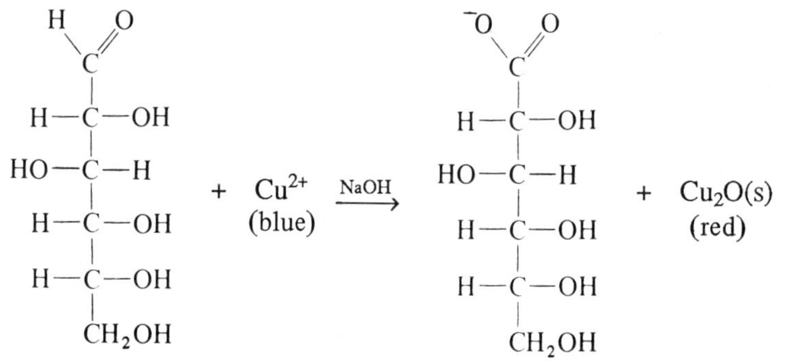

(d)

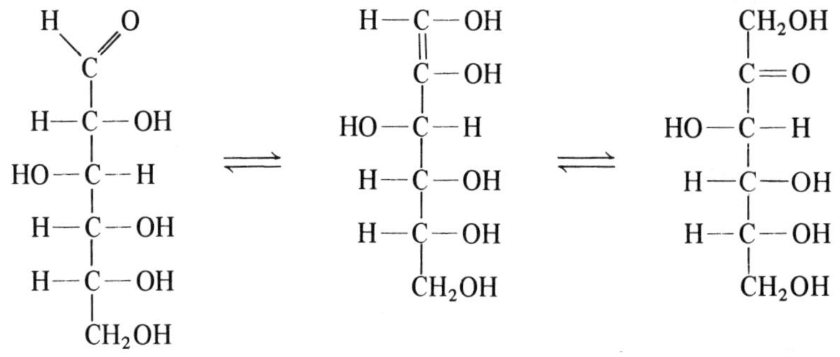

D-glucose   enediol   D-fructose

(e)

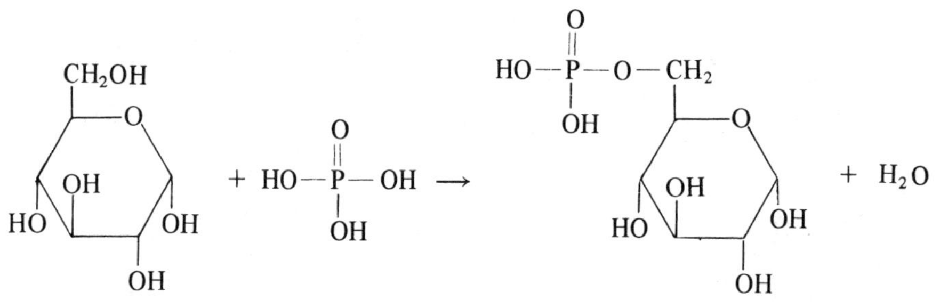

α-D-glucose   α-D-glucose 6-phosphate

11.36 Let x = percent of β-isomer expressed as a decimal
Let y = percent of α-isomer expressed as a decimal

$$x(-133°) + y(-21°) = -92°$$
$$x + y = 1.00$$
$$x = 1.00 - y$$
$$(1.00 - y)(-133) - 21y = -92$$
$$-133 + 133y - 21y = -92$$
$$112y = 41$$
$$y = .366 \text{ or } 36.6\% \text{ of the } \alpha\text{-isomer}$$
$$x = .634 \text{ or } 63.4\% \text{ of the } \beta\text{-isomer}$$

CHAPTER 12

Lipids

12.1 **(a)** An ester composed of three fatty acids joined to glycerol.
  (b) A simple triacylglycerol comprised of 3 moles of oleic acid esterified to glycerol.
  (c) A saturated fatty acid containing 16 carbons ($C_{15}H_{31}COOH$).
  (d) An 18-carbon unsaturated fatty acid containing three double bonds.
  (e) Hydrolytic enzymes that catalyze the digestion of triacylglycerols into diacylglycerols, monoacylglycerols, fatty acids, and glycerol.
  (f) A measure of the degree of unsaturation of a fat or oil.
  (g) The process of converting oils to fats by means of hydrogenation.
  (h) A compound that is added to foods, in very small amounts, to suppress rancidity.
  (i) The spontaneous start of a fire (i.e., without the aid of an ignition agent).
  (j) Aggregation of molecules that contain both polar and nonpolar groups.
  (k) Detergents with polar covalent structures that provide the water solubility. These detergents lack charged groups.
  (l) Any substance that can be broken down by microorganisms.
  (m) Substances that are added to detergents to increase their efficacy as cleansing agents.
  (n) An overabundance of plant life, and a reduction of oxygen, in a body of water.
  (o) The parent compound of the phosphoglycerides.
  (p) Any compound that contains the perhydrocyclopentanophenanthrene structure.
  (q) A tissue response to injury characterized by swelling, redness, and pain.
  (r) Steroids or synthetic derivatives that serve to stimulate protein synthesis.
  (s) A formulation containing a mixture of two compounds that are analogs of progesterone and estradiol. These drugs function to prevent ovulation.
  (t) The administration of daily doses of estrogen or estrogen substitutes to treat menopausal women who are experiencing unpleasant emotions.
  (u) A family of (20-carbon) unsaturated fatty acids having the same basic skeleton as prostanoic acid.
  (v) Recommended daily allowance.

12.2 **(a)** A fat is a lipid that is a solid at room temperature (20°C), whereas an oil is a liquid at the same temperature.
  (b) A fat is a lipid comprised of three fatty acids esterified to glycerol, whereas a wax is a lipid comprised of a fatty acid esterified to a long-chain alcohol.
  (c) A saponifiable lipid can be hydrolyzed under alkaline conditions, whereas a nonsaponifiable lipid cannot undergo hydrolysis because there are no ester linkages in the molecule.
  (d) The H atoms lie on the same side of the double bond (*cis*-configuration) in the unsaturated fatty acids that comprise lipid molecules. In the *trans*-configuration, the H atoms lie on opposite sides of the double bond.
  (e) In a simple triacylglycerol, the lipid molecule contains three identical fatty acids, whereas a mixed triacylglycerol is composed of two or three different fatty acid components.
  (f) Butter is an animal fat that contains a relatively high percentage of low molecular weight fatty acids, whereas margarine is a butter-like substance obtained from the partial hydrogenation of vegetable oil.

(g) Oxidative rancidity arises from a complex process in which oxygen reacts with triacylglycerols that are rich in polyunsaturated fatty acids. Bond cleavage results in the production of short-chain carboxylic acids that are responsible for the disagreeable odor. Hydrolytic rancidity occurs under moist, warm conditions whereby microorganisms secrete lipases that catalyze the hydrolysis of the ester linkages present in lipid molecules. The eventual release of low molecular weight acids accounts for the offensive odor associated with rancidity.

(h) Hydrophobic generally refers to a substance that lacks an affinity for water, whereas hydrophilic substances are soluble in water.

(i) Saponification is the alkaline hydrolysis of an ester, whereas emulsification is the conversion of a large, water-insoluble lipid glob into a suspension of small lipid globules.

(j) A rancid oil is an oil that has developed a disagreeable odor, whereas a drying oil is an oil that causes paint or varnish to develop a hard, protective coating.

(k) Hard water contains certain metal ions such as calcium and magnesium. These ions form complexes with the carboxylate ions of soaps and form insoluble precipitates. Soft water lacks the metal ions that form such precipitates.

(l) Hard soaps are sodium salts of long-chain carboxylic acids (fatty acids), whereas soft soaps are potassium salts of long-chain carboxylic acids.

(m) Soaps are salts of fatty acids. They tend to form precipitates in hard water. Syndets are synthetic detergents that do not form precipitates in hard water.

(n) Anionic detergents are sulfates of long-chain alcohols or sulfonate salts of aromatic hydrocarbons. In solution, the hydrophilic portion of the soap molecule carries a net negative charge. A cationic detergent (invert soap) carries a net positive charge on the water-soluble portion of the molecule.

(o) A prokaryote is a simple cell that lacks a nuclear membrane and membranated organelles. A eukaryote is a complex cell whose genetic material is enclosed within a nuclear membrane.

(p) Phosphoglycerides are phospholipids consisting of a parent compound, phosphatidic acid, and complex organic groups that are esterified to the phosphate portion of the molecule. Sphingolipids are phospholipids consisting of a sphingosine backbone (long-chain unsaturated amino alcohol), fatty acids, phosphate, and a polar alcohol component. Phospholipids are the main structural component of cell membranes.

(q) Lecithin (phosphatidyl choline) is a phosphoglyceride comprised of the quaternary ammonium salt choline esterified to the phosphoric acid portion of a phosphatidic acid parent compound. The cephalins are phosphoglycerides comprised of either ethanolamine or serine esterified to the phosphate group of the parent compound.

(r) Peripheral proteins are those proteins that form loose associations at the surface of cell membranes, whereas integral proteins are deeply embedded within the bilayer membrane structure.

(s) Arteriosclerosis or hardening of the arteries produces degenerative heart disease, stroke, and other arterial diseases. Atherosclerosis is a form of arteriosclerosis that results from the deposition of excess cholesterol within the blood vessels that acts to constrict normal blood flow and hence elevate blood pressure.

(t) Bile is synthesized in the liver and concentrated within the gallbladder. It consists of water, cholesterol, and a variety of bile salts, acids, and pigments. The most important components of bile are the bile salts. They are derived from cholesterol and aid in the emulsification and absorption of dietary lipids.

(u) Hormones are chemical compounds that are synthesized and secreted by one tissue and exert some biological effect on a target tissue. Vitamins are complex organic compounds that are essential for the maintenance of normal metabolism yet cannot by synthesized by the organism.

(v) Androgens are the male sex hormones that are responsible for the development of the male genitalia and secondary sexual characteristics. Estrogens are female sex hormones that serve to regulate the ovulation cycle.

(w) Thromboxanes are compounds related to prostaglandins that induce blood clotting by stimulating blood platelet aggregation. Prostacyclins oppose the action of thromboxanes by preventing blood platelet aggregation.

    (x) Keratomalacia is a form of blindness that results from a perforation of the cornea brought about by a vitamin A deficiency. Osteomalacia is a disease associated with a tendency for brittle bones to fracture. The disease is a direct result of vitamin D deficiency.

    (y) Ergocalciferol (vitamin D_2) is produced by irradiating the cholesterol derivative, ergosterol, with ultraviolet light. Cholecalciferol (vitamin D_3) is formed in the skin of animals by the reaction of sunlight with a cholesterol derivative called 7-dehydrocholesterol.

    (z) Low density lipoproteins are aggregates of lipid and protein that function to transport water-insoluble lipids in the bloodstream. Low density lipoproteins are comprised of approximately 80% lipid and 20% protein. (Remember that lipids are less dense than protein). High density lipoproteins are comprised of approximately 50% lipid and 50% protein. Researchers contend that the relative proportion of these lipoproteins to one another plays an underlying role in the inappropriate deposition of cholesterol in the blood vessels (atherosclerosis).

12.3 Fats are the body's primary energy reserve. In addition, they perform a number of vital functions in the body including maintenance of body temperature and insulation of organs and tissues against mechanical and electrical shocks and are integral components of biological membranes.

12.4 All fats are solids. Saturated fatty acids are solids, but unsaturated fatty acids are liquids. Both fats and fatty acids are insoluble in water. Fats and oils are soluble in nonpolar organic solvents, e.g., benzene, carbon tetrachloride.

12.5 Unsaturated fatty acids have low melting points due to the stereochemical nature of their double bonds. In unsaturated fatty acids, the *cis*-configuration of the double bonds causes severe kinks or bends in the long hydrocarbon tail of the molecule. This prevents molecules from packing tightly together and thus weakens the hydrophobic interactions that exist between adjacent fatty acids.

12.6 The melting point of elaidic acid ($C_{17}H_{33}COOH$) is lower than the melting point of the 18-carbon stearic acid (saturated), yet higher than the melting point of its *cis*-isomer, oleic acid. Although oleic acid and elaidic acid are both 18-carbon fatty acids containing one double bond, the H atoms in elaidic acid are oriented on opposite sides of the double bond (*trans*-configuration). The *trans*-configuration does not distort the linearity of the hydrocarbon tail and thus adjacent molecules experience strong intermolecular attractions. The tight packing of the molecules accounts for the elevated melting point of elaidic acid as opposed to its *cis*-isomer (oleic acid). The m.p. of *trans*-hexadecenoic acid would be lower than the m.p. of elaidic acid. The former has a lower molecular weight (fewer carbon and hydrogen atoms), and hence weaker van der Waals forces of attraction between molecules.

12.7 (a) glyceryl oleate (triolein) (b) glyceryl oleomyristostearate (c) sodium stearate
(d) glyceryl oleomyristopalmitate (e) myricyl palmitate (f) sphingosine

12.8 Two mixed triacylglycerols are possible.

12.9 (a)

(e)

$$C_{15}H_{31}O\overset{\overset{O}{\|}}{\underset{\underset{O}{\|}}{S}}-O^-\ Na^+$$

(f)

$$
\begin{array}{l}
\text{RCOCH}_2 \\
\text{RCOCH} \quad \text{OH} \\
\text{HO-P OCH}_2 \\
\text{OH}
\end{array}
$$

12.10

$$
\begin{array}{l}
\text{RCOCH}_2 \\
\text{RCOCH} \\
\text{H}_2\text{COPO} \\
\text{OH}
\end{array}
$$

12.11 (a) yes (3 moles of hydrogen gas would react)
(b) yes (1 mole of hydrogen gas would react)
(c) no (hydrogen gas would not react)
(d) yes (1 mole of hydrogen gas would react)
(e) no (hydrogen gas would not react)
(f) yes (1 mole of hydrogen gas would react)

12.12 (a) fatty acids and glycerol   (b) salts of fatty acids (soaps) and glycerol

12.13 (a) The odor of butyric acid.   (b) The hydrolysis of the ester linkages.
(c) Storing the butter covered in a refrigerator will protect the butter from water vapor in the air and the cold temperature will inhibit the growth of microorganisms that produce hydrolytic enzymes.

12.14 Yes, waxes can be converted into soaps. Waxes are esters of fatty acids and long-chain alcohols. Therefore, they may undergo alkaline hydrolysis (saponification) to yield salts of fatty acids (soaps) and alcohols.

12.15 (1) Carnauba wax (myricyl cerotate): plant wax; used as a floor and automobile wax and as a coating on carbon paper and mimeograph stencils.
(2) Spermaceti wax (cetyl palmitate): animal wax; used in the manufacture of candles, cosmetics, and ointments.
(3) Lanolin (fatty acid esters of lanosterol and agnosterol): animal wax; used as a base for creams, ointments, and salves.
(4) Beeswax (myricyl palmitate): insect wax; used in the production of candles, wax paper, cosmetics, and medicinals.

12.16 The cleansing action of soap is largely dependent upon the dual nature of the fatty acid salts. The hydrophilic carboxylate group of the soap molecule is attracted to water, whereas the hydrophobic hydrocarbon tail is repelled from the aqueous surface. The thin layer of suds that develops serves to lower the surface tension of the water. When in contact with grease, the hydrophobic groups dissolve in the lipid and the hydrophilic groups remain oriented toward the aqueous phase. Mechanical agitation serves to emulsify the grease into many water-soluble micelles that are easily washed away.

12.17 **(a)** A detergent must have both a hydrophobic portion and a hydrophilic portion within the same molecule.

(b) The advantage of detergents is that they do not precipitate out of hard water.

(c) The disadvantage of detergents is that they contribute to the eutrophication of lakes and streams.

(d) An anionic detergent.

12.18 **(a)**

$$CH_3(CH_2)_7-CH=CH-(CH_2)_7-\overset{O}{\overset{\|}{C}}-O-CH_2$$
$$CH_3(CH_2)_7-CH=CH-(CH_2)_7-\overset{O}{\overset{\|}{C}}-O-CH \quad + \quad 3\,H_2 \quad \xrightarrow[\Delta]{Ni}$$
$$CH_3(CH_2)_7-CH=CH-(CH_2)_7-\overset{O}{\overset{\|}{C}}-O-CH_2$$

$$CH_3(CH_2)_{16}-\overset{O}{\overset{\|}{C}}-O-CH_2$$
$$CH_3(CH_2)_{16}-\overset{O}{\overset{\|}{C}}-O-CH_2$$
$$CH_3(CH_2)_{16}-\overset{O}{\overset{\|}{C}}-O-CH_2$$

(b)

$$C_{17}H_{33}\overset{O}{\overset{\|}{C}}-O-CH_2$$
$$C_{17}H_{33}\overset{O}{\overset{\|}{C}}-O-CH \quad + \quad 3\,I_2 \quad \longrightarrow$$
$$C_{17}H_{33}\overset{O}{\overset{\|}{C}}-O-CH_2$$

$$CH_3(CH_2)_7\overset{I}{\overset{|}{C}}H-\overset{I}{\overset{|}{C}}H(CH_2)_7-\overset{O}{\overset{\|}{C}}-O-CH_2$$
$$CH_3(CH_2)_7\overset{I}{\overset{|}{C}}H-\overset{I}{\overset{|}{C}}H(CH_2)_7-\overset{O}{\overset{\|}{C}}-O-CH$$
$$CH_3(CH_2)_7\overset{I}{\overset{|}{C}}H-\overset{I}{\overset{|}{C}}H(CH_2)_7-\overset{O}{\overset{\|}{C}}-O-CH_2$$

(c)

$$CH_3(CH_2)_{16}\overset{O}{\overset{\|}{C}}-O-CH_2$$
$$CH_3(CH_2)_{16}\overset{O}{\overset{\|}{C}}-O-CH \quad + \quad 3\,KOH \quad \longrightarrow \quad 3\left[CH_3(CH_2)_{16}\overset{O}{C}\diagdown_{O^-K^+}\right] \quad + \quad \begin{matrix} CH_2OH \\ | \\ CHOH \\ | \\ CH_2OH \end{matrix}$$
$$CH_3(CH_2)_{16}\overset{O}{\overset{\|}{C}}-O-CH_2$$

(d)

$$C_{17}H_{31}\overset{O}{\overset{\|}{C}}-O-CH_2$$
$$C_{17}H_{31}\overset{O}{\overset{\|}{C}}-O-CH \quad + \quad 3\,HOH \quad \xrightarrow{H^+} \quad 3\left[C_{17}H_{31}\overset{O}{C}\diagdown_{OH}\right] \quad + \quad \begin{matrix} CH_2OH \\ | \\ HCOH \\ | \\ CH_2OH \end{matrix}$$
$$C_{17}H_{31}\overset{O}{\overset{\|}{C}}-O-CH_2$$

(e)

$$2\,C_{17}H_{35}\overset{O}{C}\diagdown_{O^-Na^+} \quad + \quad Mg^{2+} \quad \longrightarrow \quad \left[C_{17}H_{35}\overset{O}{C}\diagdown_{O^-}\right]_2 Mg^{2+} \quad + \quad 2\,Na^+$$

(f)

$$C_{17}H_{33}\overset{O}{C}\diagdown_{O^-K^+} \quad + \quad HCl \quad \longrightarrow \quad C_{17}H_{33}\overset{O}{C}\diagdown_{OH} \quad + \quad KCl$$

(g)

$$\begin{array}{l} R-\overset{\overset{\displaystyle O}{\|}}{C}-O-CH_2 \\ R'-\overset{\overset{\displaystyle O}{\|}}{C}-O-CH \quad + \quad HOCH_2CH_2NH_2 \quad \longrightarrow \\ HO-\overset{\overset{\displaystyle O}{\|}}{P}-O-CH_2 \\ \qquad \overset{|}{OH} \end{array}$$

$$\begin{array}{l} R-\overset{\overset{\displaystyle O}{\|}}{C}-O-CH_2 \\ R'-\overset{\overset{\displaystyle O}{\|}}{C}-O-CH \\ \qquad CH_2-O-\overset{\overset{\displaystyle O}{\|}}{P}-CH_2CH_2\overset{+}{N}H_3 \\ \qquad\qquad\qquad \overset{|}{O^-} \end{array}$$

(h)

$$\begin{array}{l} R-\overset{\overset{\displaystyle O}{\|}}{C}-O-CH_2 \\ R'-\overset{\overset{\displaystyle O}{\|}}{C}-O-CH \\ \qquad H_2C-O-\overset{\overset{\displaystyle O}{\|}}{P}-O-CH_2CH\overset{+}{N}H_3COO^- \\ \qquad\qquad\qquad \overset{|}{O^-} \end{array}$$

$$\xrightarrow[H^+]{4\ HOH}$$

$$R-\overset{\overset{\displaystyle O}{\|}}{C}\underset{OH}{} \quad + \quad R'-\overset{\overset{\displaystyle O}{\|}}{C}\underset{OH}{} \quad + \quad \begin{array}{l}CH_2OH\\ HCOH\\ CH_2OH\end{array} \quad + \quad HO-\overset{\overset{\displaystyle O}{\|}}{P}-OH \atop \overset{|}{OH} \quad + \quad HOCH_2CHNH_2COOH$$

12.19 Phospholipids are referred to as polar lipids because the phospholipid molecule contains both a polar and a nonpolar component. An oxygen of the phosphate group is negatively charged, and the nitrogen (of the nitrogenous component) is positively charged.

12.20

$$\begin{array}{l} C_{15}H_{31}\overset{\overset{\displaystyle O}{\|}}{\boxed{C-O}}-CH_2 \\ C_{17}H_{33}\overset{\overset{\displaystyle O}{\|}}{\boxed{C-O}}-CH \\ \qquad H_2C\boxed{O}\overset{\overset{\displaystyle O}{\|}}{\boxed{P}}\boxed{O}CH_2CH_2\overset{+}{N}(CH_3)_3 \\ \qquad\qquad \overset{|}{O^-} \end{array}$$

12.21 (a) an amide linkage   (b) an acetal linkage

12.22 (a) The three types of lipids found in cell membranes are phospholipids, glycolipids, and cholesterol.
(b) The proteins found in cell membranes may function as receptors for messenger molecules, as transport molecules, as enzymes, or as structural entities.

12.23 (a)

(b)

Carbons number 3, 8, 9, 10, 13, 14, 17, and 20 are chiral.

12.24 The do not contain an ester linkage and thus cannot be hydrolyzed in an alkaline solution.

12.25 Only a portion of the cholesterol molecule will react with these reagents.

(a)

(b)

12.26 (a) A hydroxyl group and an alkene group.
(b) Hydroxyl groups and a carboxyl group.
(c) Hydroxyl groups, ketone groups, and an alkene group.
(d) Hydroxyl groups, ketone groups, an aldehyde group, and an alkene group.
(e) Hydroxyl groups and an aromatic ring.
(f) Ketone group and an alkene group.

12.27 The mineralocorticoids serve to regulate the exchange of sodium, potassium, and hydrogen ions across biological membranes. The glucocorticoids regulate many biological activities including the production of glucose, the mobilization of fatty acids and amino acids, and inhibition of the in-flammatory response.

12.28 The synthetic analogs contain an acetylenic group at carbon-17.

12.29 (a) Arachidonic acid serves as the precursor for the synthesis of the prostaglandins.
(b) Potential therapeutic uses of the prostaglandins include the regulation of blood pressure, inhibition of gastric secretions, and relief of asthma and nasal congestion.

12.30 Aspirin functions to relieve pain, reduce fever, and reduce inflammation by inhibiting an enzyme, cyclooxygenase, that participates in the synthesis of prostaglandins.

12.31 (a) Functions to stimulate fluid secretion by the epithelial cells of the eye. Keratomalacia.
(b) Functions to promote the absorption of calcium from the intestine into the bloodstream. Osteomalacia.
(c) Functions in controlling oxidation–reduction reactions in a variety of body tissues. Anemia (in infants).
(d) Functions as an antioxidant in the aqueous regions of the body. Scurvy.

12.32 The inappropriate deposition of cholesterol in the arteries is responsible for a constriction of blood flow that acts to elevate blood pressure and may ultimately result in the complete blockage of an artery (atherosclerosis).

CHAPTER 13

Proteins

13.1 (a) A compound that contains an amino group alpha to a carboxylic acid group.

(b) The collection of atoms (R-group) that bestows uniqueness upon a particular amino acid.

(c) The dipolar ion form of an amino acid (electrically neutral).

(d) The pH at which an amino acid exists in a zwitterionic form (maintains a net charge = 0).

(e) Any substance that is capable of donating or accepting protons (H^+).

(f) The amide linkage that joins two adjacent amino acids.

(g) Three amino acids linked by peptide bonds.

(h) Substances that increase the volume of urine excreted.

(i) A disorder characterized by the excessive excretion of urine.

(j) An inherited disease resulting from a single amino acid substitution whereby glutamic acid is replaced by a valine residue at position-6 within two of the β-polypeptide chains of the hemoglobin molecule. The resulting abnormal blood cells have a decreased affinity for oxygen and frequently become trapped within small capillaries because of their crescent shape.

(k) Proteins contain two or more polypeptide chains (e.g., hemoglobin).

(l) The nonprotein portion of a conjugated protein.

(m) A process used to separate and identify specific proteins from a complex mixture of proteins by subjecting them to an electric field.

(n) Any change that alters the three-dimensional conformation of a protein without disrupting the integrity of the primary structure (peptide bonds).

13.2 (a) A polypeptide is an amino acid polymer whose molecular weight is generally below 5000 daltons (5–35 amino acids). A protein is a large amino acid polymer consisting of one or more polypeptide chains (MW/greater then 5000 daltons).

(b) An N-terminal amino acid is the amino acid at the end of the peptide chain that has a free amino group, whereas the C-terminal amino acid is at the opposite end of the chain and has a free carboxyl group.

(c) Oxytocin is a peptide hormone that controls lactation and promotes smooth muscle contraction. Vasopressin is a peptide hormone that induces a rise in blood pressure and serves to increase the retention of fluids by the kidney.

(d) The α-helical conformation of a polypeptide is a secondary structure maintained by intramolecular hydrogen bonding between the amide hydrogen of one peptide bond and a carbonyl oxygen of another peptide bond further along the chain. The β-pleated sheet is a secondary structure maintained by intermolecular hydrogen bonds between adjacent polypeptide chains.

(e) Primary structure is the actual number and sequence of amino acids in a polypeptide chain, whereas secondary structure describes the fixed conformation of a polypeptide backbone.

(f) Tertiary structure is a unique three-dimensional conformation produced by the folding and bending of a polypeptide backbone. Quaternary structure describes the conformation of proteins that consist of two or more polypeptide chains.

(g) Interchain H-bonds occur between components of two individual polypeptide chains, whereas intrachain H-bonds exist between components of a single polypeptide chain.

(h) Hemoglobin is a conjugated protein that consists of four polypeptide chains. (Each chain contains the prosthetic group heme.) Myoglobin is a conjugated protein that consists of a single polypeptide chain complexed to a heme group.

(i) The α-keratins are the proteins present in hair and wool. They maintain an α-helical conformation and are rich in the amino acid cysteine (disulfide bonds). β-Keratins exist in a pleated-sheet conformation and do not contain the amino acid cysteine (no disulfide bonding).

(j) A simple protein is a polymer comprised of only amino acids. A conjugated protein contains both a protein (amino acids) and a nonprotein component (prosthetic group).

(k) Globular proteins are spherical, compact proteins that are soluble in water, whereas fibrous proteins are stringy, elongated proteins that are water insoluble.

(l) Albumins are an abundant class of simple proteins that are soluble in both water and dilute salt solutions. Globulins are a class of simple proteins that are widely distributed throughout the body. Although they are insoluble in water, they are soluble in dilute salt solutions. The γ-globulins constitute the antibodies present in blood serum.

(m) Glycoproteins are conjugated proteins consisting of a sugar and a protein component. They are frequently found in the outer surface of cell membranes where they have a variety of receptor functions. Lipoproteins are conjugated proteins consisting of triacylglycerols, cholesterol esters, and phospholipids attached to protein molecules. They are found in blood serum, where they function in the transport of lipids, in egg yolk, cell nuclei, ribosomes, and in many varieties of cell membranes.

(n) The ninhydrin test is a general test for proteins that exploits the existence of the free amino group at the N-terminus (positive test: blue color). The biuret test is a general test that gives a positive reaction (violet color) with peptides that contain two or more peptide bonds.

13.3 Structural; transportation; catalytic (enzymes); communication (hormones); defense (antibodies); protection.

13.4 In addition to carbon, hydrogen, and oxygen, all proteins contain nitrogen and, to a lesser extent, sulfur and phosphorus plus traces of metal ions.

13.5

4-hydroxyproline

5-hydroxylysine

ε-*N*-methyllysine

3-methylhistidine

γ-carboxyglutamic acid

13.6 (1) Nonpolar R-groups.   (2) Polar but neutral R-group.

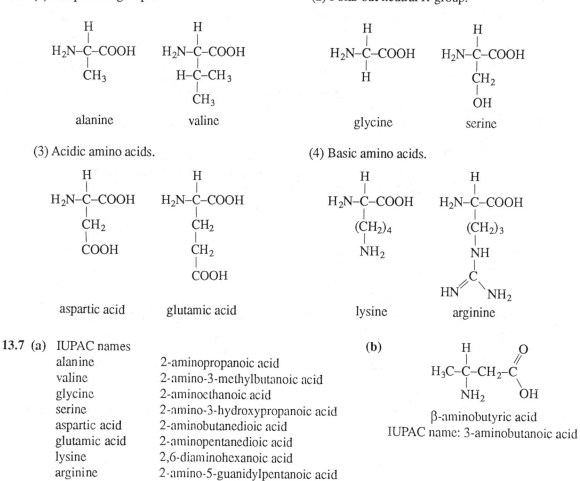

13.7 (a) IUPAC names
  alanine   2-aminopropanoic acid
  valine   2-amino-3-methylbutanoic acid
  glycine   2-aminoethanoic acid
  serine   2-amino-3-hydroxypropanoic acid
  aspartic acid   2-aminobutanedioic acid
  glutamic acid   2-aminopentanedioic acid
  lysine   2,6-diaminohexanoic acid
  arginine   2-amino-5-guanidylpentanoic acid

(b)

β-aminobutyric acid
IUPAC name: 3-aminobutanoic acid

13.8 Asparagine and glutamine are the amide derivatives of aspartic acid and glutamic acid. These amino acids have polar character, yet are electrically neutral at physiological pH. The second carboxyl group of both aspartic acid and glutamic acid is negatively charged at physiological pH and is responsible for the acidity of these latter compounds.

13.9 Two amino acids that contain more than one chiral carbon are isoleucine and threonine.

(d)

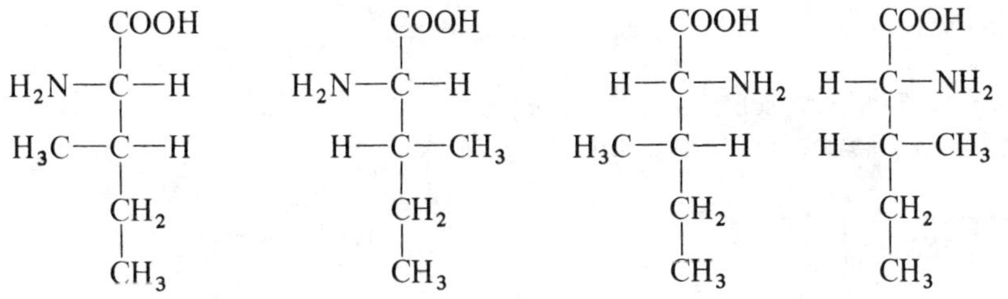

| L | L | D | D |

13.10 **(a)** proline   **(b)** histidine, tryptophan   **(c)** phenylalanine, tyrosine, tryptophan
(d) citrulline, ornithine, dihydroxyphenylalanine, thyroxine, homocysteine, homoserine, β-alanine, γ-aminobutyric acid
(e) glycine   **(f)** hydroxyproline   **(g)** asparagine   **(h)** arginine   **(i)** cysteine
(j) tyrosine   **(k)** valine, leucine, isoleucine   **(l)** proline

13.11

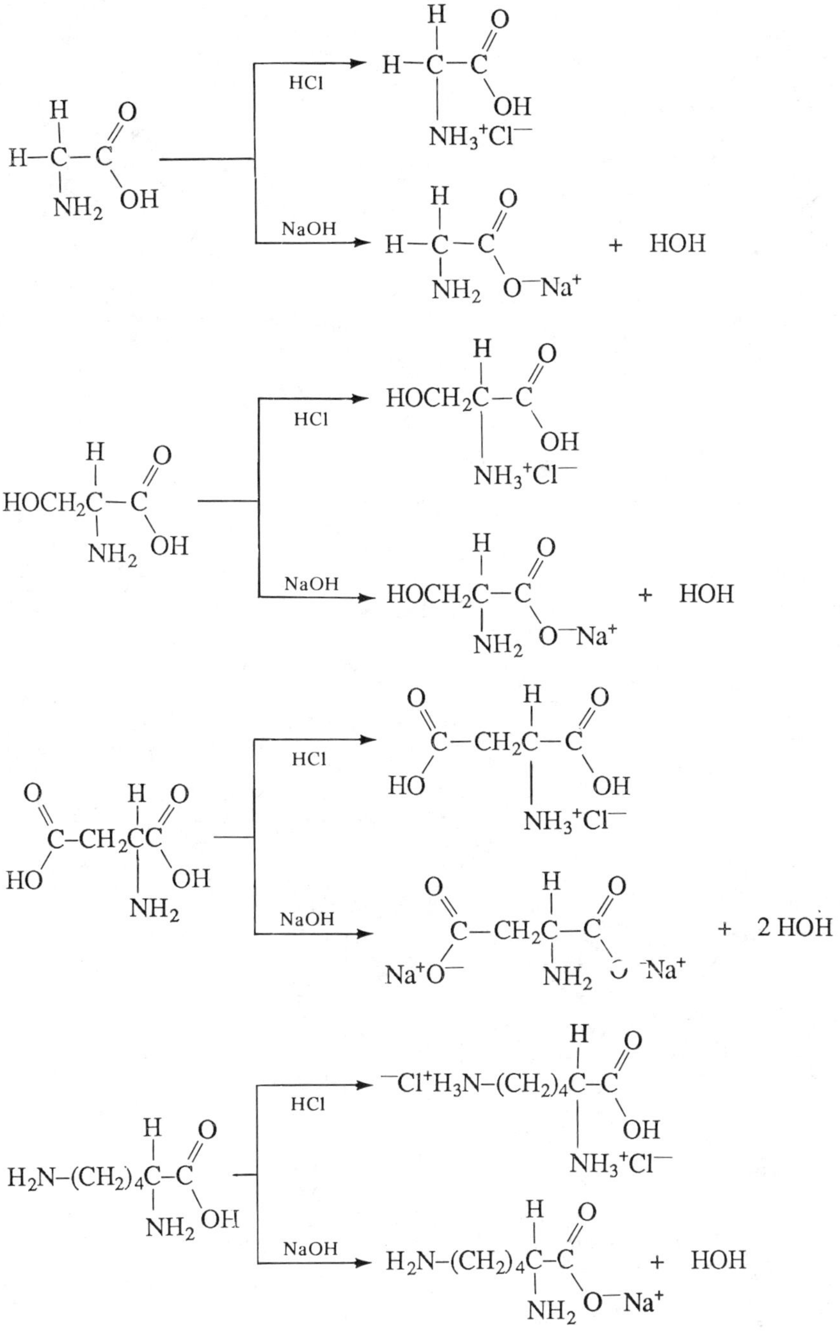

13.12 Alanine, stationary; histidine, to cathode; aspartic acid, to anode.

13.13 (a) (2.36 + 9.60)/2 = 5.98 (isoelectric pH of leucine)

   (b) (6.0 + 9.2)/2 = 7.6 (isoelectric pH of histidine)

13.14 (a)

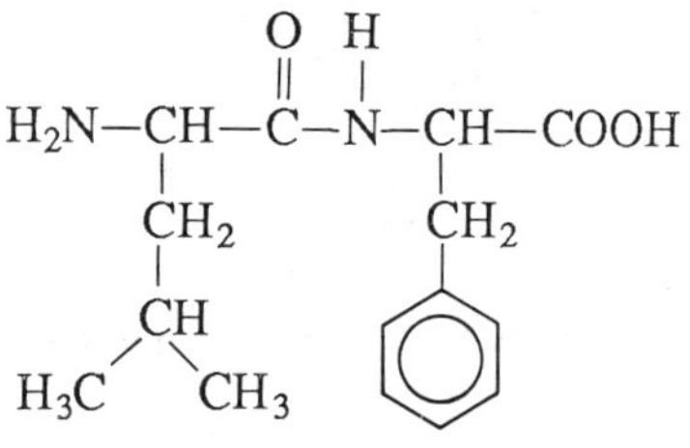

leucylphenylalanine
(Leu-Phe)

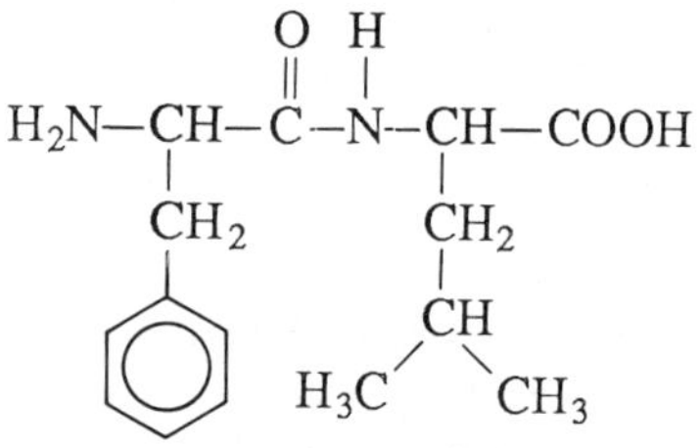

phenylalanylleucine
(Phe-Leu)

(b)

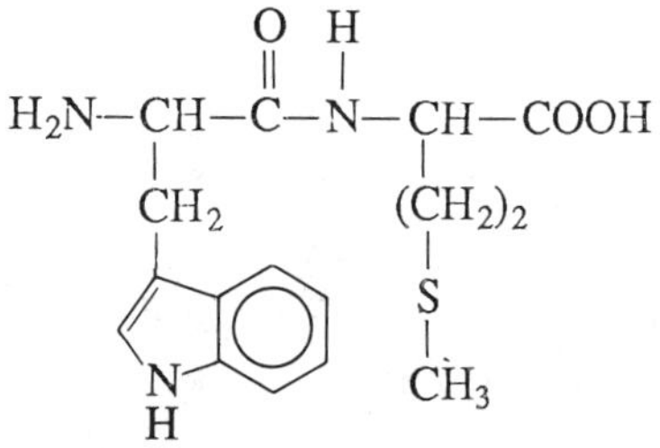

tryptophanylmethionine
(Trp-Met)

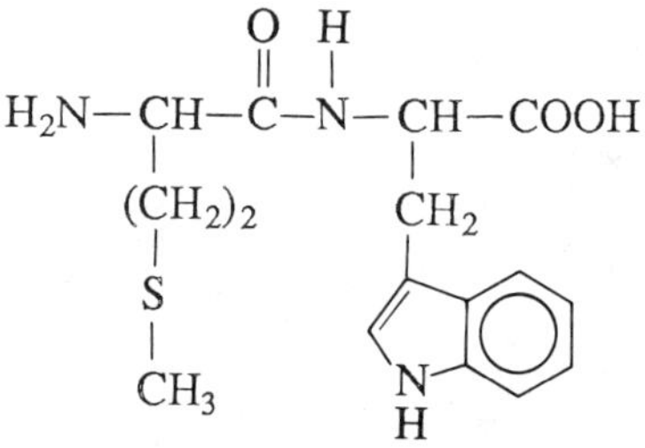

methionyltryptophan
(Met-Trp)

(c)

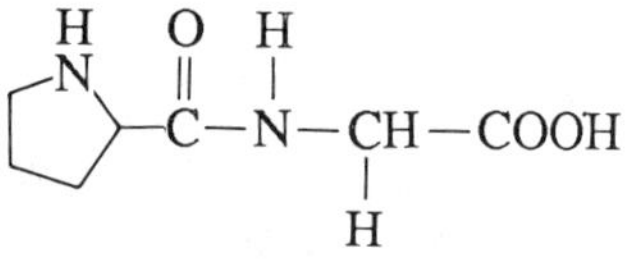

prolylglycine
(Pro-Gly)

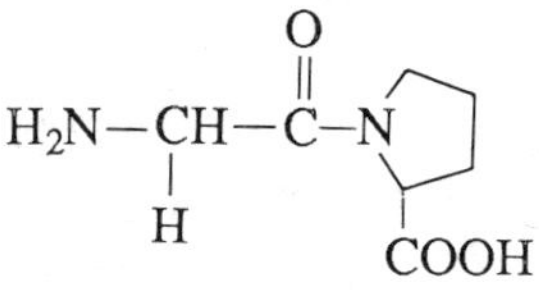

glycylproline
(Gly-Pro)

(d)

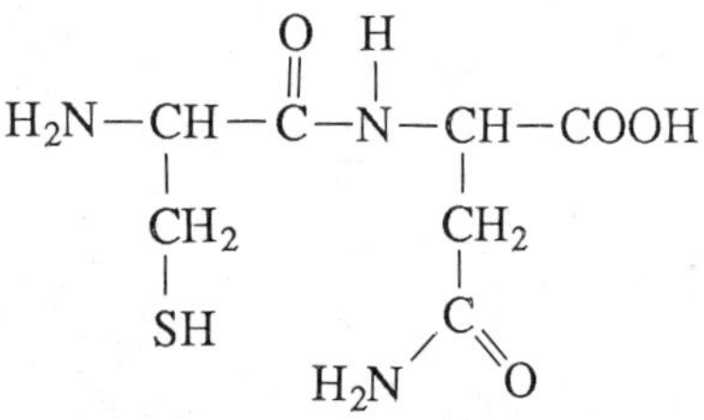

cysteinylasparagine
(Cys-Asn)

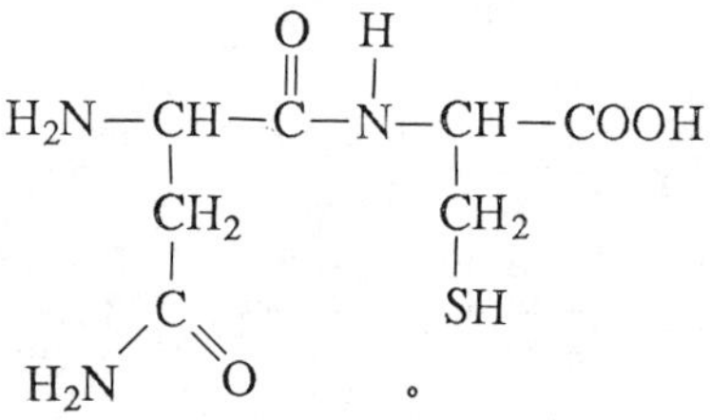

asparagylcysteine
(Asn-Cys)

(e)

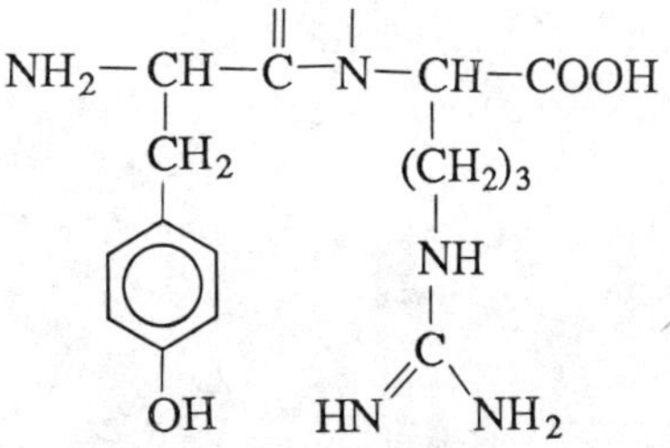

tyrosylarginine
(Tyr-Arg)

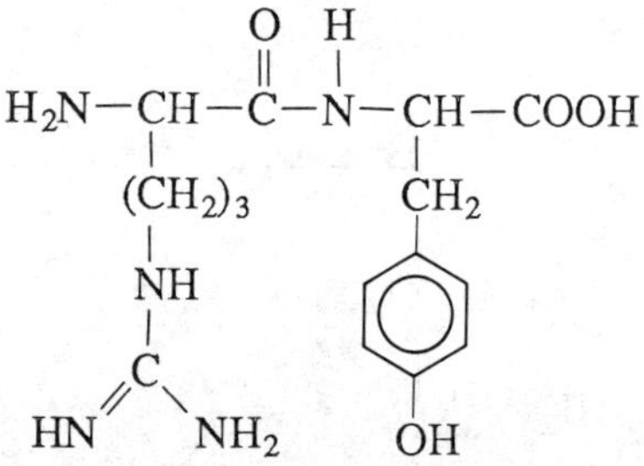

arginyltyrosine
(Arg-Tyr)

(f)

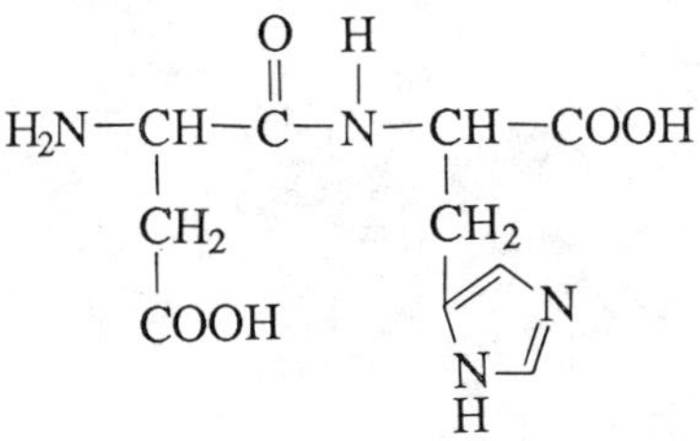

$$H_2N-CH-\underset{\underset{\text{O}}{\overset{\text{O}}{\parallel}}}{C}-N-CH-COOH$$

aspartylhistidine
(Asp-His)

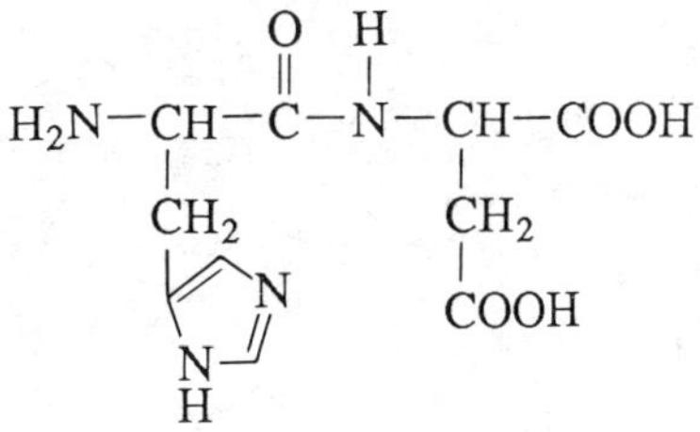

histidylaspartic acid
(His-Asp)

13.15 ABCD, ACDB, ADBC, ABDC, ACBD, ADCB, BCDA, BDAC, BACD, BADC, BCAD, BDCA, CDAB, CDBA, CBDA, CBAD, CADB, CABD, DABC, DACB, DCBA, DCAB, DBAC, DBCA

13.16 six: AAVV, AVAV, AVVA, VVAA, VAVA, VAAV.

13.17 aspartic acid

13.18

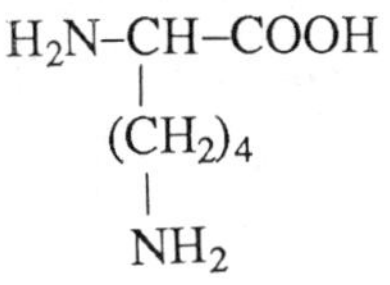

lysine 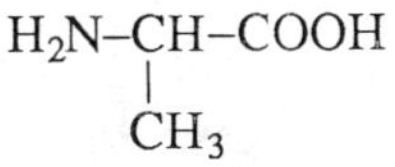   alanine 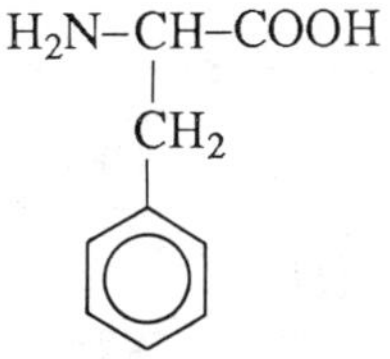   phenylalanine 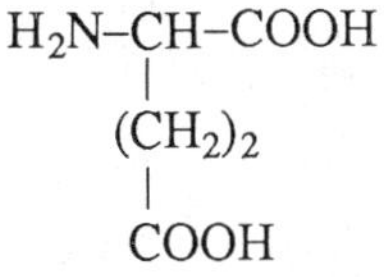   glutamic acid

13.19

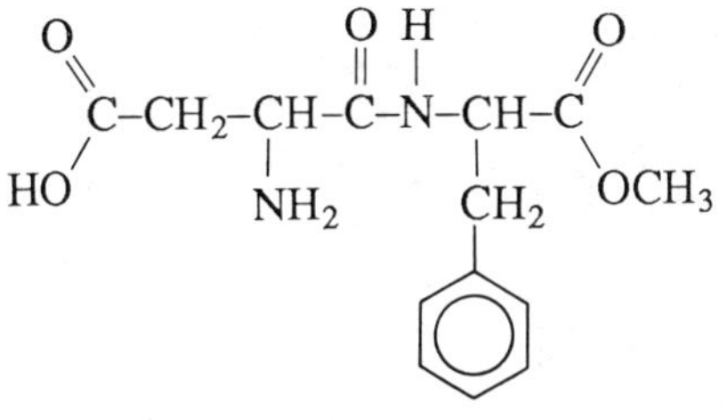

aspartame

13.20

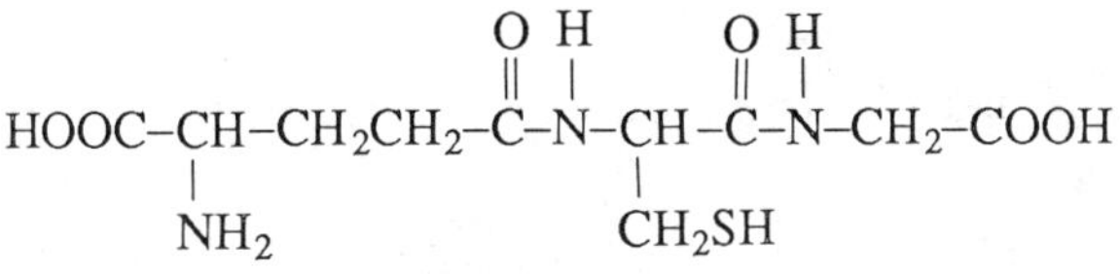

glutathione

13.21

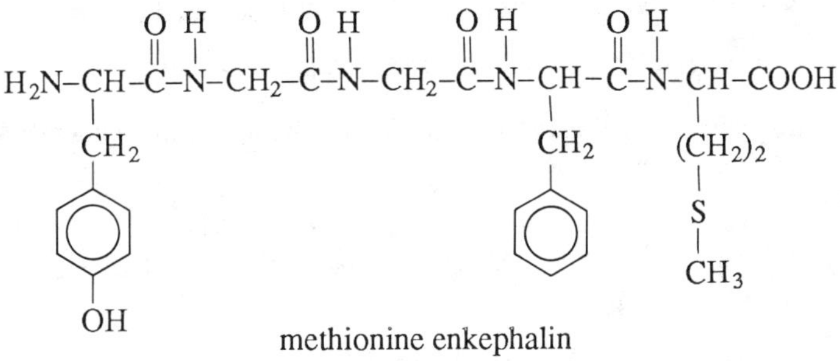

methionine enkephalin

leucine enkephalin

13.22 Leu-Val-Gly-Cys-Glu-Met-Cys-Ala

13.23 Glucagon: histidine–serine–glutamic acid–glycine–threonine–phenylalanine–threonine–serine–aspartic acid–tyrosine–serine–lysine–tyrosine–leucine–aspartic acid–serine–arginine–arginine–alanine–glutamine–aspartic acid–phenylalanine–valine–glutamine–tryptophan–leucine–methionine–asparagine–threonine

13.25

angiotensin

13.24

bradykinin

The secondary structure of bradykinin is not in an α-helical conformation because the presence of proline disrupts that orientation.

13.26 The secondary structure of a protein is stabilized by hydrogen bonds between amino hydrogens and carbonyl oxygens (of the polypeptide backbone) either within the same molecule or between separate polypeptide chains. The hydrogen bonding that serves to stabilize the tertiary structure of proteins takes place between the amino acid side chains that project from the polypeptide backbone.

13.27 carboxyl, hydroxyl, amino, sulfhydryl, imidazole, guanidyl, indole, phenol, phenyl

13.28 Peptide bonds (covalent); intrachain and/or interchain hydrogen bonds; salt linkages; disulfide linkages (covalent); hydrophobic interactions.

13.29 (a) hydrophobic interaction   (b) salt linkage   (c) hydrogen bond   (d) hydrogen bond
(e) hydrogen bond   (f) disulfide linkage   (g) hydrophobic interaction   (h) salt linkage
(i) hydrophobic interaction   (j) hydrogen bond

13.30 (a) interchain hydrogen bonds   (b) interchain disulfide bonds
(c) ionic interactions between polar side chains, interchain hydrogen bonds (and some hydrophobic interactions between nonpolar side chains)

13.31 Casein—phosphate, phosphoprotein; (2) mucin—carbohydrate, glycoprotein; (3) hemoglobin—heme, chromoprotein; (4) RNA—nucleic acid, nucleoprotein.

13.32 **(a)** albumin, hemoglobin, fibrinogen   **(b)** casein   **(c)** albumin, vitellin   **(d)** immunoglobulin G   **(e)** α-keratin   **(f)** mucin   **(g)** collagen   **(h)** lysozyme, ribonuclease   **(i)** insulin, oxytocin, vasopressin   **(k)** pepsin, trypsin   **(k)** myoglobin   **(l)** hemoglobin

13.33 The acidic and basic groups in the side chains of the protein will react with bases and acids, respectively. This removes any excess acid or base, and thus the pH is maintained nearly constant.

13.34 **(a)** low pH   **(b)** high pH   **(c)** isoelectric pH

13.35 At the isoelectric pH, the numbers of positive and negative charges on the protein are equal. The protein will not migrate in an electric field, will be least stable, and will most readily precipitate out of solution.

13.36 Bacteria in the milk produce lactic acid, which lowers the pH of the milk, causing the protein casein to precipitate out in the form of white curds.

13.37 The isoelectric pH of glucagon would be about neutral because the positive and negative charges of the constituent amphoteric R-groups are nearly equal in number at a pH = 7. Therefore, the net charge of the molecule at physiological pH is nearly zero.

13.38 **(a)** +   **(b)** +   **(c)** −

13.39 **(a)** The heat acts as a denaturing agent, disrupting the hydrogen bonds and hydrophobic interactions that stabilize the proteins present in the egg. This process results in the coagulation of the denatured proteins.
   (b) Ultraviolet light and heat that are utilized as sterilization agents serve to disrupt the integrity of bacterial enzymes by denaturing the proteins that constitute those enzymes.
   (c) A reducing agent is applied to hair in order to disrupt the disulfide bonds present within hair proteins (α-keratin). The hair is set in the desired shape or pattern, and an oxidizing agent is applied to establish disulfide bonds between different amino acids.
   (d) The silver nitrate acts as an antiseptic in that it denatures bacterial enzymes and thus serves to prevent infection in the eyes of newborn infants.

13.40 A 95% alcohol solution only coagulates the proteins at the surface of the bacteria without killing them. A 70% solution is able to pass into the bacteria and denature enzymes vital to the survival of the bacteria.

13.41 **(a)** Tannic acid disrupts salt linkages by combining with the positively charged amino groups. Mercuric chloride disrupts salt linkages by combining with the negatively charged carboxyl groups. Thus both denature the protein, causing it to precipitate out of solution.
   (b) Acids denature proteins by disrupting salt linkages. (They bind to negatively charged carboxyl groups.) Alcohols denature proteins by disrupting existing hydrogen bonds and forming their own hydrogen bonds with amino acid side chains.

13.42 Egg white is administered as an antidote for heavy metal poisoning because the proteins present within the egg white readily react with the heavy metal ions to form insoluble precipitates. The insoluble precipitate is then removed from the stomach in order to prevent the release of the poisonous ions upon digestion of the coagulated protein.

13.43 Ninhydrin, free amino group. Xanthoproteic, amino acids containing a benzene ring. Millon, tyrosine residue. Hopkins-Cole, tryptophan residue. Biuret, two or more peptide bonds present.

CHAPTER 14

Enzymes

14.1 **(a)** A complex organic catalyst produced by living cells.
   (b) The substance upon which an enzyme acts.
   (c) A conjugated protein that acts as a biological catalyst.
   (d) The metal ion of a metalloenzyme.
   (e) An apoenzyme complexed to a metal ion.
   (f) An enzyme that is secreted in an inactive form and subsequently activated by covalent modification.
   (g) A compound that cannot be synthesized by an organism but is needed for the maintenance of normal metabolism.
   (h) The intermediate compound formed by the combination of an enzyme with its substrate.
   (i) The portion of the enzyme that binds to the substrate and at which transformation from substrate to product occurs.
   (j) The concept that an enzyme will catalyze only a reaction involving a specific substrate or only a certain type of reaction.
   (k) The concentration of an enzyme in a cell at any particular time is a function of the rate of its synthesis and·the rate of its degradation.
   (l) Any factor that suppresses the catalytic activity of an enzyme molecule.
   (m) The concept whereby the end product of a multistep metabolic pathway serves to inhibit the enzyme that catalyzes the initial step of that pathway (end-product inhibition).
   (n) In the case of end-product inhibition, it is the site on the enzyme at which the inhibitory molecule binds.
   (o) The use of chemicals (drugs) to destroy infectious microorganisms or cancer cells without damaging the cells of the host.
   (p) Substances that inhibit the growth of cancer cells by interfering with DNA replication and protein synthesis.
   (q) A controversial compound extracted from apricot pits that was thought to be an effective agent against cancer. In 1981, the FDA concluded that the drug had no cancer-curing properties.
   (r) Derivative of sulfanilamide; an antibacterial agent.
   (s) Ingredient in suntan lotions that functions as a sunscreen because it absorbs the short-wavelength ultraviolet rays that produce sunburn.
   (t) A disruption in the functioning of the heart muscle tissue (heart attack).

14.2 **(a)** Experiments performed in vitro are conducted in a test tube, whereas in vivo reactions take place within a living organism.
   (b) Intracellular enzymes are synthesized in cells and carry out their catalytic activities within those cells. Extracellular enzymes are synthesized within cells, then secreted to catalyze reactions outside the cell.
   (c) Sucrose is the substrate molecule (disaccharide) that is hydrolyzed by the enzyme sucrase into its constituent monosaccharides (glucose and fructose).
   (d) An apoenzyme is the protein portion of a holoenzyme (conjugated protein enzyme), whereas a coenzyme is the nonprotein portion of a holoenzyme.

(e) Peptidases catalyze the hydrolysis of peptide linkages in proteins, whereas lipases hydolyze ester linkages in lipid molecules.

(f) Trypsin is a proteolytic enzyme that catalyzes the hydrolysis of peptides in the small intestine. Trypsinogen is the inactive precursor form (zymogen) of the enzyme trypsin.

(g) The induced-fit theory depicts the active site of an enzyme as a flexible, dynamic structure. The contact groups are a cluster of amino acids responsible for the proper orientation of the substrate within the active site, whereas the catalytic groups are those amino acids that directly participate in the transformation of the substrate into product.

(h) Enzymes that exhibit absolute specificity catalyze a particular reaction for one particular substrate and do not react with structurally similar substrates. Those enzymes that exhibit stereochemical specificity react only with a particular stereoisomeric form of a particular substrate molecule (e.g., an enzyme that reacts with the D-form of a particular sugar but not with the L-form of that sugar).

(i) Group specificity refers to the enzymes that react with structurally similar molecules that have the same functional group. Enzymes that exhibit linkage specificity react with a particular type of chemical bond regardless of the structural features surrounding that linkage; they are the least specific of all the enzyme classes.

(j) Optimum temperature is the temperature at which a given enzyme exhibits maximum catalytic activity, whereas optimum pH is the pH value at which an enzyme exhibits maximum activity.

(k) A competitive inhibitor is a compound that is structurally similar to an enzyme's natural substrate and competes with that substrate for occupation of the same active site. A noncompetitive inhibitor is a compound that is structurally dissimilar from an enzyme's normal substrate and can combine with either the free enzyme or the enzyme–substrate complex.

(l) An irreversible inhibitor is a compound that does not resemble an enzyme's natural substrate yet functions to inactivate the enzyme by forming a strong bond with a particular group at the active site. An end-product inhibitor is the final product of a sequential series of enzyme-catalyzed reactions that inhibits the enzyme that functioned to catalyze the initial reaction in the pathway.

(m) An antimetabolite is a compound that structurally resembles an enzyme's normal substrate and acts to competitively inhibit a significant metabolic reaction. An antibiotic is a substance that is produced by one microorganism (or synthesized in the laboratory) that functions to inhibit the growth of another microorganism.

(n) Bacteriostatic agents function to inhibit the growth of bacteria, whereas bacteriocidal agents are capable of killing bacteria.

14.3 Louis Pasteur studied such enzyme-catalyzed reactions as alcohol fermentation and the souring of milk. Although he did not have a conception of enzymes, he did note that such catalysis took place in the presence of microorganisms. In 1897, Edward Buchner prepared a cell-free filtrate from yeast cells and observed the conversion of glucose into alcohol (fermentation) in the absence of the intact yeast cells. This experiment is thought to mark the beginning of modern enzyme chemistry. In 1926, James Sumner isolated the first enzyme (urease) in a pure, crystalline form. He proceeded to characterize the enzyme as being proteinaceous in nature.

14.4 Enzymes and ordinary catalysts both incease the rates of chemical reactions. Enzymes are much more specific than ordinary catalysts. Although in an enzyme-catalyzed reaction equilibrium is reached more quickly, the enzyme does not alter the position of equilibrium in a reversible reaction.

14.5 Because of the unique amino acid R-groups present at the active site and the unique geometric conformation of the active site.

14.6 Animals lack the enzymes necessary for the hydrolysis of the β-1,4-glucosidic linkage found in cellulose. Animals do possess the enzymes needed to hyrolyze the α-1,4 and α-1,6-glucosidic linkages that characterize the starch polymer.

14.7 Intracellular enzymes: lactic acid dehydrogenase, fumarase, succinic acid dehydrogenase.
Extracellular enzymes: ptyalin, pepsin, trypsin.

14.8 Peptidases are degradative enzymes that hydrolyze the peptide bonds present in protein molecules. If peptidases were synthesized within cells in an active form, they would immediately begin to degrade important cellular protein constituents. Therefore, they are synthesized in an inactive form in order to safeguard the existence of intracellular proteins.

14.9 (a) the hydrolysis of carbohydrates   (b) the hydrolysis of lipids   (c) the hydrolysis of esters   (d) the hydrolysis of dipeptides   (e) an isomerization reaction   (f) an oxidation–reduction reaction   (g) a decarboxylation reaction   (h) a reaction that involves the transfer of chemical groups   (i) any hydrolysis reaction   (j) the hydrolysis of penicillin

14.10 (a) hydrolase (sucrase/invertase)   (b) hydrolase (lactase)   (c) oxido-reductase (ascorbic acid dehydrogenase)   (d) oxido-reductase (alcohol dehydrogenase)   (e) hydrolase (peptidase)   (f) lyase (urea decarboxylase)   (g) isomerase (glucose isomerase)   (h) lyase (lysine decarboxylase)   (i) hydrolase (lipase)   (j) hydrolase (maltase)

14.11 Thiamine (Vitamin B_1): Converted to an important coenzyme that participates in a variety of metabolic reactions. (Reactions that remove and/or transfer aldehyde groups.) Riboflavin (Vitamin B_2): Converted in the body to the coenzymes FAD and FMN, which participate in metabolic oxidation–reduction reactions. Pyridoxine (Vitamin B_6): Enhances the decarboxylation of L-dopa. Cyanocobalamin (Vitamin B_{12}): Converted to a coenzyme that participates in transmethylation reactions and carbon skeleton rearrangements. Meats, eggs, milk, fish, and whole grain cereals are all good dietary sources of B vitamins.

14.12 (a) riboflavin (B_2); niacin   (b) biotin   (c) thiamine (B_1)   (d) pyridoxine (B_6)   (e) folic acid   (f) cyanocobalamin (B_{12})

14.13 Enzymes function by first combining with a substrate molecule to form an intermediate compound referred to as an enzyme–substrate complex. (This intermediate is more reactive than the substrate alone.) In a subsequent step this intermediate reacts further (usually with another reactant) to form products and to regenerate the enzyme.

14.14 Enzymes are highly efficient catalysts that convert substrate molecules into product at amazing rates. Some enzymatic reactions occur in milliseconds (msec), whereas others are more rapid with rates measured in microseconds (μsec) (e.g., the enzyme catalase can convert 50,000 molecules of hydrogen peroxide into water and oxygen per second.) The enzyme molecules are constantly regenerated.

14.15 The functional groups at the active site responsible for the formation and maintenance of the enzyme–substrate complex could be any of the following: carboxyl, amino, hydroxyl, sulfhydryl, imidazole, phenol, metal ions, phosphate.

14.16 The lock and key theory is a static model. The substrate must have the correct conformation and the complementary groups to match the conformation and groups at the active site of the enzyme. Only those molecules that can bind to the active site will undergo reaction. The induced-fit theory, on the other hand, proposes that several different molecules can bind to the active site, but only those in which there is proper alignment of the catalytic groups will undergo further reaction.

14.17 Although two amino acids are far apart in the primary structure of a protein chain, they may be brought within proximity at the active site during the unique folding and bending that characterizes the three-dimensional globular nature of enzyme molecules (tertiary structure).

14.18 To maintain the structure of the active site in the proper conformation so that it will be readily available for binding to the substrate molecule.

14.19 (a) hydrogen bonding groups   (b) positively charged groups (salt linkages)

(c) hydrogen bonding groups   (d) negatively charged groups (salt linkages)
(e) hydrogen bonding groups   (f) sulfhydryl groups (disulfide linkages)
(g) nonpolar groups (hydrophobic interactions)   (h) hydrogen bonding groups

14.20 (1) Absolute specificity. Enzyme: urease; substrate: urea; products: carbon dioxide and ammonia.

(2) Stereochemical specificity. Enzyme: L-lactic acid dehydrogenase; substrate: L-lactic acid; product: pyruvic acid.

(3) Group specificity. Enzyme: pepsin; substrate: peptides; products: smaller peptides and amino acids.

(4) Linkage specificity. Enzyme: lipases; substrate: lipids; products: fatty acids and glycerol.

14.21 The digestive enzymes in moths can recognize and bind to the protein in wool, but they do not recognize and hence bind to the synthetic polymer. Nylon and most of the man-made plastics are non-biodegradable because evolution has still not produced microorganisms that can recognize synthetic polymers.

14.22 No. Enzymes are extremely specific catalysts. Often that specificity is dependent upon unique conformational changes that are induced when an apoenzyme binds to a specific coenzyme. More often, coenzymes actually participate in the specific reaction that the enzyme catalyzes by either donating or accepting a particular functional group.

14.23 The rate of an enzyme-catalyzed reaction increases as the substrate concentration increases until a rate-limiting point is reached. At that point, the enzyme molecules are saturated with substrate. After the saturation point is reached, any subsequent addition of substrate will not change the overall rate of reaction.

For all practical cases, the rate of an enzyme-catalyzed reaction increases as the concentration of enzyme increases.

The rate of an enzyme-catalyzed reaction increases with increasing temperature until a point of optimum activity is reached. Thereafter, the rate of reaction decreases with further increases of temperature.

The rate of an enzyme-catalyzed reaction is very sensitive to pH. The rate will maximize at a certain pH value and will decrease at pH values above and below this optimum value.

14.24 (a) Above the optimum temperature, excessive heat causes the denaturation of the enzyme. Heat, as a denaturing agent, disrupts those interactions responsible for maintaining the unique conformation of the enzyme. Below the optimum temperature, chemical reactions occur much more slowly because of a decrease in the kinetic energy of the molecules. Slow-moving molecules collide less frequently and therefore the probability of a favorable reaction (product formation) decreases.

(b) Above and below the optimum pH, the enzyme will become denatured.

14.25 No. The amylase present in saliva is an effective catalyst at pH values between 6.8 and 7.2. Because the pH of the stomach is low (pH = 2), amylase would be denatured.

14.26 The denaturation of an enzyme disrupts the interactions that are responsible for maintaining the unique three-dimensional conformation of the polypeptide backbone. Any structural change in the active site of an enzyme will severely hamper its ability to bind and react with the substrate. Extensive structural changes induced by a denaturing agent will render an enzyme inactive.

14.27 A competitive inhibitor is a compound that structurally resembles the substrate and competes with the substrate for the active site of an enzyme. The inhibitory effect is reversible by increasing substrate concentration.

A noncompetitive inhibitor is a compound that forms strong bonds with either the free enzyme or the enzyme–substrate complex. It binds to the enzyme at a site remote from the active site, but in so doing, alters the conformation of the active site.

An irreversible inhibitor is structurally dissimilar from the substrate and binds strongly to the active site. It thereby prevents the substrate from interacting with the active site. Both noncompetitive and irreversible inhibition cannot be reversed by increasing the substrate concentration.

End-product inhibition occurs when the final product of a sequential series of enzyme-catalyzed reactions binds to the allosteric site of the enzyme that initiates the pathway. In the process the conformation of the active site is altered and the substrate is prevented from binding. The inhibition cannot be reversed until the concentration of inhibitor decreases.

14.28 Add excess substrate to distinguish between competitive and noncompetitive inhibition. If enzyme activity increases, then it was competitive inhibition. If the addition of substrate has no effect on the inhibited enzyme system, then the inhibition was noncompetitive.

14.29 It may act as an irreversible inhibitor by blocking one or more hydroxyl groups at the active site of a crucial enzyme. Since it is an acid, it could also exert its poisoning effect by lowering the pH so that enzymes become inactivated by denaturation.

14.30 Nerve gas irreversibly inhibits the functioning of the enzyme acetylcholinesterase. The active agent in nerve gas binds to the hydroxyl group of serine at the active site and thereby prevents the interaction of the enzyme with its normal substrate. The addition of excess acetylcholine cannot reverse the inhibition because the bond between the inhibitor and the enzyme is very strong.

14.31 Yes. It would act as a competitive inhibitor.

14.32 Salts of heavy metals are poisonous to the body because they can denature enzymes that are components of vital metabolic reactions.

14.33 Allosteric enzymes have two binding sites. The active site binds to a substrate molecule, whereas the allosteric site binds to a biological modulator. A biological modulator may function to inhibit enzyme activity (negative feedback) or enhance enzyme activity (positive feedback) by inducing conformational changes at the active site.

14.34 Because they are both effective inhibitors of a wide variety of enzymes that are essential to the growth of microorganisms.

14.35

$$H_2N-\langle\bigcirc\rangle-\overset{\overset{O}{\|}}{C}\overset{}{\underset{OCH_2CH_3}{\diagdown}}$$

benzocaine

14.36 penicillin; tetracycline; streptomycin; chloramphenicol; gramicidin S

CHAPTER 15

The Nucleic Acids

15.1 **(a)** The bond that joins the pentose sugar (ribose or deoxyribose) to the nitrogenous base (purine or pyrimidine).

(b) They are nucleosides (pentose linked to a nitrogenous base) bonded to a triphosphate group. Examples are ATP, GTP, CTP, UTP, and the deoxynucleoside triphosphates.

(c) The bond that joins adjacent nucleotides in nucleic acids.

(d) The stable base pairs held together by hydrogen bonds in the double-stranded helix of DNA. Adenine pairs with thymine, and guanine pairs with cytosine.

(e) In DNA, it is the sum of the total number of adenine and thymine bases divided by the sum of the guanine and cytosine bases.

(f) The shape of DNA. It is maintained by hydrogen bonds between complementary bases.

(g) Chromosomal proteins complexed with the DNA molecule that play a functional role in the regulation of gene expression.

(h) Heating the double-stranded helix of DNA results in hydrogen bond cleavage and hence the dissociation of the complementary strands.

(i) During replication, each strand of DNA serves as a pattern for the biosynthesis of its complementary strand.

(j) The process whereby the DNA that results from replication consists of one old strand and one new strand.

(k) Small fragments of DNA that are formed during the discontinuous replication of the lagging strand of DNA.

(l) DNA sequence to which RNA polymerase binds in order to initiate gene transcription.

(m) The term applied to the situation in which several tRNAs can bind to and transfer the same amino acid.

(n) A cellular substructure that serves as the site for protein synthesis.

(o) The concept that the sequence of bases in DNA serves to direct the synthesis of proteins within the cell. 61 of the 64 possible three-base sequences signify a particular amino acid.

(p) The movement of the ribosome along the mRNA strand a distance of 1 codon in the $5' \longrightarrow 3'$ direction.

(q) The concept that more than one nucleotide triplet (codon) can code for the same amino acid.

(r) Chemical or physical agents that can alter the base sequence of DNA.

(s) Diseases that arise from nonlethal mutations in the DNA of the germ cells.

(t) A genetic disease in which the afflicted individual is unable to convert phenylalanine to tyrosine; instead phenylalanine is converted to phenylpyruvate.

(u) A diagnostic assay of amniotic fluid.

(v) Viruses that cause cancer.

(w) RNA viruses.

(x) A DNA molecule that is formed by splicing a segment of DNA from one organism into the DNA of a second organism.

(y) Small, closed loops of DNA, found in *E. Coli*, that are used as the vectors in recombinant DNA research.

(z) Enzymes that cleave phosphodiester bonds at sites along a DNA molecule containing a particular nucleotide sequence.

15.2 **(a)** The pyrimidine bases are derivatives of pyrimidine; purine bases are derivatives of purine (a pyrimidine ring fused to an imidazole ring).

(b) The major bases are adenine, guanine, thymine, cytosine, and uracil. The minor bases are modified derivatives of the major bases.

(c) Ribose is the pentose sugar found in RNA. It has an OH group on carbon-2. Deoxyribose is the pentose found in DNA, and it has an H instead of OH on carbon-2.

(d) A nucleoside consists of a purine or pyrimidine base joined to either ribose or deoxyribose. A nucleotide is a nucleoside bonded to a phosphate group.

(e) Nucleic acids are polymers of nucleotides.

(f) Germ cells are the sex cells, sperm and egg. Somatic cells are all the cells in the body that are not germ cells and that are not involved in the production of the germ cells.

(g) DNA polymerase is the enzyme that catalyzes the biosynthesis of DNA by linking nucleotides together according to the pattern established by a DNA template. DNA ligase is an enzyme that catalyzes the linkage of short fragments of DNA during replication and repairs broken strands.

(h) The T loop is that portion of the cloverleaf structure of a tRNA molecule that contains the base thymine. The D loop contains the minor base dihydrouracil.

(i) The codon is the specific three-base sequence of nucleotides on the mRNA strand that codes for a particular amino acid. The anticodon is a complementary three-base sequence on the tRNA.

(j) Chromosomes are composed of DNA molecules. A gene is a segment of the DNA molecule that codes for the synthesis of a particular polypeptide.

(k) Transcription is the process of synthesizing a mRNA molecule from one strand of DNA that serves as the template. Translation is the synthesis of a polypeptide at the ribosome from the information encoded in the mRNA.

(l) The cytoplasm refers to all the contents of the cell with the exception of the nucleus. The cytosol consists of all the fluid outside the membrane-bound organelles.

(m) The aminoacyl (A) site is the portion of the ribosome to which an aminoacyl tRNA molecule initially binds. The peptidyl (P) site is the portion of the ribosome where the growing polypeptide chain is attached.

(n) A mutation is any chemical or physical change that alters the sequence of bases in the DNA. If the alteration in the base sequence occurs at just one nucleotide in the DNA polymer, then it is called a point mutation.

15.3 (a) adenine, guanine, thymine, cytosine, deoxyribose, phosphate
(b) adenine, guanine, uracil, cytosine, ribose, phosphate

15.4   PYRIMIDINES   PURINES

uracil   thymine   cytosine   adenine   guanine

15.5 **(a)**   **(b)**   **(c)**   **(d)**

5-methylcytosine   5-hydroxymethylcytosine   6-methyladenine   2-methylguanine

(e)

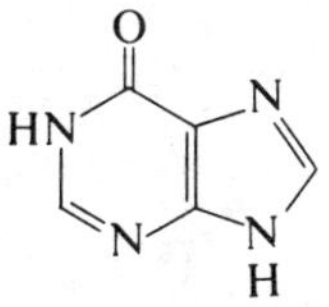

(f)

(g)

(h)

6-oxypurine   4-thiouracil   5,6-dihydrouracil   1-methylguanine

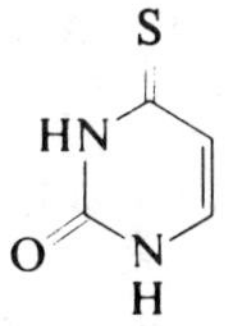

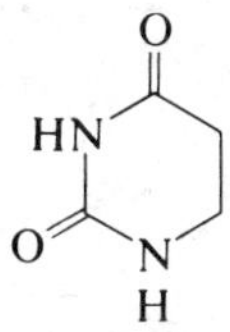

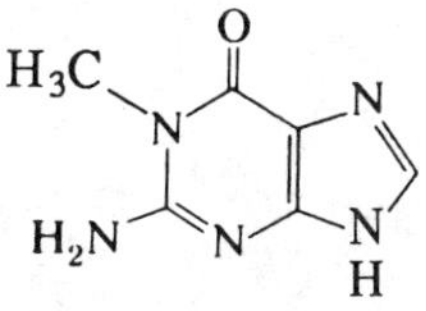

15.6 (a)

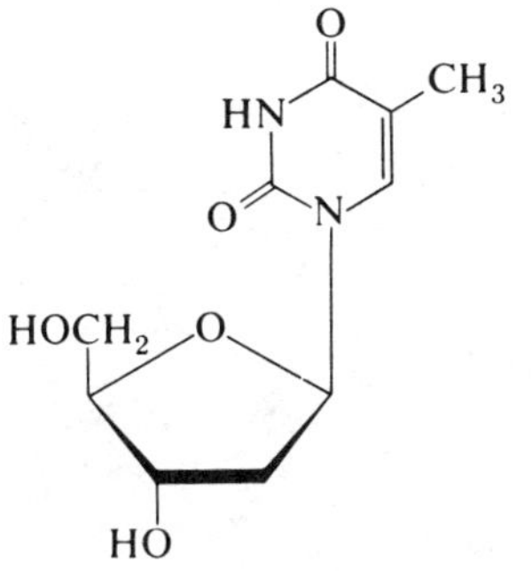

(b)

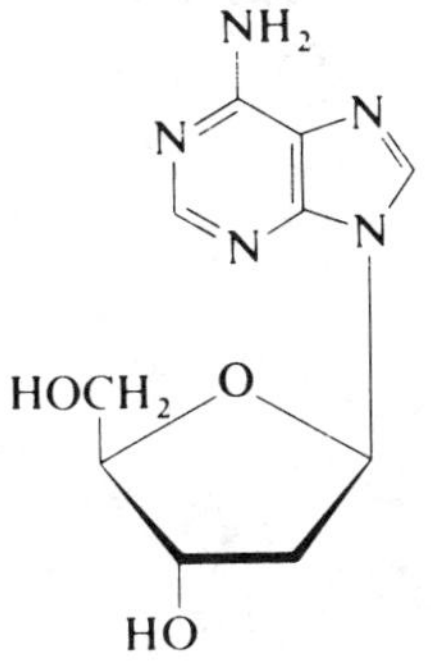

deoxythymidine   deoxyadenosine

15.7

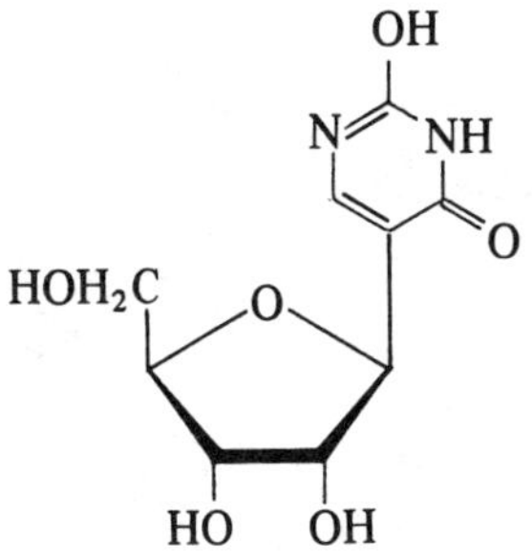

pseudouridine

15.8 Thymine differs from uracil only in having a methyl group in place of a hydrogen on carbon-5. This portion of the molecule does not participate in hydrogen bonding.

15.9 (a)

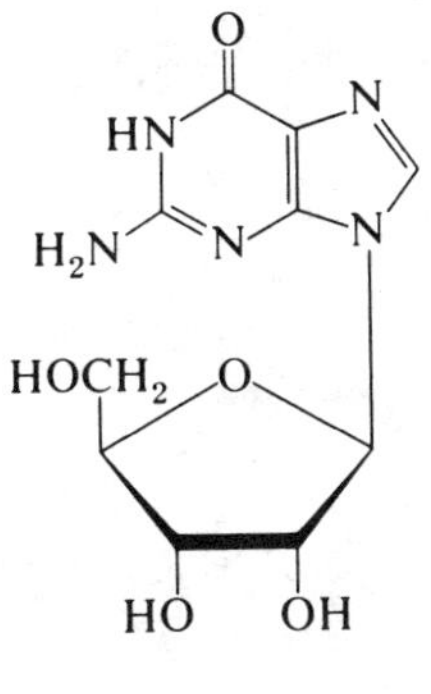

(b)

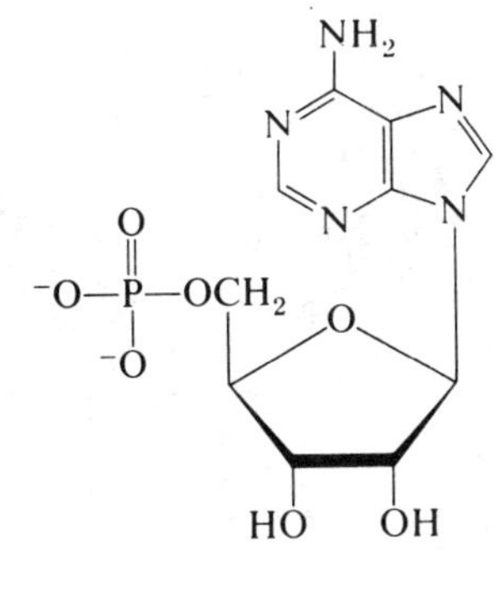

guanosine   AMP

(c)

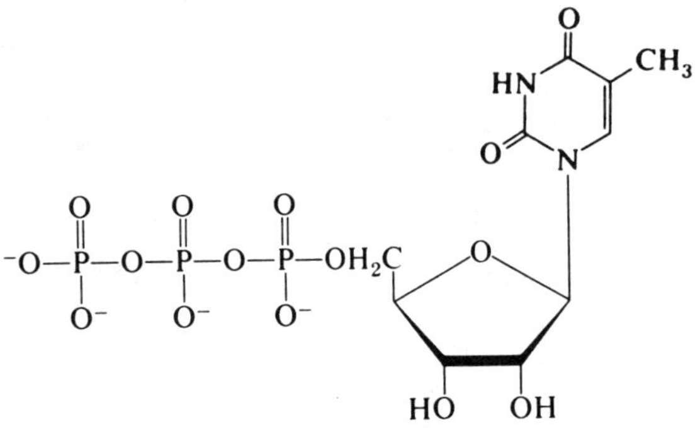

TTP

(d)

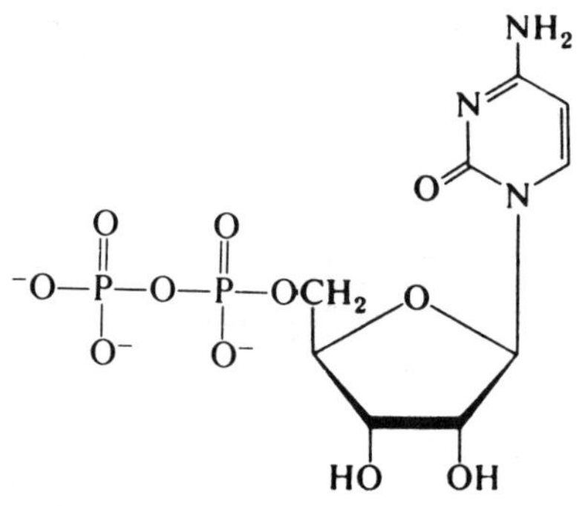

CDP

(e)

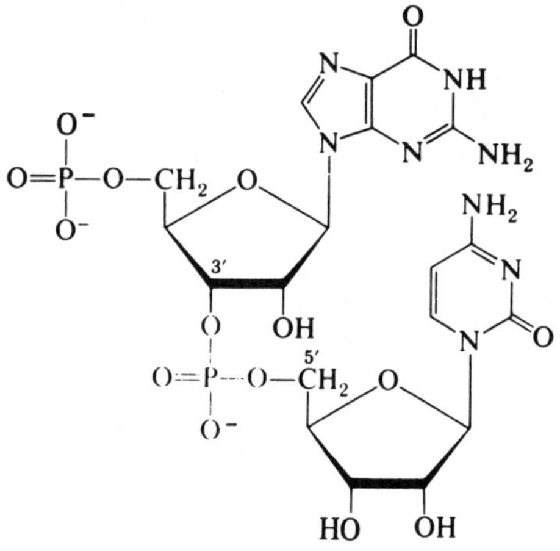

$^{5\prime}$G–C$^{3\prime}$

(f)

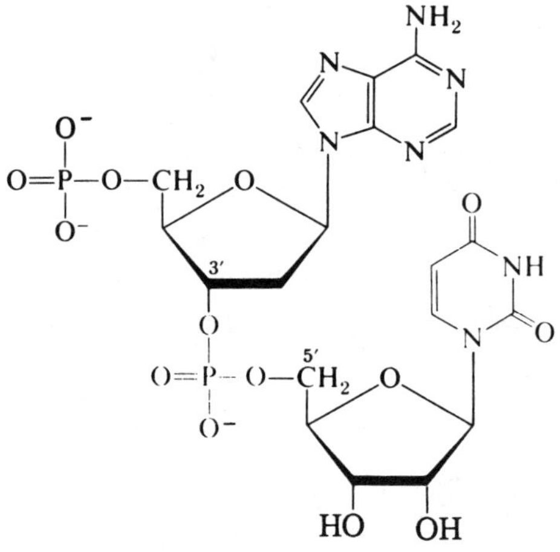

$^{3\prime}$U–A$^{5\prime}$

15.10

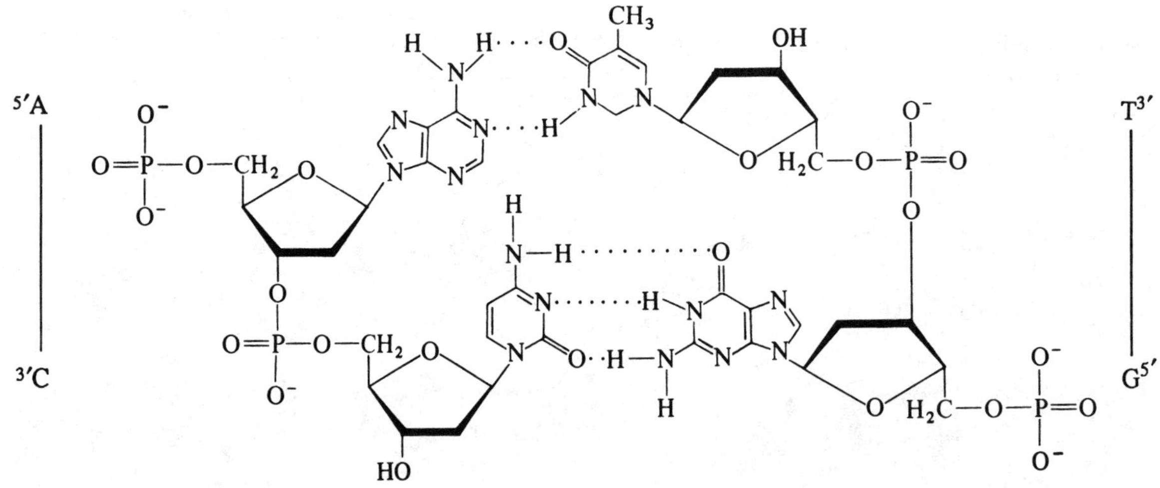

$^{5\prime}$A

$^{3\prime}$C

T$^{3\prime}$

G$^{5\prime}$

(a)   (b)

15.11

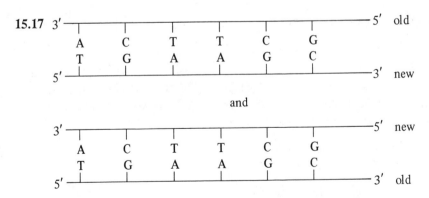

cAMP

15.12 (a) The backbone of the DNA strand is comprised of deoxyribose and phosphate groups. Adjacent nucleotides are linked by a 3′, 5′-phosphodiester bond that characterizes the sugar–phosphate backbone.

(b) One end of the DNA strand is characterized by the presence of a free 5′-phosphate group, whereas the opposite end of the same strand is characterized by the presence of a free 3′-hydroxyl group.

(c) The two strands are held together by hydrogen bonds.

15.13 Primary structure of DNA refers to the actual number and sequence of nucleotides in the strand. Secondary structure of DNA refers to the double-stranded helical conformation of that molecule.

15.14 Phosphoric acid is a major structural component of the backbone of nucleic acid molecules. At physiological pH, the hydrogens of the acid are dissociated, and the phosphate group is negatively charged.

15.15 Antiparallel structure of DNA means that the phosphodiester linkages of one strand are aligned in a 5′ ⟶ 3′ direction, whereas the phosphodiester linkages of the complementary strand are aligned in a 3′ ⟶ 5′ direction.

15.16 Inherited characteristics such as hair, eye, and skin color depend upon the proteins that are synthesized by the respective cells. Upon cell division, an exact copy of the parent cell's DNA is passed on to each daughter cell. Within this DNA is encoded the message for the synthesis of proteins needed by that cell.

15.17

```
3' ─┬────┬────┬────┬────┬────┬──── 5'  old
    A    C    T    T    C    G
    T    G    A    A    G    C
5' ─┴────┴────┴────┴────┴────┴──── 3'  new
```

and

```
3' ─┬────┬────┬────┬────┬────┬──── 5'  new
    A    C    T    T    C    G
    T    G    A    A    G    C
5' ─┴────┴────┴────┴────┴────┴──── 3'  old
```

15.18 (a) DNA is a double-stranded molecule containing the pentose deoxyribose. RNA is mainly a single-stranded molecule containing the pentose ribose.

(b) DNA contains adenine, guanine, cytosine, and thymine. RNA contains adenine, guanine, cytosine, and uracil.

15.19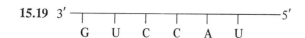
3′ ⊢────┬────┬────┬────┬────┬────⊣ 5′
         G   U   C   C   A   U

15.20 (a) DNA contains the information for the number and sequence of amino acids to be incorporated into proteins.
(b) mRNA transmits the DNA message to the area of protein synthesis.
(c) tRNA binds to particular amino acids and brings them to the ribosomes.
(d) rRNA forms part of the structure of the ribosomes.

15.21 (a) Three nucleotide units are present in a codon.
(b) By taking the bases in groups of three, it is possible to code for 4^3 or 64 different amino acids. This is more than sufficient to code for all the known amino acids found in proteins.

15.22 The site to which the amino acid binds (3′-position of the terminal adenosine nucleoside) and the three nucleotide residues that constitute the anticodon.

15.23 Highly specific enzymes (aminoacyl–tRNA synthetases) and tRNA molecules are responsible for assuring the specific interaction between particular amino acids and tRNAs.

15.24 (a) 60,000/120 = 500 amino acids present in the protein.
(b) 500 codons specifying amino acid sequence and 1 termination codon.
(c) 500 codons = 1,500 nucleotide bases; termination codon = 3 nucleotide bases.

15.25 Met–Tyr–His–Gly–Thr–Arg–Val–Leu–Leu–Ala–Asp–Gly (termination)

15.26 Met–Leu–Arg–Ser–Tyr–Glu–Gly–Phe–His–Lys–Thr–Met (termination)

15.27 (a) AUGGCUGGUUGUAAAAAUUUUUUCUGGAAGACUUUUACCAGUUGCUAG
(b) TACCGACCAACATTTTTAAAAAAGACCTTCTGAAAATGGTCAACGATC

15.28 (a) Prevents transcription by blocking the movement of RNA polymerase.
(b) Prevents the attachment of aminoacyl–tRNAs at the ribosome.
(c) Inhibits peptide bond formation at the ribosome.
(d) Prevents the movement of the ribosome (translocation) along the mRNA strand.

15.29 Physical mutagens: ultraviolet and gamma radiation
Chemical mutagens: 5-bromouracil, 2-aminopurine, hydroxylamine, nitrous acid

15.30 Phenylketonuria—phenylalanine hydroxylase; (2) histidinemia—histidase; (3) goiter—iodotyrosine dehalogenase

15.31 Codons for valine are GUU, GUC, GUA, and GUG; codons for glutamic acid are GAA and GAG. Therefore, a single base substitution in the mRNA—GUA instead of GAA and/or GUG instead of GAG—could cause the incorporation of valine instead of glutamic acid. This would result from a base sequence of CAT instead of CTT and/or CAC instead of CTC in the DNA molecule.

15.32

Type	Codon for Normal	Codon for Mutant
J	GCU, GCC	GAU, GAC
I	AAA, AAG	GAA, GAG
M	CAU, CAC	UAU, UAC
D	GAA, GAG	CAA, CAG
K	GGU, GGC	GAU, GAC

15.33 (a) UGCAAUCGGGGUCGA   (b) Cys–Asn–Arg–Gly–Arg
(c) (i) Cys–Asn–Trp–Gly–Arg;   (ii) Cys–Asn–Gln–Gly–Ser;   (iii) Cys–Asn–Gly–Val

15.34 (a) The point mutation would result in a new codon (UUG instead of CUG), yet the amino acid at position-2 of the polypeptide would remain unchanged since both of those codons encode the incorporation of leucine.
(b) The mutation would result in a new codon for the amino acid at position-4 (UGA instead of UGU). Thus a termination codon replaces the codon for cysteine and a very much shortened, incomplete polypeptide is released. No functional enzyme is produced.

15.35 Viruses contain all the necessary information for their own reproduction, yet they are not capable of metabolism or energy transformation. They must depend upon living cells to supply the necessary enzymes and molecules (amino acids, nucleotides) to synthesize viral proteins and viral nucleic acids. Viruses are intracellular parasites. They cannot reproduce except inside a living host cell.

CHAPTER 16

Carbohydrate Metabolism

16.1 **(a)** Metabolite is any compound that is involved in a metabolic reaction.

(b) Villi are small, finger-like appendages that extend from the intestinal epithelial cells into the lumen of the digestive tract. They function to increase the surface area for the absorption of dietary foodstuffs.

(c) Blood-sugar level is the concentration of glucose in the blood.

(d) Renal threshold is 160–170 mg of glucose/100 mL of blood. When the blood-sugar level exceeds this value, glucose passes into the kidneys and is excreted in the urine.

(e) Glucosuria is the presence of excess glucose in the urine.

(f) Glucose tolerance test is a diagnostic test for diabetes mellitus.

(g) Gluconeogenesis is the formation of glucose from a noncarbohydrate source.

(h) Adenyl cyclase is the enzyme that catalyzes the formation of cyclic adenosine 3′,5′-monophosphate (cAMP) from ATP.

(i) Cyclic AMP is an important intracellular messenger molecule that transmits messages received at the cell membrane to enzymes within the cell.

(j) Metabolic pathway is a sequential series of metabolic reactions (steps) requiring the formation of several intermediates before the overall reaction is completed.

(k) (P) is the symbol for the phosphite group, PO_3^{2-}.

(l) Kinase is an enzyme that catalyzes phosphorylation–dephosphorylation reactions involving ATP.

(m) Oxygen debt is a condition whereby the respiratory and circulatory systems cannot meet the body's oxygen demand for metabolic activities. As a result, muscle cells are forced to degrade glucose via the anaerobic pathway (glycolysis).

(n) Cori cycle is the anaerobic catabolism of glucose to lactic acid in the muscle cells and the subsequent reconversion of that lactic acid to glucose in the liver.

(o) Fetal alcohol syndrome is a disorder that results when pregnant women regularly consume alcohol. Alcohol that crosses the placental barrier builds up in the fetus whose liver is incapable of detoxifying the alcohol. FAS consists of facial deformities, growth deficiency, and mental retardation.

(p) Antabuse is a drug (Disulfiram) used in the treatment of alcoholism. It inhibits the second step of alcohol metabolism in the liver by competing with the intermediate acetaldehyde for the active site of acetaldehyde dehydrogenase. The resulting accumulation of acetaldehyde produces severe discomfort in the patient.

(q) Carbohydrate loading is a method used by athletes to boost performace by eating a high carbohydrate diet in order to maximize their storage of glycogen.

(r) Krebs cycle is a cyclic series of reactions that represent the aerobic continuation of the degradation of glucose (and other metabolites) to CO_2 and H_2O.

(s) Oxidative decarboxylation is an oxidation–reduction reaction accompanied by the removal of CO_2 from the substrate molecule.

(t) Mitochondria are cell organelles that are termed the "power plants" of the cell because they carry out reactions that generate energy for the cell.

(u) Respiratory chain is also called the electron transport chain. The sequence of reactions whereby the reduced forms of the coenzymes are reoxidized and energy is released.

(v) Cytochromes are enzymes containing a heme group that function in the respiratory chain.

16.2 (a) Anabolism is the biosynthesis of large molecules from smaller molecules. Catabolism is the degradation of large molecules into smaller molecules.

(b) Photosynthesis describes the series of reactions by which glucose is formed from CO_2, H_2O, and energy from the sun. Respiration is the process whereby oxygen combines with food molecules (chiefly carbohydrates) to form CO_2 and H_2O and to release energy.

(c) Digestion is the hydrolytic process whereby food molecules are broken down into smaller molecules so they can be absorbed. Metabolism describes the processes whereby the absorbed molecules from food are utilized by an organism to provide energy, growth, maintenance, and repair.

(d) Passive transport is the passage of a substance across an inert membrane by osmosis or diffusion. Active transport requires that energy be expended in order for a substance to pass across a selective cell membrane.

(e) Hypoglycemia is a condition resulting from a lower than normal concentration of sugar in the blood. Hyperglycemia results from a higher than normal blood-sugar level.

(f) Glycogenesis is the formation of glycogen from glucose. Glycogenolysis is the hydrolytic breakdown of glycogen to glucose.

(g) Anaerobic reactions are those that do not require the presence of oxygen. Aerobic reactions are those that do require the presence of oxygen.

(h) In glycolysis, muscle cells and other animal cells respiring anaerobically convert pyruvic acid into lactic acid. Under similar conditions, the enzymes in yeast convert the pyruvic acid into ethyl alcohol and CO_2 (fermentation).

(i) A mutase catalyzes the intramolecular transfer of a particular group (e.g., phosphate group in glycolysis). An isomerase catalyzes the interconversion of one form of a sugar into another form (e.g., glucose 6-phosphate to fructose 6-phosphate).

(j) A phosphatase is an enzyme that catalyzes the transfer of a phosphate group from one molecule to another. A phosphorylase is an enzyme that catalyzes the phosphorolytic cleavage of a molecule by a reaction that is analogous to a hydrolytic cleavage (i.e., inorganic phosphate is used instead of water).

(k) Facultative anaerobes may function with or without oxygen. Strict anaerobes cannot function in the presence of oxygen.

(l) Substrate level phosphorylation is the direct transfer of a phosphate group from a metabolite to ADP to form ATP. Oxidative phosphorylation is a term for the processes involved in the formation of ATP from ADP and P_i using the energy released via the respiratory chain.

16.3 Carbohydrate digestion begins in the mouth where large polysaccharides (starch) are hydrolyzed by ptyalin (an amylase) to produce many smaller polysaccharides (dextrins). The primary site of carbohydrate digestion is the small intestine. The pancreas secretes a battery of hydrolytic enzymes into the intestinal lumen in order to complete the digestion of the polysaccharides and disaccharides into their constituent monosaccharides. Amylopsin, sucrase, lactase, and maltase are the pancreatic enzymes that hydrolyze the glycosidic linkages of starch, sucrose, lactose, and maltose, respectively. The resultant monosaccharides are absorbed across the intestinal lumen and transported into the blood.

16.4 The hormonal regulation of the synthesis and breakdown of glycogen in the liver is the major factor responsible for the maintenance of a constant blood-sugar level.

16.5 The three hormones insulin, glucagon, and epinephrine play a major role in the regulation of the blood-sugar level by regulating the rates of glycogenesis and glycogenolysis. Insulin makes the cell membrane more permeable to the transport of glucose from the blood into the cell, and it enhances glycogenesis and inhibits glycogenolysis. Both glucagon and epinephrine activate the enzyme phosphorylase, which catalyzes the breakdown of glycogen to glucose. Epinephrine promotes glycogen breakdown in both the liver and in muscles, whereas glucagon acts primarily on the liver. Insulin and glucagon are synthesized and secreted from the pancreas whereas epinephrine is synthesized and secreted from the adrenal glands.

16.6   Type I: Juvenile-onset diabetes is characterized by the inability of the pancreas to produce sufficient amounts of insulin. Patients suffering from Type I diabetes mellitus must be treated with daily injections of insulin. Type II: Adult-onset diabetes is characterized by the pancreatic production of insulin but there is failure of that gland to secrete sufficient amounts of the vital hormone, or there is a lack of insulin receptors on the target cells. This form of diabetes mellitus can be controlled with a combination of diet and exercise therapy.

16.7   The major role of the glycogen stores in the muscle cells is to provide energy for muscle contraction. In those cells glucose 6-phosphate is immediately converted to fructose 6-phosphate. The glycogen that is stored in the liver serves as a reservoir for glucose molecules that may be transported by the blood to other cells. Glucose 6-phosphate in the liver is converted to glucose by glucose 6-phosphatase. The dephosphorylated glucose molecules are then free to enter the circulation. Muscle cells lack the enzyme glucose 6-phosphatase and thus cannot export glucose into the bloodstream.

16.8   (a)

(b)

16.9   Iodoacetic acid noncompetitively inhibits the enzyme glyceraldehyde 3-phosphate dehydrogenase by covalently binding to free sulfhydryl groups at the enzyme's active site. Fluoride ions inhibit the enzyme enolase by binding to magnesium ions that are required for enzyme activity.

16.10   70–80% of the lactic acid diffuses out of the muscle into the bloodstream and is transported to the liver. There it may be oxidized to pyruvic acid and hence to CO_2 and H_2O (via the Krebs cycle) or it may be converted back into glucose (gluconeogenesis). The remaining 20–30% remains in the muscle cells where it can be reoxidized to pyruvic acid, which then enters the Krebs cycle and is further oxidized to CO_2 and H_2O (assuming the availability of oxygen in the muscle cells).

16.11   (a) both   (b) both in CO_2, neither in ethanol

16.12   (a) Step 6, glyceraldehyde 3-phosphate to 1,3-diphosphoglyceric acid.
           (b) In yeast cells, NADH is reoxidized by reacting with acetaldehyde to form ethanol.
           (c) In muscle cells, NADH is reoxidized by reacting with pyruvic acid to form lactic acid (anaerobic conditions).

16.13

This is not surprising. Here is an example of an enzyme that exhibits stereochemical specificity but not absolute specificity.

16.14   (a) $C_6H_{12}O_6 \longrightarrow 2\,CH_3COCOOH + 4\,H^+ + 4\,e^-$

           (b) $4\,e^- + 4\,H^+ + 2\,CH_3COCOOH \longrightarrow 2\,CH_3CHOHCOOH$

           (c) $4\,e^- + 4\,H^+ + 2\,CH_3COCOOH \longrightarrow 2\,C_2H_5OH + 2\,CO_2$

16.15   Lactic acid, pyruvic acid, acetaldehyde, and ethanol are the only nonphosphorylated metabolites of the Embden–Meyerhof pathway. All the other metabolites are phosphorylated.

16.16 (a) acetaldehyde   (b) dihydroxyacetone phosphate + glyceraldehyde 3-phosphate   (c) enolase
     (d) glucose 6-phosphate   (e) glucose 1-phosphate   (f) triose phosphate isomerase
     (g) glucose 6-phosphate   (h) pyruvic acid   (i) phosphofructokinase   (j) glucose 6-phosphate
     (k) 1,3-diphosphoglyceric acid   (l) phosphoglycerokinase   (m) 2-phosphoglyceric acid
     (n) pyruvic acid decarboxylase   (o) phosphoenolpyruvic acid   (p) glucose 6-phosphate

16.17 (a) g; i   (b) l; o   (c) f; j   (d) b   (e) k   (f) a, h   (g) c   (h) n

16.18

16.19

(a) acetaldehyde dehydrogenase
(b) The detoxification of alcohol in the liver results in the production of acetic acid. The acetic acid is
    released from the liver and transported by the blood to cells where it is converted into acetyl-CoA.
    Acetyl-CoA may be further oxidized via the Krebs cycle to produce energy for cellular activities.

16.20 The glycolytic pathway results in the production of 4 ATP when 1 mole of glucose is converted to lactic acid. The pathway requires the expenditure of 2 moles of ATP in converting a free glucose
molecule to fructose 1,6-diphosphate [net 2 ATP]. The cleavage of a terminal glucose residue from a
glycogen molecule by a phosphorylase does not require the expenditure of phosphate bond energy
(ATP). The conversion of the resultant glucose 1-phosphate into fructose 1,6-diphosphate requires the
expenditure of a single molecule of ATP [net 3 ATP]. Therefore, it is more efficient for anaerobic cells
to utilize glycogen as a source of energy since it requires less energy to initiate the glycolytic pathway.

16.21

It serves as the storage form of high energy phosphate groups in nerve tissue and in muscle cells of
vertebrates. It easily transfers its phosphate group to ADP to form ATP when the cell requires ATP for
energy-demanding processes.

16.22 Trained athletes have larger muscle cells and a greater number of mitochondria in each muscle cell. Their muscles contain a greater amount of actin and myosin (contractile proteins) and myoglobin. Physical training also stimulates the growth of additional capillaries within the muscle fibers which bring more oxygenated blood to the muscles.

16.23 The labeled carbon atom would be the α-carbon ($-CH_2$) in the molecule of oxaloacetic acid that results after one turn of the Krebs cycle.

16.24 (a) citric acid   (b) oxaloacetic acid + acetyl-CoA   (c) L-malic acid
(d) isocitric acid dehydrogenase   (e) α-ketoglutaric acid dehydrogenase complex
(f) oxaloacetic acid   (g) GDP + P_i (h) pyruvic acid dehydrogenase complex
(i) succinic acid   (j) fumaric acid

16.25 (a) a   (b) a; b; c; i   (c) a   (d) d; e; f; h; j   (e) d; e; h   (f) g

16.26

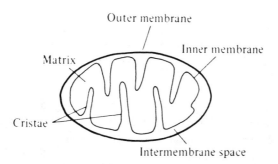

Fluorocitric acid competitively inhibits the enzyme aconitase.

16.27

Outer membrane

Inner membrane

Matrix

Cristae

Intermembrane space

16.28 See Figure 16.9.

16.29

$$CH_3\overset{O}{\overset{\|}{C}}COOH + CoASH \rightarrow CH_3\overset{O}{\overset{\|}{C}}SCoA + CO_2 + 2H^+ + 2e^-$$

$$HOOCCH_2CH(COOH)CHOHCOOH \rightarrow HOOCCH_2CH(COOH)\overset{O}{\overset{\|}{C}}COOH + 2H^+ + 2e^-$$

$$HOOCCH_2CH_2\overset{O}{\overset{\|}{C}}COOH + CoASH \rightarrow HOOCCH_2CH_2\overset{O}{\overset{\|}{C}}SCoA + CO_2 + 2H^+ + 2e^-$$

$$HOOCCH_2CH_2COOH \rightarrow HOOCCH=CHCOOH + 2H^+ + 2e^-$$

$$HOOCCH_2CHOHCOOH \rightarrow HOOCCH_2\overset{O}{\overset{\|}{C}}COOH + 2H^+ + 2e^-$$

$$NAD^+ + 2\,H^+ + 2\,e^- \rightleftharpoons NADH + H^+$$

$$FMN + 2\,H^+ + 2\,e^- \rightleftharpoons FMNH_2$$

$$2\,Fe(III)\cdot S + 2\,e^- \rightleftharpoons 2Fe(II)\cdot S$$

$$CoQ_{10} + 2\,H^+ + 2\,e^- \rightleftharpoons CoQ_{10}H_2$$

$$2\,Cyt\,b - Fe(III) + 2\,e^- \rightleftharpoons 2\,cyt\,b - Fe\,(II)$$

$$2\,Fe\,(III)\cdot S + 2\,e^- \rightleftharpoons 2\,Fe(II)\cdot S$$

$$2\,Cyt\,c_1 - Fe(III) + 2\,e^- \rightleftharpoons 2Cyt\,c_1 - Fe(II)$$

$$2\,Cyt\,c - Fe\,(III) + 2\,e^- \rightleftharpoons 2\,Cyt\,c - Fe(II)$$

$$2\,Cyt\,aa_3 - Fe(III) + 2\,e^- \rightleftharpoons 2Cyt\,aa_3 - Fe(II)$$

$$\tfrac{1}{2}O_2 + 2\,H^+ + 2\,e^- \rightleftharpoons H_2O$$

16.30 By way of the electron transport chain. The reoxidation of reduced coenzymes generates energy that is harnessed to drive the phosphorylation of ADP to ATP. ATP is the high-energy molecule that is utilized by the cell to supply energy for a variety of cellular processes.

16.31 (a) One mole of acetyl-CoA metabolized via the Krebs cycle will result in the production of 12 moles of ATP.
   (b) (12) (7300) = 87,600 cal
   Efficiency = (87,600/200,000) 100% = 43.8%

16.32 Each mole of lactic acid is converted to pyruvic acid. This occurs in the cytosol of muscle cells and one mole of NADH is formed. The NADH enters the mitochondria via the glycerol phosphate shuttle and two moles of ATP are obtained by oxidative phosphorylation (and the respiratory chain). Each mole of pyruvic acid that enters the Krebs cycle is oxidized to produce 15 moles of ATP. Therefore, a total of 17 moles of ATP is obtained from each mole of lactic acid.

16.33 The glycerol phosphate shuttle and the malic acid–aspartic acid shuttle.

16.34 (a) The oxidizing agent for this reaction is FAD, which is reduced to $FADH_2$. When $FADH_2$ is reoxidized by means of the electron transport chain, only two ATPs are formed.
   (b) The oxidation of glyceraldehyde 3-phosphate to 1,3-diphosphoglyceric acid occurs in the cytosol. Under aerobic conditions, the NADH must be reoxidized by a mechanism that involves the passage of some metabolite into the mitochondria. The process in muscle and nerve cells utilizes the glycerol phosphate shuttle, and as a result of this shuttle a molecule of $FADH_2$ is reoxidized by the electron transport chain. Hence only two ATPs are produced.

16.35 1,3-diphosphoglyceric acid $\longrightarrow$ 3-phosphoglyceric acid
   phosphoenolpyruvic acid $\longrightarrow$ pyruvic acid
   succinyl-CoA $\longrightarrow$ succinic acid

16.36 (a) 38 ATPs (recall the malic acid–aspartic acid shuttle)
   (b) [(38) (7300)/686,000] 100% = 40.4%

16.37 $C_6H_{12}O_6 + 6\,O_2 + 36\,ADP + 36\,P_i \rightarrow 6\,CO_2 + 42\,H_2O + 36\,ATP + 423,000$ cal

16.38 Carbohydrates are the chief fuels of biological systems—they provide energy. Carbohydrate metabolism and the burning of table sugar are similiar in that both processes release energy. When sugar is burned, all the energy given off is released into the environment as heat. When carbohydrates are metabolized, 38–40% of the energy released is conserved in the form of chemical energy (ATP).

CHAPTER 17

Lipid Metabolism

17.1 (a) An essential fatty acid is a fatty acid that cannot be synthesized by an organism and therefore must be supplied in the diet.

(b) Eczema is an inflammatory skin disease characterized by the development of scales and crusts. It can often be relieved by the dietary intake of essential fatty acids.

(c) Bile salts are important components of bile that aid in lipid digestion by emulsifying the lipids.

(d) Lipases is the general name for a class of enzymes that catalyze the hydrolysis of lipids.

(e) Chylomicrons are a class of lipoproteins (density of less than 0.94 g/mL) that are produced in the intestine.

(f) Lymph is tissue fluid (water and dissolved substances) that has entered a lymphatic capillary.

(g) Fat deposits are specialized cells (adipose tissue) that can store a high percentage of lipids within their cytoplasm.

(h) Obesity is the condition in which an excessive amount of fat is deposited in adipose tissue.

(i) Lipolysis is the hydrolysis of triacylglycerols.

(j) Mobilization is the release of fatty acids from adipose tissue.

(k) Fatty acyl-CoA is a complex formed between a fatty acid molecule and coenzyme A.

(l) β-oxidation is a sequence of reactions whereby the carbon atom beta to the carboxyl group of a fatty acid undergoes successive oxidations.

(m) Ketonemia is the accumulation of excess ketone bodies in the blood at concentrations greater than 3 mg/100 mL.

(n) Ketonuria is the accumulation of ketone bodies in the urine.

17.2 The oxidation of 1.0 g of lipid liberates about 9500 cal. The oxidation of 1.0 g of carbohydrate liberates about 4200 cal. The oxidation of 1.0 g of alcohol liberates about 7100 cal.

17.3 Because of the presence of many hydroxyl groups, glycogen is extremely hydrated. Therefore, 1 g of glycogen contains a higher percentage of water and a lower percentage of fuel mass than 1 g of fat.

17.4 The essential fatty acids are those that contain more than one double bond.

17.5 Lipid digestion begins in the upper portion of the small intestine. The secretion of bile from the gallbladder serves to emulsify the large lipid globules in order to increase the surface area exposed to lipolytic attack by pancreatic lipases. These degradative enzymes hydrolyze the triacylglycerols into their constituent diacylglycerols, monoacylglycerols, fatty acids, and glycerol. The products of digestion are then absorbed into the intestinal epithelial cells.

17.6

17.7 $C_{15}H_{31}COOH$ $C_{17}H_{35}COOH$ $C_{17}H_{33}COOH$ $C_{17}H_{31}COOH$

     palmitic acid   stearic acid   oleic acid   linoleic acid

17.8   After the products of lipid digestion pass across the intestinal wall, they are immediately resynthesized into triacylglycerols, phospholipids, or cholesterol esters. These are then conjugated to proteins and transported to the blood via the lymphatic system.

17.9   (a) The triacylglycerols are transported in the blood as complexes with specialized proteins (lipoproteins).

    (b) Some of the triacylglycerals are transported to the liver where they are utilized to produce energy. The remainder of the lipids either are deposited within fat storage cells (adipose tissue) or continue to circulate within the blood bound to proteins.

17.10   cholesterol esters > phospholipids > triacylglycerols.

17.11   In some cases, obesity is due to an abnormality in the functioning of one or more endocrine glands leading to a hormonal imbalance. Most individuals, however, are overweight from eating more food than is required to meet the needs of the body and/or from a lack of exercise.

17.12   To protect vital organs from mechanical injury, to insulate the organism against loss of termperature, to store reserve energy.

17.13   Epinephrine binds to receptor proteins in the membranes of adipose tissue. This binding prompts the stimulation of the enzyme adenyl cyclase, which converts ATP to cyclic AMP. cAMP stimulates the hydrolysis of lipids within the adipose cells and the subsequent release of fatty acids (mobilization) from those cells. Insulin, on the other hand, enhances the synthesis of triacylglycerols and the storage of fats in adipose tissue.

17.14   α- and β-Globulins transport lipids in the blood. Globulins are insoluble in water but are soluble in dilute salt solutions. Albumins transport fatty acids and while they are likewise soluble in dilute salt solutions, they are also soluble in water.

17.15   (a) glucose   (b) fatty acids   (c) fatty acids   (d) glucose and fatty acids   (e) fatty acids
    (f) glucose   (g) fatty acids   (h) ketone bodies

17.16   (a) Each turn of the spiral produces a fatty acyl-CoA containing two fewer carbon atoms rather than reproducing the identical fatty acyl-CoA.

    (b) The carbon atom beta to the carboxyl group of the fatty acid undergoes successive oxidations.

17.17

Step 1: $CH_3(CH_2)_9CH_2\overset{O}{\underset{OH}{C}}$ + ATP $\xrightarrow{\text{synthetase}}$ $CH_3(CH_2)_9CH_2\overset{O}{\underset{AMP}{C}}$ + PP_i

$CH_3(CH_2)_9CH_2\overset{O}{\underset{AMP}{C}}$ + HSCoA $\xrightarrow{\text{synthetase}}$ $CH_3(CH_2)_9CH_2\overset{O}{\underset{SCoA}{C}}$ + AMP

Step 2: $CH_3(CH_2)_8\overset{H}{\underset{H}{C}}-\overset{H}{\underset{H}{C}}-\overset{O}{\underset{SCoA}{C}}$ + FAD $\xrightarrow{\text{dehydrogenase}}$ $CH_3(CH_2)_8C=\overset{H}{\underset{H}{C}}-\overset{O}{\underset{SCoA}{C}}$ + $FADH_2$

fatty acyl-CoA

Step 3: $CH_3(CH_2)_8C=\overset{H}{\underset{H}{C}}-\overset{O}{\underset{SCoA}{C}}$ + HOH $\xrightarrow{\text{hydratase}}$ $CH_3(CH_2)_8\overset{HO}{\underset{H}{C}}-\overset{H}{\underset{H}{C}}-\overset{O}{\underset{SCoA}{C}}$

Step 4: $CH_3(CH_2)_8\overset{OH}{\underset{H}{C}}CH_2\overset{O}{\underset{SCoA}{C}}$ + NAD^+ $\xrightarrow{\text{dehydrogenase}}$ $CH_3(CH_2)_8\overset{O}{C}CH_2\overset{O}{\underset{SCoA}{C}}$ + NADH + H^+

Step 5: $CH_3(CH_2)_8\overset{O}{C}CH_2\overset{O}{\underset{SCoA}{C}}$ + HSCoA $\xrightarrow{\text{thiolase}}$ $CH_3(CH_2)_8\overset{O}{\underset{SCoA}{C}}$ + $CH_3\overset{O}{\underset{SCoA}{C}}$

shortened fatty acyl-CoA   acetyl-CoA

*shortened fatty acyl-CoA is further degraded to acetyl-CoA by successive repetition of steps (2–5).

17.18 Carbon-2 would appear in acetyl-CoA. Carbons 4 and 6 would appear in the molecule of butyryl-CoA.

17.19 (a) 6 turns   (b) 7 turns   (c) 12 turns

17.20

$R(CH_2)_5\overset{O}{\underset{SCoA}{C}}$ + $H_3C\overset{+}{\underset{CH_3}{N}}-CH_2-\overset{CH_3}{\underset{OH}{C}}-CH_2-\overset{H}{\underset{OH}{C}}\overset{O}{}$ $\underset{\text{acyltransferase}}{\rightleftharpoons}$ $H_3C\overset{+}{\underset{CH_3}{N}}-CH_2-\overset{CH_3}{\underset{O}{C}}-CH_2-\overset{H}{\underset{OH}{C}}\overset{O}{}$ + CoASH

fatty acyl-CoA   carnitine

$\underset{R}{\underset{|}{\underset{(CH_2)_5}{\underset{|}{C=O}}}}$

acyl carnitine

17.21 Fatty acid biosynthesis proceeds by the addition of two carbon units at a time to the growing hydrocarbon chain.

17.22 The amount of glucose that can be stored as glycogen is limited. Therefore, ingestion of excess carbohydrates results in the formation of excess acetyl-CoA. The excess acetyl-CoA molecules are utilized for the production of fats, which are then stored in adipose tissue.

17.23 Acetyl-CoA is the molecule that links carbohydrate and lipid metabolism.

17.24 The segment in which the acetyl-CoA, produced from β-oxidation of fatty acids, is metabolized via the Krebs cycle.

17.25

$$
\begin{array}{l}
\quad\quad\quad O \\
\quad\quad\quad \| \\
C_{17}H_{35}COCH_2 \\
\quad\quad\quad O \quad | \\
\quad\quad\quad \| \quad | \\
C_{17}H_{35}COCH \xrightarrow{\text{lipases}} 3\ C_{17}H_{35}COOH\ +\ \begin{array}{l} CH_2OH \\ | \\ CHOH \\ | \\ CH_2OH \end{array} \\
\quad\quad\quad O \quad | \\
\quad\quad\quad \| \quad | \\
C_{17}H_{35}COCH_2
\end{array}
$$

Each mole of stearic acid yields 146 moles of ATP as follows.

1 mole of ATP used	−2
9 moles of acetyl-CoA formed (9 × 12)	108
8 moles of FADH$_2$ formed (8 × 2)	16
8 moles of NADH formed (8 × 3)	24
	146

Therefore, 3 moles of stearic acid (3 × 146) yields 438 moles of ATP. One mole of glycerol, converted to dihydroxyacetone phosphate and metabolized via glycolysis and the Krebs cycle, yields an additional 20 moles of ATP.

$$\text{total yield of ATP} = 438 + 20 = 458 \text{ moles.}$$

The 20 moles of ATP from 1 mole of glycerol come about as follows.

$$
\begin{array}{ccc}
CH_2OH & CH_2OH & CH_2OH \\
| & \xrightarrow[\text{-1 ATP}]{\text{ATP} \quad \text{ADP}} \quad | & \xrightarrow[\text{+2 ATP}]{\text{NAD}^+ \quad \text{NADH}+H^+} \quad | \\
CHOH & CHOH & C{=}O \\
| & | & | \\
CH_2OH & CH_2O\textcircled{P} & CH_2O\textcircled{P}
\end{array}
$$

An additional 4 moles of ATP are formed by the subsequent conversion of dihydroxyacetone phosphate to pyruvic acid via the Embden–Meyerhof pathway. Then 15 moles of ATP are produced by the oxidation of pyruvic acid by the Krebs cycle and electron transport chain.

17.26 Acetoacetic acid, β-hydroxybutyric acid, acetone

17.27 Diabetics lack insulin and therefore glucose does not readily pass into the cells to be metabolized. Since glucose is unavailable, diabetics must rely on their fat reserves to supply their energy needs. The accelerated rate of lipid metabolism produces more acetyl-CoA than can be efficiently oxidized by the Krebs cycle. The excess acetyl-CoA molecules condense to produce the ketone bodies.

17.28 Two of the three ketone bodies also contain carboxyl groups, and their presence in the blood tends to lower the pH. Initially, the excess acidity is neutralized by the blood buffers. If sufficient ketone bodies are produced, however, the blood buffers will eventually be unable to prevent a decrease in blood pH. This condition is called acidosis.

17.29

$$
\begin{array}{l}
CH_2OH \\
| \\
CHOH \\
| \\
CH_2OH \\
\text{glycerol}
\end{array}
\xrightarrow[\text{kinase}]{ATP \quad ADP}
\begin{array}{l}
CH_2OH \\
| \\
CHOH \\
| \\
CH_2O\textcircled{P} \\
\text{glycerol} \\
\text{phosphate}
\end{array}
\xrightarrow[\text{dehydrogenase}]{NAD^+ \quad NADH+H^+}
\begin{array}{l}
CH_2OH \\
| \\
C=O \\
| \\
CH_2O\textcircled{P} \\
\text{dihydroxyacetone} \\
\text{phosphate}
\end{array}
$$

Dihydroxyacetone phosphate can be degraded to produce energy by being incorporated into the glycolytic pathway.

CHAPTER 18

Protein Metabolism

18.1 (a) Peptidase is an enzyme that catalyzes the hydrolysis of proteins.
   (b) Autocatalysis is the catalysis of a reaction by one of the products of the reaction.
   (c) Protein turnover is a measure that is a function of the rate at which body proteins are synthesized and degraded.
   (d) Nitrogen balance is the state in which an individual's intake of dietary nitrogen is equal to the amount of nitrogen in the excrement.
   (e) Kwashiorkor is an emaciating disease that results from protein deficiency.
   (f) Pyridoxal phosphate is a coenzyme, derived from vitamin B_6, that participates in transamination reactions.
   (g) GABA (γ-aminobutyric acid) is an inhibitory neurotransmitter that is formed from the decarboxylation of glutamic acid.
   (h) Parkinson's disease is a disease in which there is a deficiency of the neurotransmitter dopamine in brain cells. It involves both a progressive paralytic rigidity and tremors of the extremities.
   (i) L-Dopa is a drug administered to patients with Parkinson's disease. It is converted to dopamine in brain cells.
   (j) Hyperammonemia is the presence of excess ammonia in the blood.
   (k) Gout is a form of arthritis that is characterized by inflammation of the joints.
   (l) Urea is the metabolic end-product of amino acids and other nitrogenous compounds.

18.2 (a) Pepsin is the active form of the proteolytic enzyme that functions in the stomach. Pepsinogen is the inactive precursor form (proenzyme) of pepsin.
   (b) Chymotrypsin is an endopeptidase that preferentially cleaves peptide bonds involving the carboxyl groups of phenylalanine, tryptophan, and tyrosine. Trypsin is an endopeptidase that cleaves peptide bonds involving the carboxyl groups of lysine and arginine.
   (c) An endopeptidase catalyzes the hydrolysis of peptide bonds at the interior of the protein molecule (e.g., pepsin). An exopeptidase hydrolyzes the peptide bonds at the terminal ends of the protein molecule (e.g., carboxypeptidase).
   (d) Positive nitrogen balance occurs when the intake of nitrogen exceeds its excretion (e.g., growth); negative nitrogen balance occurs when the excretion of nitrogen exceeds intake (e.g., fasting).
   (e) An essential amino acid cannot be synthesized by an organism, or at least not at a rate rapid enough to meet the needs of the organism (e.g., lysine). A nonessential amino acid can readily be synthesized by an organism (e.g., glycine).
   (f) A complete protein contains an adequate amount of the essential amino acids. An incomplete protein is deficient in one or more of the essential amino acids.
   (g) In transamination, the amino group is transferred from an amino acid to an α-keto acid and a new amino acid and a new α-keto acid are formed. In oxidative deamination, a hydrogen and an amino group are replaced by oxygen; an amino acid is converted to an α-keto acid.
   (h) GPT and GOT are initials for specific transaminase enzymes. GPT is glutamic–pyruvic transaminase, and it is abundant in the liver. GOT is glutamic–oxaloacetic transaminase and it is abundant in heart muscle.
   (i) A glucogenic amino acid can be converted into any one of the metabolites of carbohydrate metabolism (e.g., alanine). A ketogenic amino acid can be converted into a ketone body (e.g., leucine).

109

(j) Histamine results from the decarboxylation of histidine. It functions to dilate the blood vessels. Antihistamines are compounds that block the action of histamine.

18.3 Protein digestion begins in the stomach. Pepsinogen present in gastric juice is converted into the active form of the enzyme, pepsin. Pepsin is an endopeptidase that cleaves internal peptide bonds of the large protein molecules in the stomach. Protein digestion is completed in the small intestine by several proteolytic enzymes present in the pancreatic secretions. The proenzymes trypsinogen and chymotrypsinogen are converted to their active forms, trypsin and chymotrypsin, respectively. Like pepsin, these endopeptidases hydrolyze internal peptide bonds of proteins and smaller polypeptides (peptones). Two exopeptidases, aminopeptidase and carboxypeptidase, cleave terminal amino acid residues from the protein chains. The complete hydrolysis of the dietary proteins results in the liberation of free amino acids which are actively transported across the intestinal epithelium.

18.4 The proteolytic enzymes are secreted in an inactive form. They are not activated until food is present in the digestive tract.

18.5 Since insulin is a protein, it would be digested (hydrolyzed) into its respective amino acids if it were taken orally. When injected directly into the blood, it is not subject to any hydrolytic attack.

18.6 (a) alanine, phenylalanine, tyrosine ⎫
       (b) isoleucine, tyrosine, serine   ⎬   (complete hydrolysis)
       (c) phenylalanine, arginine, leucine ⎭
       (d) threonine–glutamic acid–lysine   (no hydrolysis)

18.7 (a) peptide linkage at the N-terminal end   (b) peptide linkage at the C-terminal end

18.8 They found that almost 60% of the ingested amino acids were incorporated into tissue proteins. Therefore, the apparent stability of the adult organism is due not to metabolic inertness, but rather to a balance between the rates of synthesis and degradation of its constituent compounds.

18.9 The mixture of amino acids, derived either from the diet or from the degradation of tissue protein, that is contained within each cell.

18.10 The turnover rate of enzymes is much more rapid than that of muscle proteins. Muscle proteins have half-lives of approximately 180–1000 days. The half-lives of enzymes may vary from approximately 10 min to 6 h depending on their metabolic importance and the cell in which they function.

18.11 Through digestion, the plant proteins are degraded to the individual amino acids. The animal then uses these amino acids for the biosynthesis of the proteins it requires.

18.12 (a) Most plant proteins are deficient in one or more essential amino acids. Therefore, vegetarians should consume a wide variety of vegetables in order to ensure their intake of the proper quantities of essential amino acids.
       (b) The best source of essential amino acids is animal proteins.

18.13 (a) The essential amino acids contain carbon chains, or aromatic rings, that are not present as intermediates of carbohydrate or lipid metabolism. The inability to synthesize these amino acids results from the animal's inability to manufacture the correct carbon skeleton.
       (b) Since all the amino acids required for the construction of a particular protein must be present at the time of its synthesis, a deficiency in one or more essential amino acids will prevent protein synthesis from proceeding.

18.14 (1) Protein synthesis.
       (2) Amino acids play an essential role in the metabolism of all nitrogenous compounds.
       (3) Amino acids can be converted to carbohydrates and fats or oxidized to produce energy.

18.15 The reversibility of transamination reactions serves to link protein metabolism with the metabolism of carbohydrates and lipids. Transamination reactions convert certain amino acids into key metabolic intermediates that may be used to synthesize carbohydrates and lipids or may be oxidized to produce energy.

18.16

$$CH_3CH_2CH(CH_3)\overset{O}{\underset{}{C}}\!-\!\overset{O}{\underset{OH}{C}} + R\!-\!\overset{H}{\underset{NH_2}{C}}\!-\!COOH \xrightarrow{\text{transaminase}} CH_3CH_2CH(CH_3)\overset{H}{\underset{NH_2}{C}}\!-\!\overset{O}{\underset{OH}{C}} + R\!-\!\overset{O}{\underset{}{C}}\!-\!COOH$$

Isoleucine

18.17 15 moles of ATP can be produced from the complete oxidation of 1 mole of pyruvic acid.

18.18 (a) Ammonia (NH_3) and hydrogen peroxide (H_2O_2) are the toxic compounds formed in oxidative deamination reactions.
(b) Ammonia is converted to urea in the liver and excreted in the urine. Hydrogen peroxide is detoxified by the enzyme catalase which converts H_2O_2 into O_2 and H_2O.

18.19 (a) Benadryl is an antihistamine that exerts its effect by binding to H_1 receptors and thus inhibits the binding of histamine, which normally serves to initiate a hypersensitive response.
(b) H_1 receptors are histamine receptors found in the walls of capillaries and in the smooth muscles of the respiratory tract. When bound to histamine, they affect vascular dilation and muscular constriction associated with hypersensitivity. H_2 receptors are found in the wall of the stomach and function to increase HCl secretions when activated.

18.20 (a) leucine   (b) α-ketoglutaric acid   (c) α-ketoglutaric acid   (d) 5-hydroxytryptophan
(e) glutamic acid   (f) glutamine

18.21 (a)

$$R\!-\!\overset{NH_2}{\underset{}{C}}H\!-\!\overset{O}{\underset{OH}{C}} + FAD + H_2O \underset{}{\overset{\text{amino acid oxidase}}{\rightleftharpoons}} R\!-\!\overset{O}{\underset{}{C}}\!-\!\overset{O}{\underset{OH}{C}} + FADH_2 + NH_3$$

(b)

$$R\!-\!\overset{NH_2}{\underset{}{C}}H\!-\!\overset{O}{\underset{OH}{C}} + R'\!-\!\overset{O}{\underset{}{C}}\!-\!\overset{O}{\underset{OH}{C}} \underset{}{\overset{\text{transaminase}}{\rightleftharpoons}} R\!-\!\overset{O}{\underset{}{C}}\!-\!\overset{O}{\underset{OH}{C}} + R'\!-\!\overset{NH_2}{\underset{}{C}}H\!-\!\overset{O}{\underset{OH}{C}}$$

(c)

$$R\!-\!\overset{NH_2}{\underset{}{C}}H\!-\!\overset{O}{\underset{OH}{C}} \xrightarrow{\text{decarboxylase}} R\!-\!\overset{NH_2}{\underset{H}{C}}\!-\!H + CO_2$$

18.22 (a)

$$CH_3\overset{NH_2}{\underset{}{C}}HCOOH + H_2O + O_2 \xrightarrow{\text{alanine oxidase}} CH_3\overset{O}{\underset{}{C}}COOH + NH_3 + H_2O_2$$

alanine   pyruvic acid

(b) (imidazole ring)$-CH_2\overset{}{\underset{NH_2}{C}}HCOOH \xrightarrow{\text{histidine decarboxylase}}$ (imidazole ring)$-CH_2CH_2\underset{NH_2}{}$ + CO_2

histidine   histamine

(c) $\underset{\text{glutamic acid}}{HOOCCH_2CH_2\overset{\overset{\displaystyle NH_2}{|}}{C}HCOOH} + H_2O + O_2 \xrightarrow{\text{oxidase}} \underset{\alpha\text{-ketoglutaric acid}}{HOOCCH_2CH_2\overset{\overset{\displaystyle O}{\|}}{C}COOH} + NH_3 + H_2O_2$

(d) $\underset{\text{glutamic acid}}{HOOCCH_2CH_2\overset{\overset{\displaystyle NH_2}{|}}{C}HCOOH} + NH_3 \underset{\text{synthetase}}{\overset{\text{ATP} \quad \text{ADP}}{\longrightarrow}} \underset{\text{glutamine}}{\underset{\underset{\displaystyle H_2N}{\diagup}}{\overset{\overset{\displaystyle O}{\|}}{C}}CH_2CH_2\overset{\overset{\displaystyle NH_2}{|}}{C}HCOOH} + H_2O + P_i$

(e) $\underset{\text{lysine}}{H_2N-(CH_2)_4\overset{\overset{\displaystyle NH_2}{|}}{C}HCOOH} \xrightarrow{\text{decarboxylase}} \underset{\text{cadaverine}}{H_2N-(CH_2)_4CH_2NH_2}$

(f) $\underset{\text{aspartic acid}}{HOOCCH_2\overset{\overset{\displaystyle NH_2}{|}}{C}HCOOH} + H_2O + O_2 \xrightarrow{\text{oxidase}} \underset{\text{oxaloacetic acid}}{HOOCCH_2\overset{\overset{\displaystyle O}{\|}}{C}COOH} + NH_3 + H_2O_2$

(g) $\underset{\text{serine}}{HOCH_2\underset{\underset{\displaystyle NH_2}{|}}{C}HCOOH} \xrightarrow{\text{serine decarboxylase}} \underset{\text{ethanolamine}}{HOCH_2CH_2\atop{|}\atop NH_2} + CO_2$

(h) 3,4-dihydroxyphenylalanine (L-dopa) $\xrightarrow{\text{dopa decarboxylase}}$ dopamine $+ CO_2$

18.23 (a) The liver is the organ responsible for the synthesis of urea in mammals.
(b) The kidney is the organ responsible for the excretion of urea in the urine.

18.24 The aerobic catabolism of carbohydrates and lipids results in the production of acetyl-CoA, which is fed into the Krebs cycle in order to produce energy by way of the respiratory chain and oxidative phosphorylation. Protein catabolism can result in the production of metabolites of carbohydrate metabolism, which may be likewise oxidized to produce energy or used in biosynthetic pathways. Acetyl-CoA is the key intermediate that links the metabolism of the three major foodstuffs. It is important to note that the conversion of pyruvic acid to acetyl-CoA is an irreversible reaction. Although the catabolism of carbohydrates may provide for the biosynthesis of lipids from acetyl-CoA, the catabolism of lipids cannot directly provide for the biosynthesis of carbohydrates. (see Figure 18.6).

CHAPTER 19

The Blood

19.1 (a) Lymph is the extracellular fluid that is circulating through the lymphatic system.
    (b) Formed elements are the blood cells.
    (c) Hematocrit value is the volume (in percent) of red blood cells in a sample that has been centrifuged under standard conditions.
    (d) Leukemia is a cancerous condition whereby the leukocytes in the blood divide uncontrollably.
    (e) Immunity is resistance to specific pathogens (infectious agents such as viruses, bacteria, fungi, and protozoans), cancer cells, or the toxins secreted by the pathogens.
    (f) Osmosis is the diffusion of water from a dilute solution (low solute concentration) through a semipermeable membrane into a more concentrated solution (high solute concentration).
    (g) Edema is the accumulation of fluids in a tissue causing swelling.
    (h) Hypertension is high blood pressure.
    (i) Hemophilia is a sex-linked hereditary disease in which one of the protein factors involved in the formation of thrombin is lacking or is inactive.
    (j) Anticoagulant is a substance that inhibits the clotting of blood.
    (k) Thrombosis is the formation of a clot within a blood vessel.
    (l) Embolism is a blood clot that has broken away from its site of origin and is carried away by the blood to be lodged in a small blood vessel elsewhere.
    (m) Methemoglobinemia is the presence of a greater-than-normal concentration of methemoglobin in the blood.
    (n) Jaundice is a yellow pigmentation that develops in the skin from the inappropriate deposition of bile pigments (bilirubin).

19.2 (a) Aplastic anemia results from the decreased production of erythrocytes. Hemolytic anemia results from the increased destruction of erythrocytes.
    (b) Anemia is a condition that is characterized by an abnormally low percentage of erythrocytes in the plasma. Conversely, polycythemia is the condition arising from an abnormally high percentage of red blood cells in the plasma.
    (c) The lymphocytes (B- and T-cells) are a class of leukocytes that comprise the humoral and cell-mediated systems of immunity. The phagocytes are a special class of leukocytes whose function is the engulfing and digestion of foreign matter.
    (d) T-cells represent a class of lymphocytes that are responsible for maintenance of the cellular immune response. B-cells are the class of lymphocytes that are responsible for the humoral immune response (antibody formation).
    (e) Antibodies are proteins produced by sensitized B-cells that can specifically bind to the antigen that stimulated their production. Antigens are molecules that stimulate antibody production in the body.
    (f) Cellular immunity involves the binding and destruction of antigens by T-cells. Humoral immunity involves the production of antibodies by B-cells in order to confer immunity against a given class of antigens.
    (g) Osmotic pressure is the pressure required to prevent the occurrence of osmosis. Blood pressure is the pressure of the blood that results from the pumping action of the heart.

(h) Systolic pressure is the maximum pressure achieved during contraction of the heart ventricles. Diastolic pressure is the lowest pressure that remains in the arteries before the next ventricular contraction.

(i) Vasodilation is an increase in the diameter of the blood vessels, whereas vasoconstriction is a decrease in the diameter of the blood vessels.

(j) Plasma is the fluid portion of the blood. Serum is the portion of the blood plasma remaining after the protein fibrinogen has been removed.

(k) Stroke is the condition that results from the death of tissue in the brain. Coronary thrombosis is the condition that results from the death of heart muscle tissue.

(l) Hypoventilation is a lower than normal rate of breathing, whereas hyperventilation is a higher than normal rate of breathing.

19.3 The three circulating fluids in the animal organism are blood, interstitial fluid, and lymph.

19.4 (1) To help maintain osmotic relationships between the tissues.
(2) To transport oxygen and nutritive materials to the cells and waste products to the excretory organs.
(3) To regulate the body temperature.
(4) To control the pH of the body.
(5) To protect the organism against infection.

19.5 Assuming a weight of 150 pounds,

$$(0.08)(150) = 12 \, lb$$

$$1 \, lb = 454 \, g$$

$$\text{volume of blood} = \frac{(454)(12)}{1.06} = 5140 \, mL = 5.1 \, L$$

19.6 Carbohydrates, lipids, amino acids, hormones (nonprotein), vitamins, inorganic ions.

19.7 $Na^+, K^+, Ca^{2+}, Mg^{2+}, HCO_3^-, Cl^-, HPO_4^{2-}, SO_4^{2-}$

19.8 (a) Albumins maintain the osmotic balance and transport fatty acids.
(b) Globulins (alpha and beta) form complexes with lipids and transport them to all parts of the body.
(c) Fibrinogen functions in blood clotting.
They are all synthesized in the liver.

19.9 (a) They function to combat infectious microorganisms, cancer cells, and chemical toxins present in the body.
(b) They are synthesized by B-lymphocytes.
(c) IgM, IgG, IgA, IgD, IgE
(d) IgG

19.10 Erythrocytes effect transportation of oxygen to the cells. Leukocytes destroy invading bacteria and other foreign substances. Thrombocytes liberate substances that are involved in blood clotting.

19.11 (a) Red blood cells are similar to other cells in that they are bound by a phospholipid bilayer and are able to produce energy via substrate level phosphorylation in the Embden–Meyerhof pathway.
(b) Red blood cells do not contain a nucleus or mitochondria. They cannot reproduce or respire aerobically, and are unable to synthesize carbohydrates, proteins, or lipids.
(c) Red blood cells are derived from stem cells in the bone marrow.
(d) Red blood cells are eliminated by special tissues in the liver and the spleen.

19.12 An immune response is initiated when free antigen or antigen-bearing cells stimulate the phagocytes and lymphocytes that are routinely patrolling the body in search of such foreign molecules. The sensitization of T-cells elicits a maturation of their killer functions in that certain T-cells bind specifically to the antigen-bearing cells and release toxic factors that are lethal to the invaders (cellular immunity). Other sensitized T-cells stimulate B-cells to produce the highly specific antibodies that bind and agglutinate antigen-bearing cells (humoral immunity). The antigen–antibody complexes cue the destruction of foreign cells by phagocytes that engulf and degrade the invaders.

19.13 **(a)** At the arterial end of the capillary, the blood pressure exceeds the osmotic pressure. There is a flow of material (nutrients, oxygen, etc.) from the capillaries into the interstitial fluid.
(b) At the venous end of the capillary, the osmotic pressure is greater than the blood pressure. There is a flow of material (metabolic waste products) from the interstitial fluid into the capillary.

19.14 Osmotic pressure is directly related to the concentration of proteins (particularly albumins) in the plasma.

19.15 Blood pressure measurements are reported as a ratio of systolic pressure to diastolic pressure in units of mmHg (e.g., 120/80).

19.16 (1) Administer diuretics and/or reduce the sodium ion intake.
(2) Negate the stimulating effects of epinephrine binding sites.
(3) Administer vasodilators in order to promote the relaxation of the smooth muscles in the arterial walls.

19.17 **(a)** Prothrombin is a globulin plasma protein produced by the liver. It is a zymogen that when activated by autoprothrombin C, catalyzes the conversion of fibrinogen into fibrin.
(b) Thrombin is the actual clotting enzyme that catalyzes the activation of fibrinogen.
(c) Fibrinogen is the soluble plasma protein that is converted by thrombin into the insoluble protein fibrin.
(d) Thromboplastin is a group of compounds released by blood platelets and damaged tissue. Thromboplastin autocatalytically converts autoprothrombin III to autoprothrombin C.
(e) Calcium ions are necessary for the catalytic activity associated with both thromboplastin and autoprothrombin C.
(f) Fibrin is the insoluble protein endproduct of the blood-clotting cascade. Fibrin monomers polymerize, resulting in the formation of needle-like threads that enmesh to seal off the area where a blood vessel has been damaged.

19.18 **(a)** Vitamin K is a coenzyme involved in the activation of the zymogen prothrombin.
(b) Heparin blocks the catalytic activity of both thromboplastin and thrombin.
(c) Dicumarol, a metabolic antagonist of vitamin K, either represses prothrombin formation or inhibits the enzyme for which vitamin K is a coenzyme.
(d) These anions have a strong affinity for calcium ions. They tie up the calcium ions, and therefore free calcium ions are not present to help effect the conversion of prothrombin to thrombin.
(e)

$$
\begin{array}{c}
COOH \\
| \\
H_2N-C-H \\
| \\
CH_2 \\
| \\
H-C-COOH \\
| \\
COOH
\end{array}
$$

19.19 Hemoglobin is a conjugated protein with a molecular weight of about 68,000 daltons. Upon hydrolysis it yields a simple protein, globin, and four heme groups. The heme groups account for about 4% of the total molecular weight.

19.20 **(a)** Heme is the prosthetic group of the conjugated protein hemoglobin.
   (b) Myoglobin is a protein consisting of 153 amino acid units arranged in a single polypeptide chain; hemoglobin contains four polypeptide chains—two identical α-chains and two identical β-chains.
   (c)–(f) Oxyhemoglobin is a hemoglobin bound to oxygen. CO-hemoglobin is a hemoglobin bound to carbon monoxide. Methemoglobin contains iron in the +3 oxidation state. Carbaminohemoglobin contains a CO_2 molecule bound to the globin portion of the hemoglobin molecule.

19.21 **(a)** II   **(b)** II   **(c)** III   **(d)** II

19.22 **(a)** The ferrous ions have a much greater affinity for CO (carbon monoxide) than for O_2. Therefore, CO molecules occupy the sites of oxygen binding, and much less oxygen is transported to the cells by the blood. The brain cells, in particular, require a continual supply of oxygen.
   (b) By greatly increasing the oxygen concentration in the blood either by artificial respiration or by breathing in pure oxygen from an oxygen tank.

19.23 hemoglobin, cytochromes, catalase

19.24 biliverdin and bilirubin

19.25 (1) Infectious hepatitis where the liver is malfunctioning and cannot remove sufficient bilirubin from the blood
   (2) The obstruction of bile ducts by gallstones
   (3) An acceleration of erythrocyte destruction in the spleen.

19.26 **(a)** Arterial blood contains dissolved nutrients, oxygen, hormones, and vitamins. Venous blood contains metabolic waste products and has a lower oxygen content.
   (b) Arterial blood, crimson; venous blood, dark red.

19.27 See Section 19.7.

$$HHb + O_2 \underset{\text{Tissue}}{\overset{\text{Lung}}{\rightleftharpoons}} HbO_2^- + H^+$$

$$H_2O + CO_2 \underset{\text{Lung}}{\overset{\text{Tissue}}{\rightleftharpoons}} H_2CO_3 \rightleftharpoons H^+ + HCO_3^-$$

$$HHbNH_2 + CO_2 \overset{\text{Tissue}}{\longrightarrow} HHbNH-\overset{\displaystyle O}{\underset{\displaystyle OH}{C}}$$

19.28 Carbonic anhydrase catalyzes the reaction of CO_2 with H_2O to form carbonic acid (H_2CO_3), and the reverse reaction, the decomposition of carbonic acid to H_2O and CO_2.

19.29 The normal pH range of the blood is 7.35–7.45.

19.30 If acidosis occurs, the pH of the organism decreases. Since hemoglobin is a protein, it is denatured by changes in pH. This denaturation interferes with its oxygen-carrying capacity, among other things.

19.31 When acids enter the blood, they are neutralized by bicarbonate ions to produce carbonic acid.

$$H^+ + HCO_3^- \rightleftharpoons H_2CO_3$$

When a base enters the blood, it reacts with carbonic acid to produce bicarbonate ions.

$$OH^- \ + \ H_2CO_3 \ \longrightarrow \ HOH \ + \ HCO_3^-$$

19.32 The bicarbonate pair, H_2CO_3/HCO_3^-, and the phosphate pair, $H_2PO_4^-/HPO_4^{2-}$. The proper buffer ratio is maintained by the decomposition of any excess carbonic acid to water and carbon dioxide. The carbon dioxide is removed from the equilibrium condition by its elimination at the lungs.

19.33 **(a)** Respiratory acidosis is brought about by hypoventilation. When the rate of breathing is too slow, carbon dioxide is not expelled from the lungs at a fast enough rate. The equilibrium is shifted to the left, increasing the concentration of hydrogen ions in the blood. The pH of the blood decreases.
   (b) Respiratory alkalosis results from hyperventilation. When the rate of breathing is too rapid, there is an increased loss of carbon dioxide from the lungs. The equilibrium is shifted to the right, decreasing the concentration of hydrogen ions in the blood. The pH of the blood increases.

PART II
Practice Questions

CHAPTER 2

The Saturated Hydrocarbons

MULTIPLE CHOICE

1. Methane, ethane, propane, and butane are

   (a) conformers   (b) homologs   (c) isomers   (d) isotopes

2. Which is a paraffin hydrocarbon?

   (a) acetylene   (b) benzene   (c) ethylene   (d) octane

3. Choose the statement that best describes hydrocarbons.

   (a) Flammable ionic compounds that are insoluble in water and are less dense than water.
   (b) Flammable covalent compounds that are insoluble in water and are less dense than water.
   (c) Flammable covalent compounds that are insoluble in water and are more dense than water.
   (d) Nonflammable covalent compounds that are insoluble in water and are more dense than water.

4. All of the members of the alkane series of hydrocarbons have the general formula

   (a) C_nH_{2n} (b) C_nH_n (c) C_nH_{2n+2} (d) C_nH_{2n-2}

5. Which of the following formulas represents a saturated hydrocarbon?

   (a) C_2H_6 (b) C_2H_4 (c) C_3H_6 (d) C_5H_{10}

6. A member of a homologous series of compounds differs from the next member by a

   (a) C atom   (b) H atom   (c) CH group   (d) CH_2 group

7. Which of the following compounds exhibits isomerism?

   (a) butane   (b) ethane   (c) methane   (d) propane

8. Two compounds that are isomers have the same

   (a) melting points   (b) boiling points   (c) structural formulas   (d) molecular formulas

9. Which of the following compounds is an isomer of $CH_3CH_2\overset{\displaystyle CH_3}{\overset{|}{C}}HCH_3$

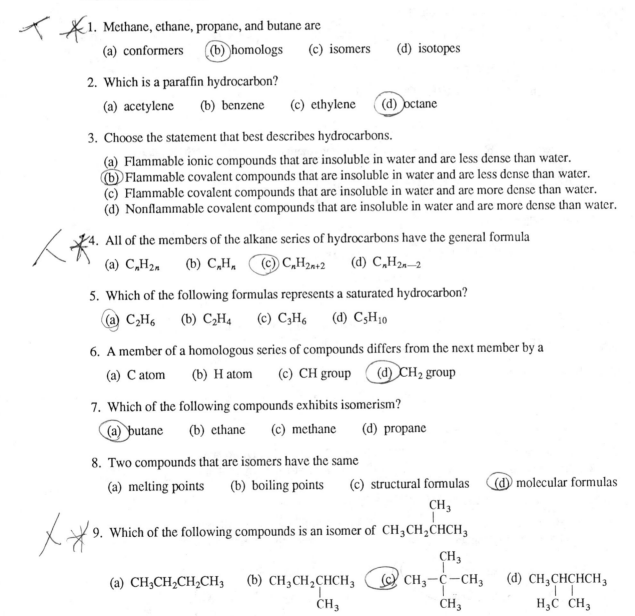

(a) $CH_3CH_2CH_2CH_3$ (b) $CH_3CH_2\underset{\displaystyle CH_3}{\underset{|}{C}}HCH_3$ (c) $CH_3-\overset{\displaystyle CH_3}{\underset{\displaystyle CH_3}{\overset{|}{\underset{|}{C}}}}-CH_3$ (d) $CH_3\underset{\displaystyle H_3C\ \ CH_3}{\underset{|\ \ \ \ |}{CHCHCH_3}}$

121

10. An isomer of $CH_3CH_2CH_2CH(CH_3)_2$ is

   (a) $(CH_3)_2CHCH_2CH_2CH_3$ (b) $CH_3(CH_2)_5CH_3$ (c) $CH_3CH_2CH_2CH_2CH_3$
   (d) $CH_3CH_2CH(CH_3)CH_2CH_3$

11. Which of the following compounds is an isomer of 2-methylbutane?

   (a) butane   (b) 2-methylpropane   (c) pentane   (d) propane

12. Which of the following formulas represent the same compound?

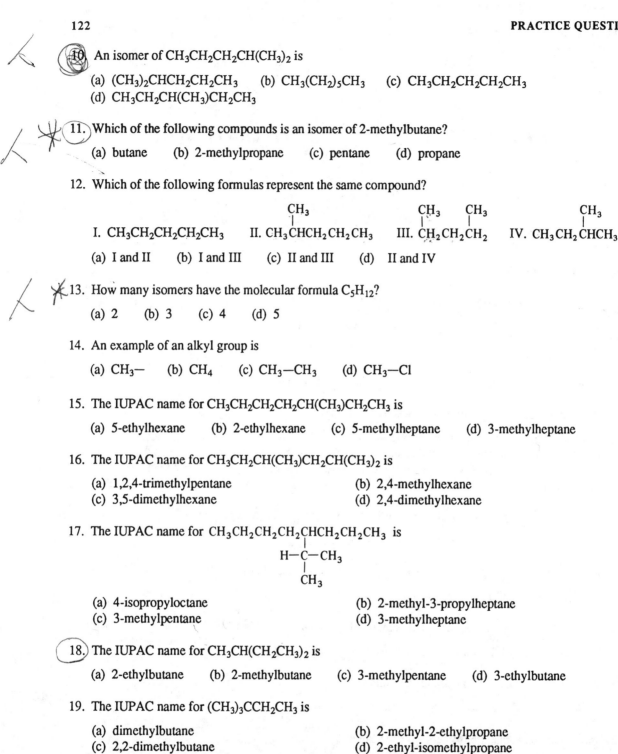

   I. $CH_3CH_2CH_2CH_2CH_3$ II. $CH_3CHCH_2CH_2CH_3$ III. $CH_2CH_2CH_2$ IV. $CH_3CH_2CHCH_3$

   (a) I and II   (b) I and III   (c) II and III   (d) II and IV

13. How many isomers have the molecular formula C_5H_{12}?

   (a) 2   (b) 3   (c) 4   (d) 5

14. An example of an alkyl group is

   (a) CH_3- (b) CH_4 (c) CH_3-CH_3 (d) CH_3-Cl

15. The IUPAC name for $CH_3CH_2CH_2CH_2CH(CH_3)CH_2CH_3$ is

   (a) 5-ethylhexane   (b) 2-ethylhexane   (c) 5-methylheptane   (d) 3-methylheptane

16. The IUPAC name for $CH_3CH_2CH(CH_3)CH_2CH(CH_3)_2$ is

   (a) 1,2,4-trimethylpentane   (b) 2,4-methylhexane
   (c) 3,5-dimethylhexane   (d) 2,4-dimethylhexane

17. The IUPAC name for $CH_3CH_2CH_2CH_2CHCH_2CH_2CH_3$ is

   (a) 4-isopropyloctane   (b) 2-methyl-3-propylheptane
   (c) 3-methylpentane   (d) 3-methylheptane

18. The IUPAC name for $CH_3CH(CH_2CH_3)_2$ is

   (a) 2-ethylbutane   (b) 2-methylbutane   (c) 3-methylpentane   (d) 3-ethylbutane

19. The IUPAC name for $(CH_3)_3CCH_2CH_3$ is

   (a) dimethylbutane   (b) 2-methyl-2-ethylpropane
   (c) 2,2-dimethylbutane   (d) 2-ethyl-isomethylpropane

20. The IUPAC name for $CH_3CH_2C(CH_3)_2CH_2CH(CH_2CH_3)_2$ is

   (a) 2-methyl-2,4-diethylhexane   (b) 3,5-diethyl-5-methylhexane
   (c) 5-ethyl-3,3-dimethylheptane   (d) 2,2-dimethyl-5-ethylheptane

21. Three of the following structures are identical. Which one is different?

   (a) $(CH_3)_2C(CH_2CH_3)CH_2CH_2CH_2CH_3$ (b) $CH_3CH_2C(CH_3)_2CH_2(CH_2)_2CH_3$
   (c) $CH_3(CH_2)_3C(CH_3)_2CH_2CH_3$ (d) $CH_3CH_2CH(CH_3)CH(CH_3)CH_2CH_2CH_3$

22. Alkanes are soluble in

    (a) dilute HCl   (b) dilute NaOH   (c) polar solvents   (d) nonpolar solvents

23. Which of the following pairs of compounds are not mutually soluble?

    (a) CH_4 and C_4H_{10} (b) C_4H_{10} and H_2O (c) $CH_3CH_2CH_2CH_3$ and $CH_3(CH_2)_4CH_3$
    (d) pentane and cyclopentane

24. Alkanes are _____ in water, and they are _____ than water.

    (a) soluble; heavier   (b) soluble; lighter   (c) insoluble; heavier   (d) insoluble; lighter

25. When an alkane burns completely, it forms

    (a) $CO + H_2O$ (b) $CO + CO_2$ (c) $CO_2 + H_2O$ (d) $CO_2 + C$

26. Which of the following compounds would react with bromine, in sunlight, to form only one monobromo compound?

    (a) butane   (b) ethane   (c) isobutane   (d) propane

27. Alkanes do not react with the following reagents except

    (a) HCl   (b) NaOH   (c) $KMnO_4$ (d) Br_2 and light

28. The burning of alkanes in the presence of air is known as

    (a) combustion   (b) polymerization   (c) pyrolysis   (d) substitution

29. Which of the following represents a free radical?

    (a) CH_3:   (b) CH_3 (c) CH_3·   (d) Cl:

30. Complete combustion of octane yields

    (a) $H_2O + CO_2$ (b) hexane and $CO_2 + CH_4$ (c) $H_2 + CO_2$ (d) none of these

31. The result of mixing pentane with concentrated sodium hydroxide is

    (a) cyclopentane   (b) cyclopentane + H_2O (c) sodium pentanol   (d) none of these

32. Gasoline is composed predominantly of hydrocarbon molecules that have a carbon-chain length of

    (a) less than 6 atoms   (b) 6–12 atoms   (c) 12–15 atoms   (d) greater than 15 atoms

33. Which of the following compounds would have the highest octane number?

    (a) $CH_3CH_2CH_2CH_2CH_2CH_2CH_3$ (b) $CH_3CH_2CH_2\underset{\underset{CH_3}{|}}{C}HCH_2CH_3$

    (c) $CH_3CH_2\underset{\underset{CH_3}{|}}{\overset{\overset{CH_3}{|}}{C}}CH_2CH_3$ (d) They all have the same octane number

34. The general formula for a cycloalkane is

    (a) C_nH_n (b) C_nH_{2n} (c) C_nH_{2n+2} (d) C_nH_{2n-2}

35. Which of the following cycloakanes contains 12 hydrogen atoms?

(a) △   (b) ☐CH$_3$ (c) ⬡CH$_2$CH$_3$ (d) ☐ CH$_3$ CH$_3$

36. An isomer of ☐ CH$_3$ CH$_3$ is

(a) ☐   (b) ☐CH$_3$ (c) ☐CH$_2$CH$_3$ (d) H$_3$C ☐ CH$_3$

37. The name of the compound ⬠ is

(a) propane   (b) pentane   (c) cyclopropane   (d) cyclopentane

38. The IUPAC name of △ CH$_2$CH$_3$ CH$_2$CH$_3$ is

(a) diethylcyclopropane   (b) 1,1-ethylcyclopropane
(c) 1,1-diethylcyclopropane   (d) diethylcyclobutane

39. Free radical chlorination of cyclohexane gives how many isomers with the formula $C_6H_{11}Cl$?

(a) 1   (b) 2   (c) 3   (d) 4

40. Which of the following is not an isomer of cyclopentane?

(a) methylcyclobutane   (b) 1,1-dimethylcyclobutane
(c) ethylcyclopropane   (d) 1,2-dimethylcyclopropane

FILL IN THE BLANKS

41. Compounds containing only carbon and hydrogen are called _____.

42. Organic compounds that contain only carbon, hydrogen, and single bonds are classed as _____.

43. The general formula for alkanes is _____.

44. The molecular formula of octane is _____.

45. Removal of one of the hydrogens of an alkane produces an _____ group.

46. The general formula for an alkyl group is _____.

47. How many structural isomers have the molecular formula C_4H_{10}? _____

48. The alkane that contains nine carbon atoms is called _____.

49. Give the IUPAC name for each of the following.

    (a) $CH_3CH_2CH(CH_3)CH(CH_3)_2$ _____

(b) $CH_3CH_2CH(CH_2CH_3)CH(CH_2CH_3)_2$ _____

(c) $CH_3CH_2CH(CH_3)CH_2CH(CH_2CH_2CH_3)_2$ _____

(d) $(CH_3)_2CHCH(CH_3)CH_2CH(CH_2CH_3)C(CH_3)_3$ _____

50. Give a condensed structural formula for each of the following.

(a) 2,4-dimethylhexane _____

(b) 3,3,4-trimethylheptane _____

(c) 2-chloro-2-methylbutane _____

(d) 1-bromo-2,2-dimethylpentane _____

(e) 4-ethyl-3-methyloctane _____

(f) 5,6-diethyl-2,3,4,7-tetramethylnonane _____

51. An alkyl group that has an odd, or unpaired, electron is called a _____.

52. Combustion involves the reaction of a compound with _____.

53. The most abundant alkane in natural gas is _____.

54. The principal component of bottled gas is _____.

55. An octane number of 100 has been assigned to _____ and a value of zero has been

assigned to _____.

56. The general formula for a cycloalkane is _____.

57. The cycloalkane that has seven carbon atoms has the molecular formula _____.

58. The formula of bromocyclobutane is _____.

59. The symbol ⬡ is used to represent _____.

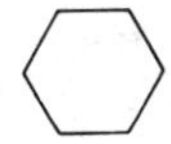

60. The name for (cyclopentane with two CH_3 groups) is _____.

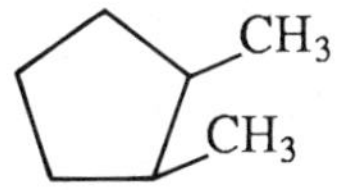

61. The molecular formula for (cyclobutane with CH_2CH_3) is _____.

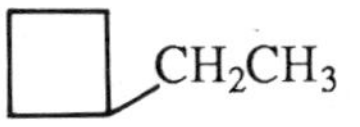

62. The two conformations of cyclohexane are called the _____ form and the _____ form.

TRUE OR FALSE

63. T   F   The general formula for the alkanes is C_nH_{2n}.

64. T   F   Isomers are compounds that have the same structural formulas but different molecular formulas.

65. T   F   The names of the alkanes containing less than five carbon atoms are derived from a Greek prefix
followed by the suffix -ane.

66. T F A straight-chain alkane and a branched-chain alkane with the same number of carbon atoms have different molecular formulas.

67. T F The compound $CH_3CH_2CH(CH_3)CH_2CH_2CH_3$ is a saturated hydrocarbon.

68. T F Alkanes are polar molecules.

69. T F The main component of natural gas is ethane.

70. T F Alkanes are insoluble in water.

71. T F Hydrocarbons are more dense than water.

72. T F Some alkanes are liquids at room temperature.

73. T F The odor of natural gas is due to substances that have been added.

74. T F Saturated hydrocarbons are generally unreactive.

75. T F Incomplete combustion of alkanes results in the formation of carbon monoxide and carbon.

76. T F Bromine does not react with alkanes at ordinary temperatures and in the dark.

77. T F In the halogenation of methane, the halogen atom replaces a hydrogen atom of methane.

78. T F The electron dot representation for a chlorine atom is $: \overset{..}{\underset{..}{Cl}} \cdot$.

79. T F The free radical chlorination of methane yields only methyl chloride.

80. T F The complete combustion of one mole of butane yields four moles of carbon dioxide.

81. T F The major use of alkanes is in the production of energy.

82. T F Petroleum resulted from the decomposition of plant and animal matter.

83. T F Crude oil is the liquid component of petroleum.

84. T F The maximum octane number that can be assigned to a fuel is 100.

85. T F Increased branching in alkanes results in a lower octane gasoline.

86. T F The carbon atoms of all cycloalkanes lie in a plane.

87. T F 1,1-Dimethylcyclopropane is an isomer of cyclopentane.

CHAPTER 3

The Unsaturated Hydrocarbons

MULTIPLE CHOICE

1. Which of the following is an alkene?

    (a) C_5H_{10} (b) C_5H_{12} (c) C_6H_{14} (d) C_6H_{10}

2. Which of the following compounds is unsaturated?

    (a) CH_4 (b) C_2H_6 (c) C_4H_6 (d) $C_{10}H_{22}$

3. The general structure of a 1-alkene is

    (a) $RCH=CH_2$ (b) $RC\equiv CH$ (c) RCH_2CH_2 (d)  △

4. How many compounds have the molecular formula C_3H_6?

    (a) one   (b) two   (c) three   (d) four

5. The IUPAC name of $CH_3CH_2CH=C(CH_3)_2$ is

    (a) 4-methyl-3-pentene   (b) 4-methyl-4-pentene
    (c) 2-methyl-2-pentene   (d) 2-methyl-3-pentene

6. The IUPAC name of $CH_3CH=CHCH_2CH_2CH_2Cl$ is

    (a) 1-chloro-2-hexene   (b) 1-chloro-4-hexene   (c) 6-chloro-2-hexene   (d) 1-chlorohexene

7. The IUPAC name of $CH_3CH_2\overset{\displaystyle |}{\underset{\displaystyle CH_2CH_3}{C}}=CH_2$ is

    (a) 2-ethyl-1-butene   (b) 3-ethyl-1-butene   (c) 3-ethyl-3-butene   (d) 2-ethyl-2-butene

8. The IUPAC name of 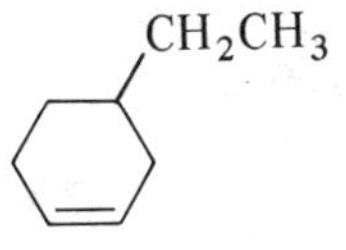 is

    (a) 1-ethylcyclohexene   (b) 2-ethylcyclohexene
    (c) 3-ethylcyclohexene   (d) 4-ethylcyclohexene

9. The IUPAC name of 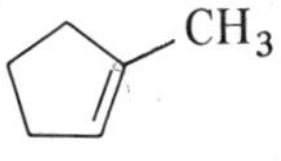 is

    (a) 1-methylcyclopentene   (b) 3-methylcyclopentene
    (c) 2-methylcyclopentene   (d) 1-methyl-2-cyclopentene

10. Which of the following is not an isomer of 2-pentene?

    (a) 1-pentene   (b) 2-methyl-1-butene   (c) 2-methyl-2-butene   (d) all are isomers of 2-pentene

Know this rule

11. Markovnikov's rule is useful in predicting the product of a reaction between an alkene and

    (a) H_2 (b) Br_2 (c) HBr   (d) O_2

12. The product of the reaction 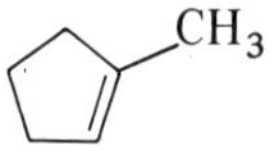 + HCl → is

    (a) 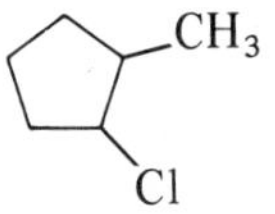   (b) 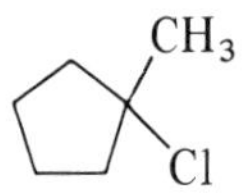   (c) 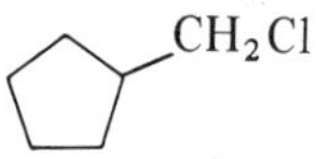   (d)

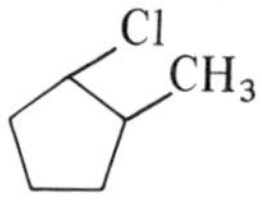

13. Which of the following compounds undergoes addition of Br_2?

    (a) 1-bromobutane   (b) cyclopentane   (c) cyclopentene   (d) pentane

14. Which of the following is the most stable carbocation?

    (a) CH^+ (b) $(CH_3)_2CH^+$ (c) $CH_3CH_2CH_2^+$ (d) $(CH_3)_3C^+$

15. A compound whose molecular formula is C_6H_{12} decolorizes a solution of bromine in carbon tetrachloride. The compound is probably

    (a) an alkene   (b) an alkane   (c) a cycloalkane   (d) a polymer

FILL IN THE BLANKS

16. Two classes of unsaturated hydrocarbons are _____ and _____ .

17. Olefins is another name for the _____ .

18. The molecular formula of a cycloalkene that contains seven carbons and two double bonds is _____

19. How many isomers are possible for C_3H_6? _____

20. Give the IUPAC name for each of the following.

    (a) $CH_3CH_2CH=CHCH(CH_3)_2$ _____

    (b) $CH_3CHClCH=CHCH_3$ _____

    (c) $CH_2=C(CH_3)CHBrCHICH_3$ _____

    (d) $CH_2FCH=CHCH_2CH(CH_3)_2$ _____

    (e)

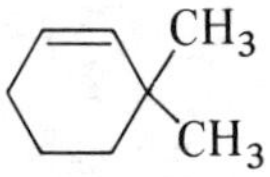

    (f)

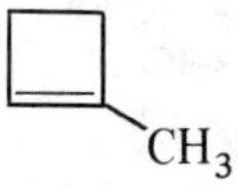

21. Give a condensed structural formula for each of the following.

   (a) 1-butene _____

   (b) 2-methyl-1-pentene _____

   (c) 2,4-dimethyl-1-hexene _____

   (d) 5-chloro-6-iodo-3-heptene _____

   (e) 1,6-dibromo-3-methyl-3-hexene _____

   (f) 4-isopropyl-2-octene _____

   (g) 4,7-diethyl-3,6,6,8,8-pentamethyl-2-nonene _____

   (h) 2,4-hexadiene _____

   (i) 1,3-cyclohexadiene _____

   (j) 3-isopropyl-4-methylcycloheptene _____

22. Cyclopentene can be prepared by the dehydration of _____ .

23. The catalytic hydrogenation of 1-pentene yields _____ .

24. Complete the following reactions.

   (a) $CH_2=CHCH_2CH_3 + HCl \rightarrow$ _____

   (b) $CH_2=CHCH_2CH_3 + Br_2 \rightarrow$ _____

   (c) $CH_3CH=CHCH_3 + H_2 \xrightarrow{\text{Ni}}$ _____

   (d) $CH_3CH=CH_2 + H_2O \xrightarrow{H_2SO_4}$ _____

   (e) 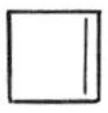 $\square\!\!\parallel + Cl_2 \rightarrow$ _____

   (f) $(CH_3)_2C=CHCH_2CH_2CH_3 + HBr \rightarrow$ _____

   (g) 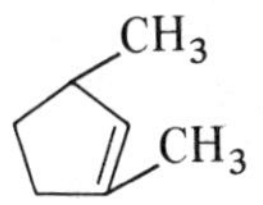 $+ HCl \rightarrow$ _____

   (h) 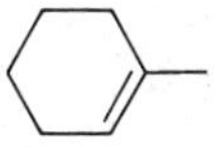 $-CH_2CH_3 + HI \rightarrow$ _____

   (i) $CH_3CH_2CH=CH_2 \xrightarrow[\text{dilute}]{KMnO_4}$ _____

   (j) $(CH_3)_2C=CHCH_3 \xrightarrow[\text{conc.}]{KMnO_4}$ _____

25. The reaction in which many alkene molecules combine to form a large molecule is called _____ .

26. A _____ is composed of many monomeric units joined together.

27. The general formula for an alkyne is _____ .

28. What is the IUPAC name for acetylene? _____

29. What is the structural formula for acetylene? _____

30. What is the IUPAC name for $CH_3C \equiv CCH(CH_3)_2$? _____

31. Write a condensed structural formula for 4-bromo-2-hexyne. _____

32. How many moles of hydrogen are required to saturate 4-octyne? _____

33. Complete the following equation.

$$CH_3C \equiv CH + 2 \, HBr \rightarrow \underline{\hspace{2cm}}$$

TRUE OR FALSE

34. T F All compounds whose general formula is C_nH_{2n} are alkenes.

35. T F All of the atoms in ethylene lie in the same plane.

36. T F Two structural isomers are possible for ethylene.

37. T F Propene and cyclopropane are isomers.

38. T F If an acyclic compound has the molecular formula $C_{10}H_{16}$, it must have three double bonds.

39. T F 3-Methyl-2-hexene is an isomer of 2-heptene.

40. T F $(CH_3)_2CHCH_2CH_3$ and $(CH_3)_2C=CHCH_3$ are isomers.

41. T F The alkenes resemble the alkanes in their physical properties.

42. T F An electrophile is a species with an unshared pair of electrons.

43. T F Alkenes and alkynes are similar to alkanes in reactivity.

44. T F Alkyl groups are electron-withdrawing.

45. T F The most characteristic general reaction of alkenes is substitution.

46. T F An alkene can be converted into an alkane through the hydrogenation reaction.

47. T F Dehydration and dehydrohalogenation require different catalysts and conditions.

48. T F An alkene undergoes dehydration in the presence of an acid to produce an alcohol.

49. T F The use of Markovnikov's rule is necessary to predict the product of HBr and $CH_3CH=CHCH_3$.

50. T F A compound with a molecular formula C_6H_{12} is unsaturated and will react with potassium permanganate.

51. T F 2-Methyl-3-hexene reacts with a solution of bromine in carbon tetrachloride.

52. T F The acetylene molecule is linear.

53. T F Two structural isomers are possible for acetylene.

54. T F Propane, propene, and propyne have similar physical properties.

55. T F Most of the chemical reactions of alkynes are similar to those of alkenes.

CHAPTER 4

The Aromatic Hydrocarbons

MULTIPLE CHOICE

1. In 1865, _____ suggested a regular hexagon structure for benzene.

   (a) Faraday   (b) Kekule   (c) Markovnikov   (d) Wöhler

2. The carbon-carbon bond distance in benzene is

   (a) longer than a C–C single bond.
   (b) longer than a C=C double bond.
   (c) shorter than a C=C double bond.
   (d) shorter than a C≡C triple bond.

3. Regarding the benzene ring, we now know that

   (a) all the bond lengths between carbon atoms are the same.
   (b) there are alternating double and single bonds between carbon atoms.
   (c) the angles between carbon atoms vary from 90° to 120°.
   (d) benzene is not a planar molecule.

4. The structure of 1,2,4-trichlorobenzene is

5. The structure of 3-chloroethylbenzene is

6. The structure of resorcinol is

7. The structure of 3-chlorophenol is

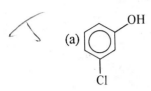

8. The structure of *p*-dinitrobenzene is

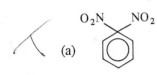

9. A correct name for is

   (a) 1-methylethylbenzene   (b) 3-methyl-2-ethylbenzene
   (c) 1-ethylmethylbenzene   (d) 1-ethyl-2-methylbenzene

10. A correct name for ⟨O⟩—CHCH=CH₂ is
                                 CH₃

   (a) 1-phenyl-1-methyl-1-propene   (b) 3-phenyl-3-methyl-1-propene
   (c) 2-phenyl-3-butene   (d) 3-phenyl-1-butene

11. A correct name for [structure with NH₂ and Cl] is

   (a) 2-chlorophenol   (b) 2-chlorotoluene   (c) 2-chloroaniline   (d) 1-chloroaniline

12. A correct name for HO—[ring with NO₂]—Cl is

   (a) 4-chloro-3-nitrophenol   (b) 4-chloro-5-nitrophenol
   (c) 4-chloro-3-nitroaniline   (d) 4-chloro-5-nitroaniline

13. A correct name for Cl—[ring with Cl] is

   (a) *o*-dichlorobenzene   (b) 1,2-dichlorobenzene
   (c) *m*-dichlorobenzene   (d) 1,5-dichlorobenzene

14. Which isomer has the highest melting point?

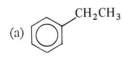 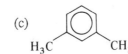 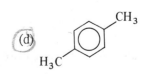

✻ Know products of SN's

134   **PRACTICE QUESTIONS**

15. An aryl group is

    (a) CH_3CH_2- (b) 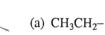   (c) $CH_2=CH-$ (d) $CH\equiv C-$

16. is called

    (a) benzyl   (b) cyclohexyl   (c) phenol   (d) phenyl

17. is called

    (a) benzylbenzene   (b) biphenyl   (c) dibenzene   (d) naphthalene

18. is called

    (a) anthracene   (b) naphthalene   (c) phenanthrene   (d) tribenzene

19. Bromobenzene can be prepared from benzene, an iron catalyst, and

    (a) Br_2 (b) HBr   (c) $AlBr_3$ (d) NaBr

20. The _____ of toluene is an aromatic reaction that is important in the manufacturing of explosives.

    (a) alkylation   (b) halogenation   (c) nitration   (d) sulfonation

21. The structure of the product of the reaction of nitric acid/sulfuric acid with benzene is

    (a) 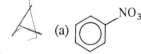   (b) 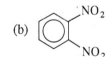   (c)   (d)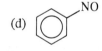

22. The reaction least likely to occur is

    (a) ⬡ + HNO_3 $\xrightarrow{H_2SO_4}$ ⬡$-NO_2$

    (b) ⬡ + Cl_2 $\xrightarrow{light}$ ⬡$-Cl$

    (c) ⬡ + H_2SO_4 $\xrightarrow{SO_3}$ ⬡$-SO_3H$

    (d) ⬡ + Br_2 $\xrightarrow{FeBr_3}$ ⬡$-Br$

FILL IN THE BLANKS

23. The molecular formula of toluene is ___C_7H_8___.

24. Disubstituted benzene can exist as three different isomers designated as ___ortho___, ___meta___, and ___para___.

25. Give a correct name for each of the following

(a) 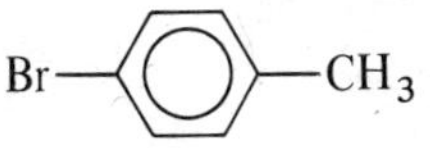 Br—⟨O⟩—CH$_3$ _____

(b) 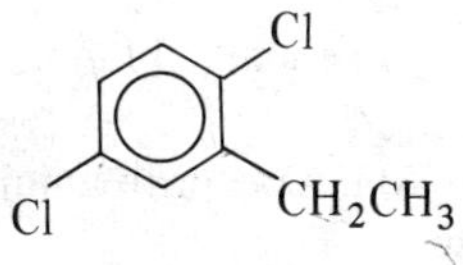 [structure with Cl, Cl, CH$_2$CH$_3$] _____

(c) 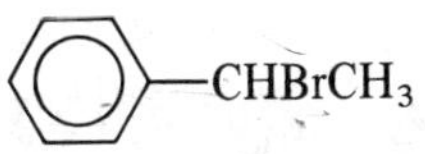 ⟨O⟩—CHBrCH$_3$ _____

(d) 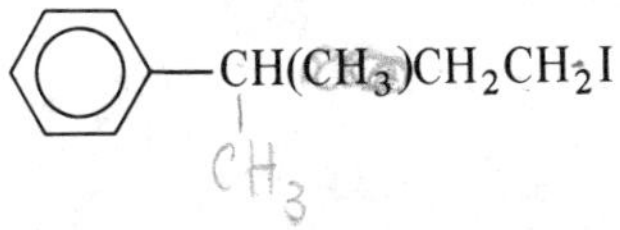 ⟨O⟩—CH(CH$_3$)CH$_2$CH$_2$I _____
            CH$_3$

26. Give a structural formula for each of the following.
     (a) the phenyl group _____   (b) benzyl bromide _____

     (c) naphthalene _____   (d) p-iodotoluene _____

     (e) o–bromochlorobenzene _____   (f) 2,5-diethylphenol _____

     (g) m-dinitrobenzene _____   (h) 2-fluoro-1-phenylpropane _____

27. Benzene is much ___less___ reactive than most alkenes.

28. The characteristic reactions of benzene are __substitution__ reactions.

29. Complete the following equation. ⟨O⟩ + Br$_2$ $\xrightarrow{\text{Fe Br}_3}$ ⟨O⟩-Br + HBr

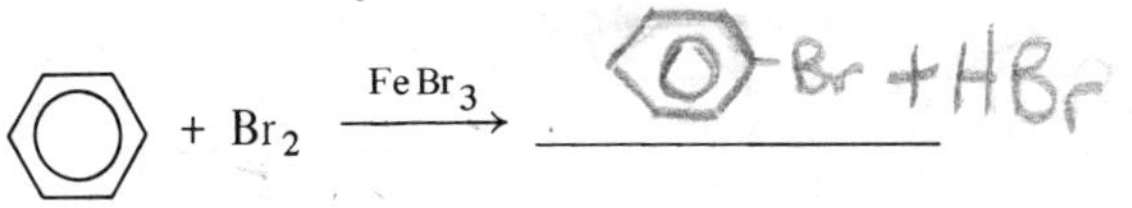

30. __benzo a pyrene__ is a carcinogenic compound that has been found in cigarette smoke and charcoal-broiled steaks.

TRUE OR FALSE

31. T F All aromatic compounds that we have studied are cyclic.

32. T F All cyclic compounds are aromatic.

33. T F Two hydrogens are bonded to each carbon atom of benzene.

34. T F All the carbon–carbon bonds in benzene are the same length.

35. T F The compound 1,3,5-trichlorobenzene can also be correctly named m-trichlorobenzene.

36. T F There are four dibromobenzene isomers.

37. T F Naphthalene is an aromatic compound.

38. T F Benzene is a liquid and naphthalene is a solid.

39. T F Aromatic compounds burn with a very sooty flame.

40. T F In an aromatic substitution reaction, a hydrogen atom attached to a benzene ring is replaced by another group.

41. T F The chemical behavior of benzene is very similar to that of alkenes.

42. T   F   Natural gas is a good source of aromatic compounds.

43. T   F   Benzene is used extensively in chemical laboratories.

44. T   F   Aromatic ring compounds do not occur in animals.

CHAPTER 5

Halogen Derivatives of Hydrocarbons

MULTIPLE CHOICE

1. Which of the following is not an alkyl halide?

   (a) CH_3CH_2F (b) $CH_3CH_2CH_2Br$ (c) $(CH_3)_3CI$ (d) 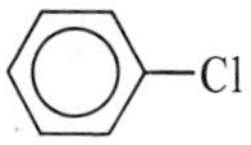 ⬡—Cl

2. Which of the following is methylene chloride?

   (a) $CHCl_3$ (b) CH_3Cl (c) CH_2Cl_2 (d) CCl_4

3. Another name for isopropyl iodide is

   (a) iodopropane   (b) 1-iodopropane   (c) 2-iodopropane   (d) 1-iodo-1-methylethane

4. The IUPAC name for $CH_3CHClCH_2CH_2CH_2Cl$ is

   (a) 1-methyl-1,4-dichlorobutane   (b) 2,5-dichloropentane
   (c) 1,4-dichloropentane   (d) 1,4-dichloro-1-methylbutane

5. What is the common name for $(CH_3)_2CHBr$?

   (a) isobutyl bromide   (b) isopropyl bromide   (c) propylene bromide   (d) propyl bromide

6. The IUPAC name for [structure showing cyclopentane with Cl groups] is

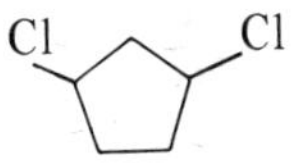

   (a) 1,3-dichloropentane   (b) 1,3-chlorocyclopentane
   (c) 1,3-dichlorocyclopentane   (d) 2,5-dichlorocyclopentane

7. The IUPAC name for $CH_2=C(CH_2CH_3)CHBrCHCH_3$ is

   (a) 3-bromo-4-ethyl-1-hexene   (b) 3-bromo-2-ethyl-1-pentene
   (c) 3-bromo-4-ethyl-1-pentene   (d) 3-bromo-2-ethyl-1-hexene

8. The structural formula for *o*-difluorobenzene is

   (a)  F—⬡—F   (b) 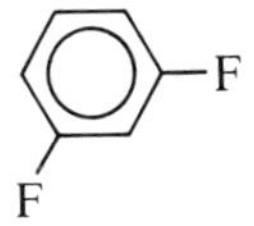 ⬡ with F groups   (c) 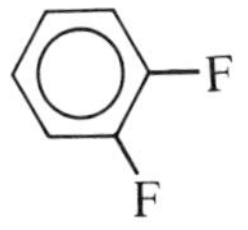 ⬡—F with F   (d) 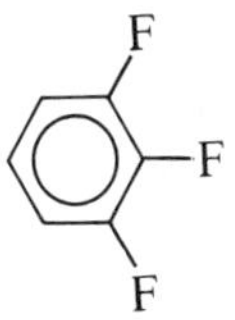 ⬡ with F groups

137

rxn. to KOH

9. The condensed structural formula for 5-bromo-2-chloroheptane is

    (a) $CH_3CH_2CHBrCH_2CH_2CHClCH_3$ (b) $CH_3CHClCH_2CH_2CHBrCH_3$

    (c) $(CH_3)_2CClCH_2CHBrCH_2CH_3$ (d)

10. Which of the following methyl halides is not a gas at room temperature?

    (a) CH_3Br (b) CH_3Cl (c) CH_3F (d) CH_3I

11. Which of the following has the highest boiling point?

    (a) $CH_3CH_2CH_2CH_2F$ (b) CH_3CH_2Cl (c) $CH_3CH_2CH_2Br$ (d) CH_3I

12. Which of the following compounds will react with HBr to yield $CH_3CHBrCH_2CH_3$?

    (a) $CH_3CH_2CH=CH_2$ (b) $CH_3CH=CHCH_2CH_3$
    (c) $CH_2=CHCH(CH_3)_2$ (d) $CH_3C\equiv CCH_3$

13. The reaction $CH_3CH = C(CH_3)CH_2CH_3 + HCl \rightarrow$ yields

    (a) 3-methylpentane   (b) 2-chloro-3-methylpentane
    (c) 3-chloro-3-methylpentane   (d) 2,3-dichloro-3-methylpentane

14. The reaction of cyclohexene with Br_2 yields

    (a) 1,2-dibromocyclohexane   (b) 1,2-dibromohexane
    (c) 1,2-dibromocyclohexene   (d) 1,2-dibromohexene

15. Most of the reactions that the alkyl halides undergo are either

    (a) addition or elimination   (b) addition or substitution
    (c) elimination or substitution   (d) decomposition or elimination

16. _____ are species that can donate a pair of electrons in a chemical reaction.

    (a) electrophilic reagents   (b) nucleophilic reagents
    (c) carbocations   (d) free radicals

17. The conversion of an alkyl bromide to an alkene occurs via a _____ reaction.

    (a) dehydration   (b) dehydrohalogenation   (c) halogenation   (d) substitution

18. The product of the reaction is

    (a)   (b)   (c)   (d)

19. Which of the following is a local anesthetic?

    (a) ethyl chloride   (b) ethylene chloride   (c) methyl chloride   (d) methylene chloride

20. Which compound was once used as an inhalation anesthetic?

    (a) CH_3Cl (b) CH_2Cl_2 (c) $CHCl_3$ (d) CCl_4

21. Vinyl chloride is still used

    (a) as a drycleaning solvent.   (b) in the production of polyvinyl chloride plastics.
    (c) as an aerosol propellant.   (d) as an antiseptic.

22. Which of the following halogenated hydrocarbons is a polymer?

    (a) fluoroacetate   (b) halothane   (c) iodoform   (d) Teflon

23. Which of the following halogenated hydrocarbons is nontoxic?

    (a) freon   (b) iodoacetate   (c) phosgene   (d) ethylene dibromide

24. Which of the following is a herbicide?

    (a) DFP   (b) PCBs   (c) TCDD   (d) 2,4,5-T

25. Which of the following is not a characteristic of DDT?

    (a) It is readily absorbed through the skin when it is in powder form.
    (b) It is one of the most persistent pesticides.
    (c) DDT concentrates in living tissue.
    (d) Numerous insects have developed a resistance to DDT.

26. DDT and other pesticides exert their effect by blocking the action of

    (a) carbohydrates   (b) enzymes   (c) lipids   (d) nucleic acids

FILL IN THE BLANKS

27. The general designation for an alkyl halide is ___R—X___.

28. Give the IUPAC name for each of the following.

    (a) $CF_2BrCHClF$ _____   (b) $(CH_3)_2CBrCH_2CBr(CH_3)_2$ _____

    (c) [cyclohexane with CH_3 and F substituents] _____   (d) [cyclohexane with two CH_3 and I substituents] _____

    (e) $CH_3CHFCH(CH_2CH_3)C(CH_3)_3$ _____

    (f) $CH_2=CHClCH_2CH_2Cl$ _____

    (g) I—⬡—NO_2 _____   (h) [cyclopentene with Br substituent] _____

29. Give a structural formula for each of the following.

    (a) 1,2-dichloroethane _____   (b) 3-bromo-4-methylheptane _____

    (c) 1,1-diethyl-3-fluorocyclobutane _____

    (d) 1-iodo-3-methylcyclopentane _____

    (e) methylene bromide _____   (f) dichlorodifluoromethane _____

    (g) *m*-dibromobenzene _____   (h) phenyl iodide _____

30. The common name for $CHCl_3$ is _____.

31. Carbon tetrachloride is _____ in water.

32. Complete the following reactions

(a) 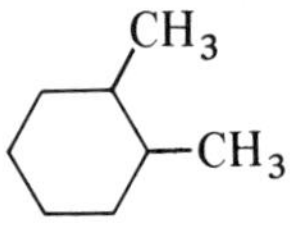 [cyclohexane with CH_3 substituent and CH_3] + HBr → _____

(b) $CH_3CH_2CH=C(CH_3)_2 + Br_2$ → _____

(c) 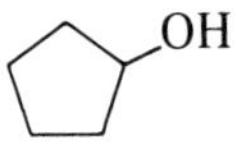 [cyclopentane with OH substituent] + PBr_3 → _____

(d) $(CH_3)_3COH + HBr$ → _____

(e) 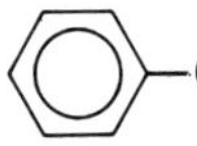 [benzene ring]$-CH_2CH_2Br + KOH \xrightarrow{HOH}$ _____

(f) $CH_3CH_2Br + KCN$ → _____

33. When 2-bromobutane undergoes an elimination reaction, _____ different alkenes are formed.

34. A _____ reagent is a species with an unshared pair of electrons.

35. The freons are still used as _____.

36. A mixture of _____ has been used as a synthetic blood substitute.

37. The initials DDT represent _____.

38. Name two other polychlorinated pesticides (other than DDT) _____.

CHAPTER 6

Alcohols, Phenols, Ethers, and Thiols

MULTIPLE CHOICE

1. The alcohol 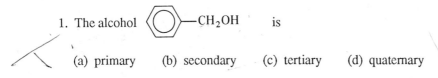 is

   (a) primary   (b) secondary   (c) tertiary   (d) quaternary

2. Which of the following is a secondary alcohol?

3. The hydroxyl groups in $HOCH_2CH_2CCH_2OH$ are

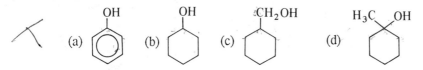

   (a) only primary   (b) only secondary   (c) only tertiary   (d) both primary and secondary

4. Which is a primary alcohol?

   (a) $CH_3CH_2CH_2CH_2OH$ (b) $CH_3CHOHCH_2CH_3$
   (c) $(CH_3)_2CHCHOHCH_3$ (d) $(CH_3)_2COHCH_2CH_3$

5. Which compound is a tertiary alcohol?

6. An example of a secondary alcohol is

   (a) $CH_3CH_2CHOHCH_3$ (b) $CH_3CH_2CH_2CH_2OH$
   (c) $CH_3CH_2COH(CH_3)_2$ (d) $(CH_3)_3COH$

7. The IUPAC name for the secondary alcohol that contains four carbon atoms is

   (a) 2-butanol   (b) 3-butanol   (c) 2-methyl-2-propanol   (d) 2-methyl-1-propanol

Lewis Acid

8. The IUPAC name for $CH_3CH_2CH_2\underset{\underset{CH_2CH_3}{|}}{\overset{\overset{CH_3}{|}}{C}}OH$ is

    (a) 1-ethyl-1-methylbutanol   (b) 3-methyl-3-hexanol
    (c) 2-ethyl-2-pentanol   (d) 3-methylheptanol

9. The IUPAC name for 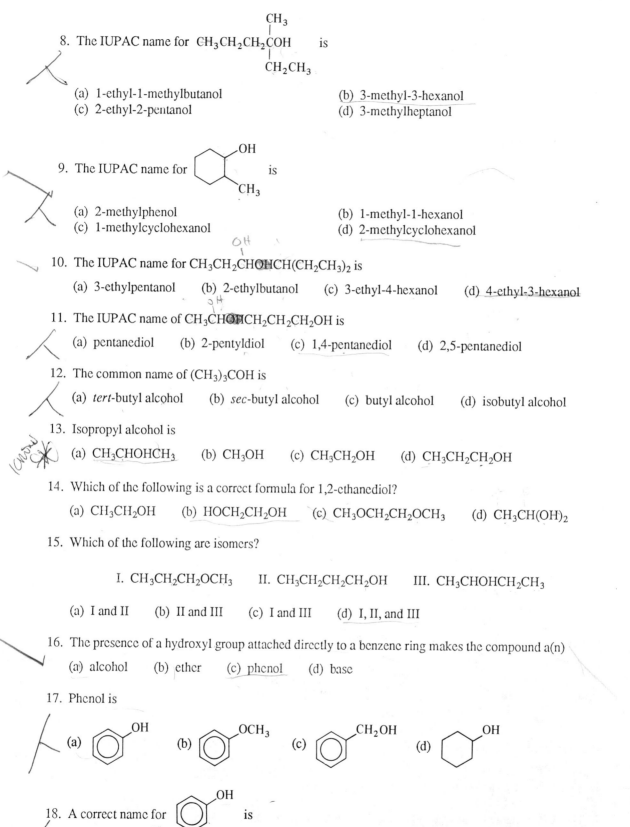 is

    (a) 2-methylphenol   (b) 1-methyl-1-hexanol
    (c) 1-methylcyclohexanol   (d) 2-methylcyclohexanol

10. The IUPAC name for $CH_3CH_2CHOHCH(CH_2CH_3)_2$ is

    (a) 3-ethylpentanol   (b) 2-ethylbutanol   (c) 3-ethyl-4-hexanol   (d) 4-ethyl-3-hexanol

11. The IUPAC name of $CH_3CHOHCH_2CH_2CH_2OH$ is

    (a) pentanediol   (b) 2-pentyldiol   (c) 1,4-pentanediol   (d) 2,5-pentanediol

12. The common name of $(CH_3)_3COH$ is

    (a) *tert*-butyl alcohol   (b) *sec*-butyl alcohol   (c) butyl alcohol   (d) isobutyl alcohol

13. Isopropyl alcohol is

    (a) $CH_3CHOHCH_3$ (b) CH_3OH (c) CH_3CH_2OH (d) $CH_3CH_2CH_2OH$

14. Which of the following is a correct formula for 1,2-ethanediol?

    (a) CH_3CH_2OH (b) $HOCH_2CH_2OH$ (c) $CH_3OCH_2CH_2OCH_3$ (d) $CH_3CH(OH)_2$

15. Which of the following are isomers?

        I. $CH_3CH_2CH_2OCH_3$ II. $CH_3CH_2CH_2CH_2OH$ III. $CH_3CHOHCH_2CH_3$

    (a) I and II   (b) II and III   (c) I and III   (d) I, II, and III

16. The presence of a hydroxyl group attached directly to a benzene ring makes the compound a(n)
    (a) alcohol   (b) ether   (c) phenol   (d) base

17. Phenol is

18. A correct name for is

    (a) 2-hydroxy-4-nitrobenzene   (b) 2-nitro-4-hydroxybenzene   (c) 3-nitrophenol   (d) 3-aminophenol

19. A correct name for OH (structure with Br at 3,5 positions) is

    (a) 1,3-dibromophenol
    (c) 2,4-dibromophenol

    (b) 4,6-dibromophenol
    (d) 3,5-dibromophenol

20. Which compound is an ether?

    (a) $CH_2OHCHOHCH_2OH$

    (b) $CH_3CHOHCH_3$

    (c) $CH_3CH_2CH_2OCH_3$

    (d) CH_3—(ring)—OH

21. The IUPAC name for methyl ethyl ether is

    (a) methoxyethane   (b) ethoxyethane   (c) 2-methoxyethane   (d) 1-ethoxyethane

22. The structure of 1-propoxypropane is

    (a) $CH_3CH_2OCH_2CH_2CH_3$
    (c) $CH_3CH_2CH_2OCH(CH_3)_2$

    (b) $CH_3CH_2OC(CH_3)_3$
    (d) $CH_3CH_2CH_2OCH_2CH_2CH_3$

23. Four compounds of similar molecular weight belong to the following classes of organic compounds. The compound with the highest boiling point is an

    (a) aromatic hydrocarbon   (b) alkene   (c) alcohol   (d) ether

24. An alcohol with a very long hydrocarbon chain is likely to be

    (a) insoluble in water but soluble in organic solvents.
    (b) insoluble in organic solvents but soluble in water.
    (c) soluble in both water and organic solvents.
    (d) insoluble in both water and organic solvents.

25. Which compound would be most soluble in water?

    (a) CH_3CH_2Cl (b) $CH_3CH_2CH_2Br$ (c) $CH_3CH_2CH_3$ (d) $CH_3CH_2CH_2OH$

26. Which compound would be most soluble in water?

    (a) $CH_3OCH_2CH_2CH_3$ (b) $CH_3CHOHCH_3$ (c) $CH_3CH_2OCH_2CH_3$ (d) $CH_3CH=CHCH_3$

27. Which compound would be most soluble in water?

    (a) $CH_2=CHCH_2CH_2OH$
    (c) $CH_3CH_2CHOHCH_2OH$

    (b) $CH_3CH_2OCH_2CH_3$
    (d) $CH_3CHOHCH_2CH_3$

28. Which compound would have the highest boiling point?

    (a) CH_3CH_2Cl (b) $CH_3CH_2CH_3$ (c) CH_3CH_2OH (d) CH_3OCH_3

29. Which compound would have the highest boiling point?

    (a) $CH_3CH_2OCH_2CH_3$
    (c) $CH_3CH_2CHOHCH_2OH$

    (b) $CH_3OCH_2CH_2CH_3$
    (d) $CH_3CHOHCH_2CH_3$

30. Which compound is the least soluble in water?

    (a) CH_3OH (b) CH_3CH_2OH (c) $CH_3CH_2CH_2OH$ (d) $CH_3CH_2CH_2CH_2OH$

31. Which compound is the least soluble in water?

    (a) CH_3CH_2OH (b) CH_3OCH_3 (c) $CH_3CH_2CH_2SH$ (d) CH_3OH

32. Which compound is the least soluble in water?

    (a) $CH_3CHOHCHOHCH_3$ (b) $CH_3CH_2CH_2CH_2OH$ (c) $(CH_3)_3COH$ (d) $CH_3OCH_2CH_3$

33. Which compound is the least soluble in water?

    (a) C_4H_9OH (b) C_3H_7OH (c) C_2H_5OH (d) $C_5H_{11}OH$

34. Which compound would have the lowest boiling point?

    (a) diethyl ether   (b) 2-propanol   (c) methoxyethane   (d) phenol

35. Alcohols boil at appreciably higher temperatures than hydrocarbons of similar molecular weight because

    (a) alcohols are more acidic.   (b) alcohols are ionic compounds.
    (c) alcohols are soluble in water.   (d) none of these.

36. Which of the following best represents hydrogen bonding in methanol?

    (a)
    $$CH_3-O \qquad\qquad O-H$$
    $$\diagdown \qquad\qquad \diagup$$
    $$H \cdots CH_3$$

    (b) $CH_3-O-H \cdots H-O-CH_3$

    (c)
    $$CH_3-O \quad H$$
    $$\diagdown \quad \diagdown$$
    $$H \cdots O-CH_3$$

    (d)
    $$CH_3-O \cdots CH_3-O$$
    $$\diagdown \qquad\qquad \diagdown$$
    $$H \qquad\qquad\quad H$$

37. An alcohol that dissolves in water yields _____ solution.

    (a) an acidic   (b) a neutral   (c) a basic   (d) a colloidal

38. Phenols differ from alcohols in that

    (a) alcohols are basic, whereas phenols are neither acidic nor basic.
    (b) phenols are acidic, whereas alcohols are neither acidic nor basic.
    (c) alcohols are acidic, whereas phenols are neither acidic nor basic.
    (d) phenols are basic, whereas alcohols are neither acidic nor basic.

39. The compound in which intermolecular hydrogen bonding is not possible is

    (a) CH_3OCH_3 (b) CH_3CH_2OH (c) $CH_3CHOHCH_3$ (d)

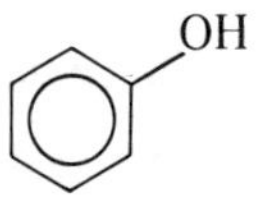

40. The double bond of an alkene can add water in the presence of a catalyst to form

    (a) an ether   (b) a phenol   (c) a thiol   (d) an alcohol

41. Dehydration of 1-pentanol produces

    (a) a ketone   (b) an acid   (c) an aldehyde   (d) an alkene

42. What is the major organic product of the following reaction?

    $$CH_3CH_2CH_2OH + NaOH \rightarrow$$

(a) $CH_3CH_2CH_2OCH_2CH_2CH_3$

(b) $CH_3CH_2CH_2ONa$

(c) $CH_3CH_2CH_3$

(d) no reaction occurs

43. Which product would be expected from the hydration of 1-butene?

(a) $CH_3CH_2CH_2CH_2OH$

(b) $CH_3CHCH_2CH_3$
       |
       OH

(c) $CH_3CH_2CH_2OCH_3$

(d) CH_3CHCH_2OH
       |
       CH_3

44. Which alkene forms upon acid-catalyzed hydration?

(a) (b) (c) (d)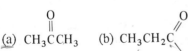

45. Which of the following products results when water is added to propene (with sulfuric acid as a catalyst)?

(a) $CH_3CH_2CH_2OH$ (b) $CH_3CHOHCH_3$ (c) $CH_3CH_2CH(OH)_2$ (d) $CH_3CHOHCH_2OH$

46. Which compound is not readily oxidized?

(a) $(CH_3)_2CHCH_2OH$

(b) $CH_3CH_2CHOHCH_3$

(c) $CH_3COH(CH_2CH_3)_2$

(d) $(CH_3)_3CCH_2OH$

47. What is the product of the following reaction?

$$CH_3CHOHCH_2CH_3 + K_2Cr_2O_7 \rightarrow$$

(a) $CH_3CH=CHCH_3$

(b) $CH_2=CHCH_2CH_3$

(c) $CH_3\overset{O}{\overset{||}{C}}CH_2CH_3$

(d) $CH_3OCH_2OCH_2CH_3$

48. What is the product of the following reaction?

$$CH_3CHOHCH_3 + KMnO_4 \rightarrow$$

(a) $CH_3\overset{O}{\overset{||}{C}}CH_3$

(b) $CH_3CH_2\overset{O}{\overset{\diagup\diagdown}{C}}_H$

(c) $CH_3\overset{CH_3}{\overset{|}{CH}}-O-\overset{CH_3}{\overset{|}{C}HCH_3}$

(d) no reaction

49. The compound that reacts most readily with the Lucas reagent is

(a) $CH_3CH_2OCH_3$ (b) $CH_3CHOHCH_3$ (c) [benzene ring with OCH₃]   (d) [cyclohexane with OCH₃]

50. Dehydration of alcohols does not produce

    (a) ethers   (b) alkenes   (c) phenols   (d) water

51. Which term is used to describe the following reaction?

$$CH_2=CH_2 + H_2O \longrightarrow CH_3CH_2OH$$

    (a) reduction   (b) dehydration   (c) hydration   (d) oxidation

52. 2-Hexanol is oxidized by a solution of potassium dichromate to

    (a) an aromatic hydrocarbon   (b) an aldehyde   (c) a ketone   (d) a phenol

53. Which of the following compounds would result from the dehydration of cyclohexanol?

    (a)  ⬡   (b) 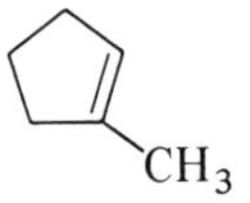 [cyclopentene with CH₃]   (c) 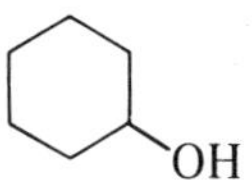 [cyclohexane with OH]   (d) 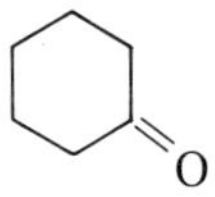 [cyclohexanone with O]

54. In chemical reactivity, ethers resemble

    (a) alcohols   (b) alkanes   (c) phenols   (d) thiols

55. Which of the following statements is incorrect?

    (a) Ethers do not react with bases such as NaOH.
    (b) Ethers do not react with acids such as HI.
    (c) Ethers do not react with oxidizing reagents such as $KMnO_4$.
    (d) Ethers do not react with metals such as Na.

56. Which of these compounds is used to make an explosive?

    (a) glycerol   (b) ethylene glycol   (c) propanol   (d) propanethiol

57. Automobile antifreeze is

    (a) CH_3CH_2OH (b) $CH_2OHCHOHCH_2OH$ (c) CH_2OHCH_2OH (d) $CH_3CHOHCH_2CH_3$

58. Rubbing alcohol is

    (a) CH_3OH (b) $CH_3CH_2CH_2OH$ (c) CH_3CH_2OH (d) $CH_3CHOHCH_3$

59. Which of the following alcohols is the most dangerous to spill on the skin?

    (a) methyl alcohol   (b) ethyl alcohol   (c) isopropyl alcohol   (d) glycerol

60. A new germicide is tested for its effectiveness against certain bacteria. It is found that a 1% solution of the new germicide is as effective as a 8% solution of phenol. The phenol coefficient of the germicide is

    (a) 8.0   (b) .80   (c) 7.2   (d) 16

61. What is the percent alcohol in 90 proof whiskey?

    (a) 9   (b) 45   (c) 90   (d) 180

62. In the U.S. ethyl alcohol is made by

    (a) fermenting grain alcohol   (b) fermenting sugars and starches
    (c) adding water to it   (d) adding water to ethane

63. Liver enzymes oxidize ethanol to

    (a) methanol   (b) formaldehyde   (c) CO_2 and H_2O (d) ethylene glycol

64. Denatured alcohol refers to

    (a) any alcohol not produced by fermentation.
    (b) grain alcohol that is highly taxed.
    (c) ethyl alcohol that has been treated with something to make it unfit to drink.
    (d) none of these.

65. The toxicity of wood alcohol results from its oxidation by liver enzymes to

    (a) carbon dioxide and water   (b) grain alcohol
    (c) formaldehyde and formic acid   (d) methanol and isopropyl alcohol

66. When ingested, ethyl alcohol has the physiological action of a(n)

    (a) stimulant   (b) depressant   (c) antiseptic   (d) none of these

67. A compound once widely used as an anesthetic is

    (a) CH_3SCH_3 (b) $CH_3CH_2CH_2OH$ (c) $CH_3CH_2OCH_2CH_3$ (d) $CH_3CH_2CH_2SH$

68. The phenol used in some mouthwashes and throat lozenges is

    (a) bakelite   (b) cresol   (c) hexachlorophene   (d) hexylresorcinol

69. Which of the following compounds does not contain a sulfhydryl group?

    (a) ethyl mercaptan   (b) hydrogen sulfide   (c) 2-propanethiol   (d) methyl ethyl sulfide

70. Which of the following names is incorrect for $CH_3CH_2SCH_2CH_2CH_3$?

    (a) 2-pentanethiol   (b) ethyl propyl sulfide
    (c) 1-ethylthiopropane   (d) they are all correct

71. _____ is used as an antidote in heavy metal poisoning.

    (a) BAL   (b) Lewisite   (c) DMSO   (d) acetyl-CoA

72. Which compound has the lowest boiling point?

    (a) CH_3CH_2SH (b) $CH_3CHSHCH_3$ (c) CH_3CH_2OH (d) $CH_3CHOHCH_3$

73. Which compound is the strongest acid?

    (a) CH_3CH_2SH (b) CH_3CH_2OH (c) CH_3SCH_3 (d) CH_3OCH_3

74. Which compound is not easily oxidized?

    (a) CH_3CH_2SH (b) CH_3CH_2OH (c) CH_3SCH_3 (d) CH_3OCH_3

75. Isopropyl mercaptan is

    (a) 2-propanesulfide   (b) 2-propyl sulfide   (c) propyl thiol   (d) 2-propanethiol

76. $CH_3CH_2SSCH_2CH_3$ is

    (a) ethyl disulfide   (b) ethyl sulfide   (c) ethyl mercaptan   (d) 3-thiobutane

77. $CH_3-\overset{\displaystyle O}{\overset{\|}{S}}-CH_3$ is

 (a) dimethylsulfonic acid   (b) dimethylthiol   (c) dimethylsulfoxide   (d) dimethylsulfone

78. Dimethyl sulfide is

 (a) $CH_3-\overset{\displaystyle O}{\underset{\displaystyle O}{\overset{\|}{\underset{\|}{S}}}}-CH_3$ (b) CH_3-S-CH_3 (c) $CH_3-\overset{\displaystyle O}{\overset{\|}{S}}-CH_3$ (d) $CH_3-S-S-CH_3$

79. [structure] does not contain a

 (a) thioether group   (b) sulfide group   (c) sulfhydryl group   (d) disulfide group

80. Arsenic and mercury ions react with

 (a) sulfide bonds   (b) disulfide bonds   (c) sulfone groups   (d) sulfhydryl groups

81. Which reagent will bring about the following conversion?

$$CH_3S-SCH_3 \rightarrow 2\,CH_3SH$$

 (a) $LiAlH_4$ (b) NaOH   (c) H_2SO_4 (d) H_2O_2

82. What is the product of the reaction of CH_3SH and I_2?

 (a) CH_3SCH_3 (b) $CH_3-S-S-CH_3$ (c) $CH_3\overset{\displaystyle O}{\overset{\|}{S}}CH_3$ (d) $CH_3\overset{\displaystyle O}{\underset{\displaystyle O}{\overset{\|}{\underset{\|}{S}}}}CH_3$

FILL IN THE BLANKS

83. Alcohols and phenols both contain the _____ functional group.

84. Cyclohexanol is classified as a _____ alcohol.

85. Write a formula of a secondary alcohol that contains four carbon atoms. _____

86. Give the IUPAC name for each of the following.

 (a) $(CH_3)_2CHCH_2CH_2OH$ _____

 (b) $CH_3\,CH_2CH_2CHOHCHOHCH_3$ _____

 (c) $(CH_3)_2COHCH_2CH_2CHClCH_2CH_3$ _____

 (d) $HO-\langle\bigcirc\rangle-CH(CH_3)_2$ _____

87. Give a structural formula for each of the following.

    (a) ethylene glycol _____

    (b) glycerol _____

    (c) 2-bromo-4-ethyl-2-hexanol _____

    (d) 2-iodo-3-methylphenol _____

88. As the carbon chain of an alcohol increases, its solubility in water _____ .

89. Primary alcohols may be oxidized to _____ .

90. The complete combustion of an alcohol yields _____ and _____ .

91. Alcohols undergo _____ in the presence of sulfuric acid to produce alkenes.

92. Complete the following equations.

    (a) $CH_3CH_2CH_2OH \xrightarrow[180^\circ C]{H_2SO_4}$ _____

    (b) $2 CH_3CH_2CH_2OH \xrightarrow[140^\circ C]{H_2SO_4}$ _____

    (c)  $\bigcirc - CHOHCH_2CH_3 \xrightarrow[180^\circ C]{H_2SO_4}$ _____

    (d)  $\bigcirc - OH + NaOH \rightarrow$ _____

    (e)  $\bigcirc - OH + PBr_3 \rightarrow$ _____

    (f)  $\bigcirc - CH_2OH + SOCl_2 \rightarrow$ _____

93. Alcohol that is rendered unfit for use in beverages is said to be _____ .

94. _____ are compounds in which an oxygen is bonded to two hydrocarbon groups.

95. Ethers have _____ boiling points than alcohols of similar molecular weights.

96. What is the common name for $CH_3OCH(CH_3)_2$? _____

97. Write a structural formula for ethoxypropane. _____

98. A danger in the use of ethers is that they can be oxidized by air to form _____ .

99. The SH group is called the _____ group.

100. The prefix thio in the name of a compound indicates that sulfur has been substituted in place of

     _____ .

101. The sulfur analogs of ethers are called _____ , whereas _____ are the sulfur analogs of alcohols.

102. Thiols are _____ acidic than alcohols.

103. Complete the following equations.

    (a) $CH_3CH_2SH \xrightarrow{[O]}$ _____

    (b) $CH_3S-SCH_3 \xrightarrow{[H]}$ _____

TRUE OR FALSE

104. T F The suffix -ol usually indicates the presence of an –OH group in an organic molecule.

105. T F $(CH_3)_2CHOH$ is a primary alcohol.

106. T F The term tertiary alcohol implies an alcohol that contains three hydroxyl groups.

107. T F The IUPAC name of $(CH_3)_2CHCH_2OH$ is 2-methyl-1-propanol.

108. T F Hydrogen bonding accounts for the water solubility of some alcohols and ethers.

109. T F CH_3CH_2OH is more soluble in water than $CH_3CH_2CH_2CH_2OH$.

110. T F Alcohols are generally less soluble in water and have lower boiling points than ethers of comparable molecular weights.

111. T F Dehydration of an alcohol is accomplished by distilling the alcohol.

112. T F Secondary alcohols may be oxidized to aldehydes and tertiary alcohols are oxidized to ketones.

113. T F Alcohol denaturation is accomplished by heating the alcohol in the presence of an acid catalyst.

114. T F Methyl alcohol is a product of the fermentation process.

115. T F Another name for 1-propanol is rubbing alcohol.

116. T F The term 50% alcohol is the same as 50 proof alcohol.

117. T F ⬡–OH is more acidic than ⬡–OH

118. T F Phenols contain the alkoxy substituent.

119. T F Methoxyethane and 2-propanol are isomers.

120. T F Ethers are generally soluble in water.

121. T F Ethers are generally inert.

122. T F Ethers are soluble in concentrated sulfuric acid.

123. T   F   Ethyl ether is a useful solvent for inorganic compounds.

124. T   F   Three different classes of organic sulfur compounds are the mercaptans, the sulfides, and the thiols.

125. T   F   The thiols are characterized by their obnoxious odors.

CHAPTER 7

Aldehydes and Ketones

MULTIPLE CHOICE

1. The name of the functional group of aldehydes and ketones is

    (a) carbonyl group   (b) carboxyl group   (c) double bond   (d) hydroxyl group

2. A group that both aldehydes and ketones have in common is

    (a) $-\overset{\overset{\textstyle O}{\|}}{C}-H$ (b) $-\overset{\overset{\textstyle O}{\|}}{C}-$ (c) $-OH$ (d) $-O-$

3. Which of the following compounds has a carbonyl group?

    (a) $H-\overset{\overset{\textstyle O}{\|}}{C}-H$ (b) $CH_3-\overset{\overset{\textstyle O}{\|}}{C}-H$ (c) $CH_3-\overset{\overset{\textstyle O}{\|}}{C}-CH_3$ (d) all of these

4. The compound $H_2C{=}O$ is

    (a) methyl alcohol   (b) formaldehyde   (c) acetone   (d) methanone

5. The compound $CH_3CH_2CH_2CH_2\overset{\overset{\textstyle O}{\diagup\!\!\!\!\|}}{\underset{\diagdown H}{C}}$ is

    (a) propanal   (b) cyclopentanal   (c) pentanal   (d) 1-pentanone

6. The compound 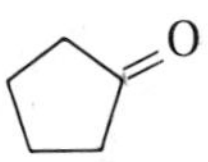 is

    (a) pentanal   (b) cyclopentanal   (c) cyclopentanol   (d) cyclopentanone

7. The formula for 3-methylcyclohexanone is

    (a) 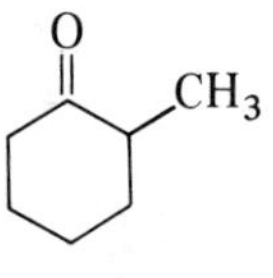   (b) 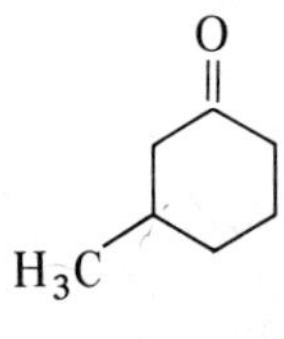   (c) 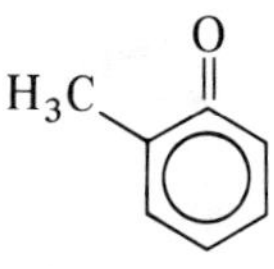   (d) none of these

8. The IUPAC name for $CH_2BrCH_2\overset{\overset{\textstyle O}{\diagup\!\!\!\!\|}}{\underset{\diagdown H}{C}}$ is

152

(a) 1-bromo-3-propanone
(c) 1-bromopropanal

(b) 3-bromopropanal
(d) 3-bromo-1-propanone

9. The IUPAC name for [structure: cyclobutanone ring with O and CH$_3$] is

(a) 2-methylcyclobutanone
(c) 2-methylcyclobutanal

(b) 1-methylcyclobutanal
(d) 1-methylcyclobutanone

10. The name of $CH_3CH_2CH_2\overset{O}{\overset{\|}{C}}CH_3$ is

(a) butyl methyl ketone   (b) 4-pentanone   (c) 2-pentanone   (d) 2-pentyl aldehyde

11. $CH_3-\overset{O}{\overset{\|}{C}}-CH_3$ is not

(a) acetone   (b) propanone   (c) dimethyl aldehyde   (d) dimethyl ketone

12. The compound butanal is also called

(a) butanol   (b) butanone   (c) butyraldehyde   (d) n-butylaldehyde

13. The IUPAC name for $CH_3CH_2CH_2CH(OH)CH(C_2H_5)CHO$ is

(a) 2-ethyl-3-hydroxyhexanal
(c) 4-hydroxy-5-ethylhexanal

(b) 4-hydroxy-3-heptanal
(d) 1-ethyl-2-hydroxypentylaldehyde

14. The common name for $CH_3COCH_2CH_3$ is

(a) diethylketone   (b) ethylethyl ketone   (c) 2-propanone   (d) ethyl methyl ketone

15. The IUPAC name for $(CH_3)_2CHCHClCH_2CHO$ is

(a) 2-chloro-3-methylpentanal
(c) 3-chloro-4-methylpentanal

(b) 2-chloro-3-methylbutanal
(d) 3-chloro-4-methyl-1-pentanone

16. The IUPAC name for [phenyl]$-CH_2\overset{O}{\overset{\|}{C}}CH_2CH_3$ is

(a) phenyl ethyl ketone
(c) benzylpropanone

(b) 4-phenyl-3-butanone
(d) 1-phenyl-2-butanone

17. The structure of 2,5,5-trimethyl-3-hexanone is

(a) $(CH_3)_2CH\overset{O}{\overset{\|}{C}}CH_2CH_2CH(CH_3)_2$
(b) $CH_3(CH_2)_3\overset{O}{\overset{\|}{C}}CH(CH_3)_2$
(c) $(CH_3)_2C\overset{O}{\overset{\|}{C}}CH_2\overset{O}{\overset{\|}{C}}CH_2CH(CH_3)_2$
(d) $(CH_3)_2CH\overset{O}{\overset{\|}{C}}CH_2C(CH_3)_3$

18. Which compound has the lowest boiling point?

(a) $CH_3CH_2CH_2OH$ (b) $CH_3CH_2\overset{O}{\overset{\|}{C}}H$ (c) $CH_3-O-CH_2CH_3$ (d) $CH_3-\overset{O}{\overset{\|}{C}}-CH_3$

19. Which of the following compounds would you expect to have the highest boiling point?

    (a) butanal   (b) diethyl ether   (c) butanone   (d) 1-butanol

20. Formaldehyde would be likly to form hydrogen bonds with

    (a) acetaldehyde   (b) acetone   (c) hexanal   (d) water

21. Dehydrogenation of isopropyl alcohol produces

    (a) an acid   (b) an aldehyde   (c) a ketone   (d) no reaction

22. Which of the following alcohols can be oxidized to 2-methylpropanal?

    (a) $(CH_3)_2CHOH$ (b) $(CH_3)_2CHCH_2OH$ (c) $(CH_3)_2CHCH_2CH_2OH$ (d) $(CH_3)_3COH$

23. The oxidation of 2-methyl-2-butanol with $K_2Cr_2O_7$ and H_2SO_4 would yield

    (a) 2-methyl-2-butanone   (b) 2-methyl-2-butanal
    (c) isopropyl alcohol and ethane   (d) none of these

24. Aldehydes react with alcohols to form

    (a) acids   (b) hemiacetals   (c) hydrates   (d) ketones

25. The general structure of a hemiacetal is

    (a) $R_2\overset{\text{OH}}{\underset{|}{C}}OR'$ (b) $R\overset{\text{OH}}{\underset{|}{C}}HOR'$ (c) $R\overset{\text{OH}}{\underset{|}{C}}HR'$ (d) $R\overset{\text{OR}}{\underset{|}{C}}HOR'$

26. Which of the following is an acetal?

    (a) $(CH_3)_2CHOCH_3$ (b) $CH_3CH(OCH_3)_2$ (c) $CH_3CH(OH)_2$ (d) none of these

27. Which of these substances can undergo an addition reaction with ethyl alcohol?

    (a) 2-propanol   (b) methyl ethyl ether   (c) propanone   (d) 1-propanol

28. What is the product of the following reaction?

$$CH_3CH_2-\overset{\text{OH}}{\underset{\underset{\text{H}}{|}}{\overset{|}{C}}}-OCH_3 + CH_3OH \xrightarrow{H^+}$$

    (a) $CH_3CH_2-\overset{\text{O}}{\overset{||}{C}}-OCH_3$ (b) $CH_3CH_2CH_2-OCH_3$ (c) $CH_3CH_2-\overset{\text{OCH}_3}{\underset{\underset{\text{H}}{|}}{\overset{|}{C}}}-OCH_3$ (d) $CH_3CH_2-\overset{\text{OH}}{\underset{\underset{\text{OCH}_3}{|}}{\overset{|}{C}}}-H$

29. In general, which is the most stable type of compound?

    (a) acetal   (b) hemiacetal   (c) hemiketal   (d) hydrate

30. The aldol condensation produces a compound that has

    (a) two hydroxyl groups   (b) two carbonyl groups
    (c) an aldehyde and a ketone group   (d) an alcohol and a carbonyl group

31. Which compound is most readily oxidized?

(a) acetaldehyde   (b) an ether   (c) an alcohol   (d) a ketone

32. Oxidation of an aldehyde produces

(a) an acid   (b) an ether   (c) an alcohol   (d) a ketone

33. Oxidation of a ketone produces

(a) an acid   (b) an ether   (c) an alcohol   (d) no reaction

34. In a positive Tollens' test

(a) silver atoms are oxidized to silver ions.
(b) the aldehyde is an oxidizing reagent.
(c) a silver mirror is formed.
(d) all of these.

35. Which of the following compounds would give a positive Tollens' or Fehling's test?

(a) CH_3COOH (b) $(CH_3)_3COH$ (c) CH_3CHO (d) CH_3COCH_3

36. Which of the following compounds could be oxidized by Benedict's reagent?

(a) $CH_3\overset{O}{\underset{\|}{C}}CH_2CH_3$ (b) $CH_3CH_2CH_2\overset{O}{\underset{\|}{C}}H$ (c) $CH_3\overset{O}{\underset{\|}{C}}CHOHCH_3$ (d) $CH_3\overset{O}{\underset{\|}{C}}CH_2CH_2OH$

37. In a reaction of substance X with substance Y, substance X gains hydrogen. Which of the following is true?

(a) Substance Y is an oxidizing agent.
(b) Substance X is reduced.
(c) Substance Y is reduced.
(d) Substance X is a reducing agent.

38. A positive Benedict's test is indicated by the formation of

(a) Cu_2O (b) Cu (c) Cu^{2+} (d) a blue solution

39. The oxidizing agent present in Tollens' reagent is

(a) OH^- (b) Cu^+ (c) Cu^{2+} (d) Ag^+

40. Which of the following products is formed when hydrogen reacts with 3-methyl-2-pentanone?

(a) a primary alcohol   (b) a secondary alcohol   (c) a tertiary alcohol   (d) a ketal

FILL IN THE BLANKS

41. The _____ group is the functional group of aldehydes and ketones.

42. It is unnecesary to designate the position of the carbonyl group when naming _____.

43. The general formula for a ketone is _____.

44. Give the IUPAC name for each of the following.

(a) $CH_3CHBrCH_2C$ $\overset{O}{\underset{H}{\diagup\diagdown}}$ _____

(b) $CH_3CH_2\overset{\overset{O}{\|}}{C}CH_2OCH_3$ _____

(c) $CH_3CHOHCH_2CH_2C\overset{O}{\underset{H}{\diagup\diagdown}}$ _____

(d) 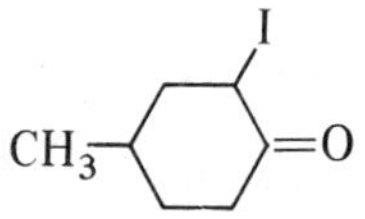 CH_3-⬡$=O$ _____

(e) $(CH_3)_2CHCH_2CH(CH_3)\overset{\overset{O}{\|}}{C}CH_3$ _____

(f) $CH_3CH_2CH_2C(CH_3)BrCH_2C\overset{O}{\underset{H}{\diagup\diagdown}}$ _____

(g) $(CH_3CH_2)_2CHCH(CH_3)CH_2\overset{\overset{O}{\|}}{C}CH_2CH_3$ _____

(h) 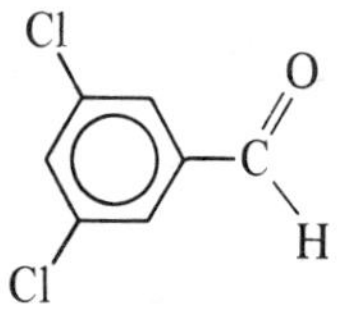 [3,5-dichlorobenzaldehyde structure] _____

45. Give a structural formula for each of the following.

(a) 2-chloro-4-methyl-3-heptanone _____

(b) 6-methoxy-2-octanone _____

(c) 3-fluorocyclopentanone _____

(d) 5-bromohexanal _____

(e) 2-phenyl-5,6-dimethylheptanal _____

(f) 2-ethyl-6-iodobenzaldehyde

46. Ketones may be prepared by the oxidation of _____ alcohols.

47. The addition of one mole of an alcohol to one mole of a ketone yields an unstable product called a
_____.

48. The addition of two moles of an alcohol to one mole of an aldehyde yields a stable product called an _____.

49. Carbonyl compounds that contain _____ hydrogens can undergo the aldol condensation.

50. The enol form of
$$\text{C}_6\text{H}_5-\overset{\overset{\displaystyle O}{\|}}{\text{C}}-\text{CH}(\text{CH}_2\text{CH}_3)_2 \quad \text{is} \ _____.$$

51. Oxidation of an aldehyde yields an _____.

52. The red precipitate formed in the Benedict's test is _____.

53. Aldehydes are reduced to _____ alcohols.

54. _____ is the most important ketone, and _____ is industrially the most important aldehyde.

55. Complete the following equations.

(a) $\text{CH}_3\text{OH} + \text{CH}_3\text{C}\overset{\displaystyle O}{\underset{\displaystyle H}{\diagup\backslash}} \ \overset{H^+}{\rightleftharpoons} \ _____$

(b) $\text{CH}_3\text{CH}_2\text{C}\overset{\displaystyle O}{\underset{\displaystyle H}{\diagup\backslash}} + 2\ \text{CH}_3\text{OH} \ \overset{H^+}{\longrightarrow} \ _____$

(c) $\text{CH}_3\overset{\overset{\displaystyle O}{\|}}{\text{C}}\text{CH}_3 + \text{CH}_3\text{CH}_2\text{OH} \ \overset{H^+}{\rightleftharpoons} \ _____$

(d) $\text{C}_6\text{H}_5-\text{C}\overset{\displaystyle O}{\underset{\displaystyle H}{\diagup\backslash}} + 2\ \text{CH}_3\text{OH} \ \overset{H^+}{\longrightarrow} \ _____$

(e) $\text{C}_6\text{H}_5-\text{C}\overset{\displaystyle O}{\underset{\displaystyle H}{\diagup\backslash}} + \text{HCN} \ \longrightarrow \ _____$

(f) $2\ \text{CH}_3\overset{\overset{\displaystyle O}{\|}}{\text{C}}\text{CH}_3 \ \overset{OH^-}{\longrightarrow} \ _____$

(g) $\text{C}_6\text{H}_5-\text{C}\overset{\displaystyle O}{\underset{\displaystyle H}{\diagup\backslash}} + \text{CH}_3\text{C}\overset{\displaystyle O}{\underset{\displaystyle H}{\diagup\backslash}} \ \overset{OH^-}{\longrightarrow} \ _____$

(h) $\text{C}_6\text{H}_{11}-\text{OH} \ \overset{K_2Cr_2O_7}{\underset{H^+}{\longrightarrow}} \ _____$

(i) $(CH_3)_2CHCH_2C\overset{\displaystyle O}{\underset{\displaystyle H}{\Vert}}$ $\xrightarrow{Ag^+(NH_3)}$ _____

(j) $CH_3\overset{\displaystyle O}{\overset{\Vert}{C}}CH_2CH_2CH_3$ $\xrightarrow[Ni]{H_2}$ _____

TRUE OR FALSE

56. T   F   The common name for CH_3CHO is acetal.

57. T   F   The IUPAC name for formaldehyde is formal.

58. T   F   Aldehydes are considerably more soluble in water than ketones of the same molecular weight.

59. T   F   Aldehydes and ketones have a lower boiling point than alcohols of approximately the same molecular weight.

60. T   F   The higher molecular weight aldehydes and ketones are soluble in water.

61. T   F   Low molecular weight aldehydes are insoluble in organic solvents.

62. T   F   In general, aldehydes and ketones exhibit greater water solubility than alcohols and ethers of similar molecular weights.

63. T   F   Pure solutions of aldehydes undergo hydrogen bonding.

64. T   F   Both aldehydes and ketones can form hydrogen bonds with water.

65. T   F   The carbonyl group is polar.

66. T   F   The hydrogen atoms on carbon atoms alpha to the carbonyl group are basic.

67. T   F   Reaction of $KMnO_4$ with propanal yields propanone.

68. T   F   The oxidation of tertiary alcohols forms ketones.

69. T   F   Propanal and propanone can be distinguished by use of the Tollens' test.

70. T   F   Ketones, like aldehydes, are easily oxidized to carboxylic acids.

71. T   F   Ketones give a positive Tollens' test and a negative Benedict's test.

72. T   F   Fehling's solution will oxidize acetaldehyde but not benzaldehyde.

73. T   F   Butanal is reduced to 1-butanol.

74. T   F   A reagent that is specific for the reduction of carbonyl groups to alcohols is sodium borohydride, $NaBH_4$.

75. T   F   Formaldehyde is a preservative.

76. T   F   Formaldehyde is a gas but is not usually handled in that form.

77. T   F   Some acetone is formed in the human body.

CHAPTER 8

Acids and Esters

MULTIPLE CHOICE

1. The general formula of a carboxylic acid is

   (a) RCOOH   (b) ROCOH   (c) RCOOR   (d) RCOH

2. The IUPAC name for $CH_3CH_2CHBrCH_2COOH$ is

   (a) β-bromovaleric acid   (b) 3-bromopentanoic acid
   (c) α-bromopentanoic acid   (d) 2-bromopentanoic acid

3. The IUPAC name for $(CH_3)_2CHCH_2CH_2COOH$ is

   (a) 2-methylhexanoic acid   (b) 2-methylpentanoic acid
   (c) 4-methylpentanoic acid   (d) 4-methylhexanoic acid

4. The IUPAC name for $CH_3CH_2CH(C_2H_5)COOH$ is

   (a) 3-ethylbutanoic acid   (b) 2-pentanoic acid
   (c) 2-ethylbutanoic acid   (d) ethylpentanoic acid

5. The IUPAC name for 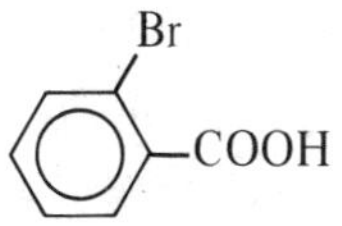 is

   (a) bromobenzoic acid   (b) 2-bromobenzoic acid
   (c) 1-bromobenzoic acid   (d) 5-bromobenzoic acid

6. The IUPAC name for $(CH_3)_2CHCOOH$ is

   (a) isobutyric acid   (c) 2-methylbutanoic acid
   (b) 2-methylpropanoic acid   (d) 1-methylpropanoic acid

7. *m*-Methylbenzoic acid is

8. The IUPAC name for $CH_3(CH_2)_3CH(CH_3)CHOHCOOH$ is

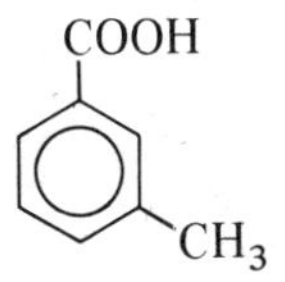

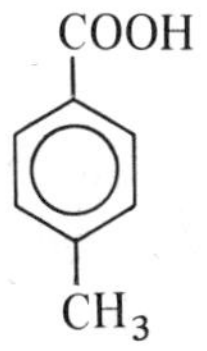

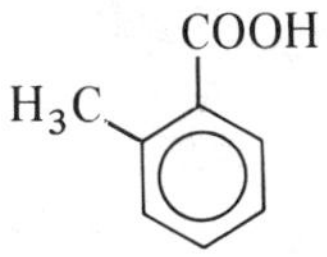

   (a) 2-hydroxy-3-methylheptanoic acid   (b) 5-methyl-6-hydroxyheptanoic acid

159

(c) 2-hydroxy-3-methylpentanoic acid   (d) 2-hydroxyoctanoic acid

9. The name for the dicarboxylic acid $HOOCCH_2CH_2COOH$ is

   (a) oxalic acid   (b) malonic acid   (c) adipic acid   (d) succinic acid

10. A carboxylic acid can be prepared by the oxidation of

    (a) a secondary alcohol   (b) a ketone   (c) a primary alcohol   (d) a tertiary alcohol

11. Which of these compounds could be oxidized by potassium permaganate to form propanoic acid?

    (a) isopropyl alcohol   (b) propanone   (c) methyl ethyl ether   (d) 1-propanol

12. The alcohol used to prepare $(CH_3)_2CHCOOH$ is

    (a) CH_3CH_2OH (b) $CH_3CH_2CH_2OH$ (c) $(CH_3)_2CHCH_2OH$ (d) $(CH_3)_2CHCH_2CH_2OH$

13. Which of the following compounds can be oxidized to $(CH_3)_3CCOOH$?

    (a) $(CH_3)_3COH$ (b) $CH_3\overset{\overset{\displaystyle O}{\|}}{C}C(CH_3)_3$ (c) $CH_3CH_2C(CH_3)_2CH_2OH$ (d) $(CH_3)_3CC\overset{\diagup O}{\underset{\diagdown H}{}}$

14. Which product is formed by the oxidation of propanal?

    (a) acetic acid   (b) formic and acetic acids   (c) propanoic acid   (d) 1-propanol

15. Which compound has the highest boiling point?

    (a) $CH_3CH_2C\overset{\diagup O}{\underset{\diagdown H}{}}$ (b) $CH_3CH_2CH_2OH$ (c) $CH_3\overset{\overset{\displaystyle O}{\|}}{C}CH_3$ (d) $CH_3CH_2C\overset{\diagup O}{\underset{\diagdown OH}{}}$

16. Carboxylic acids have relatively high boiling and melting points because

    (a) they are generally unassociated.   (b) they have a high solubility in water.
    (c) they form dimers.   (d) they have a nonpolar end.

17. Which term correctly describes the following reaction?

    $$CH_3COOH \rightleftharpoons CH_3COO^- + H^+$$

    (a) dehydration   (b) dissociation   (c) hydrolysis   (d) neutralization

18. Carboxylic acids are weak acids because they

    (a) are always found in low concentrations.
    (b) turn litmus red.
    (c) do not completely dissociate in water.
    (d) are found in food.

19. Indicate the order of increasing acidity of the following compounds.

    A. $CH_3(CH_2)_5OH$ B. 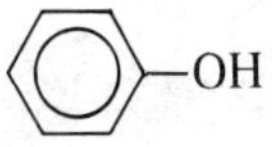 ⬡—OH   C. $CH_3(CH_2)_4COOH$

    (a) A, B, C   (b) C, B, A   (c) B, C, A   (d) C, A, B

20. Which is the strongest acid?

    (a) hydrochloric acid   (b) benzoic acid   (c) carbonic acid   (d) formic acid

21. Which is not a stronger acid than water?

    (a) CH_3CH_2OH (b) 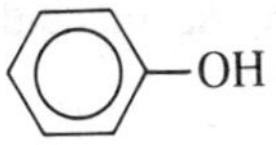 ⬡—OH   (c) CH_3COOH (d) $CH_3(CH_2)_5COOH$

22. Which of the following compounds is the strongest acid?

    (a) CH_3COOH (b) CH_3CCl_2COOH (c) CCl_3COOH (d) $(CH_3)_3CCOOH$

23. Which of the following compounds is the weakest acid?

    (a) CH_2ClCH_2COOH (b) CH_3COOH (c) 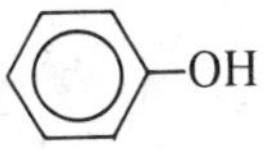 ⬡—OH   (d) 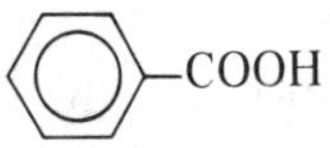 ⬡—COOH

24. Which of the following compounds is the strongest acid?

    (a) $CH_3CHOHCH_3$ (b) $CH_2ClCOOH$ (c) CH_3COOH (d) 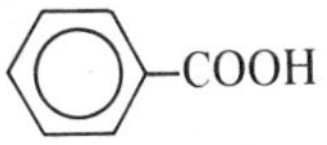 ⬡—COOH

25. Which of the following statements is true of carboxylic acids and their salts in water?

    (a) The acids are generally insoluble, but their salts are soluble.
    (b) The acids are generally soluble, but their salts are insoluble.
    (c) Both the acids and their salts are generally soluble.
    (d) Neither the acids nor their salts are generally soluble.

26. What reagent is needed to complete the following reaction?

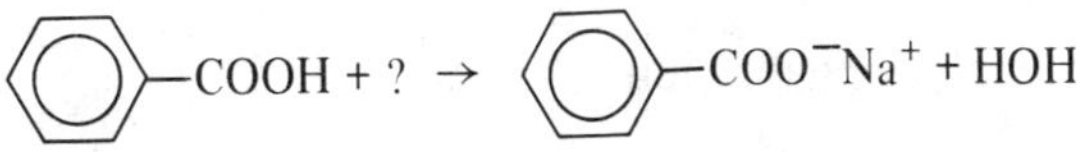

    $$⬡—COOH + ? \rightarrow ⬡—COO^-Na^+ + HOH$$

    (a) Na   (b) Na^+ (c) NaOH   (d) NaO_2

27. Which compound will produce bubbles of carbon dioxide gas when mixed with sodium bicarbonate?

    (a) 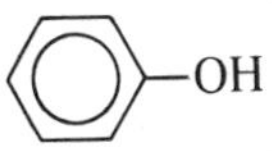 ⬡—OH   (b) 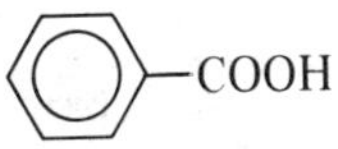 ⬡—COOH   (c) 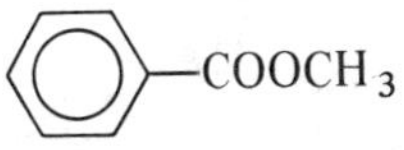 ⬡—COOCH$_3$ (d) CH_3CH_2OH

28. The name of the salt formed when formic acid reacts with sodium hydroxide is

    (a) sodium formate   (b) formic hydroxide   (c) methyl formate   (d) none of these

29. The structure of the acetyl group is

    (a) $CH_3-C\overset{O}{\diagup}$ (b) $CH_3CH_2-C\overset{O}{\diagup}$ (c) $H-C\overset{O}{\diagup}$ (d) $(CH_3)_2CH-C\overset{O}{\diagup}$

30. The acid found in rancid butter is

    (a) formic acid   (b) butyric acid   (c) tartaric acid   (d) oxalic acid

31. The acid found in sour milk is

    (a) benzoic acid   (b) citric acid   (c) lactic acid   (d) salicylic acid

32. Which is the acid found in vinegar?

    (a) formic acid   (b) lactic acid   (c) tartaric acid   (d) acetic acid

33. Glacial acetic acid is

    (a) a frozen solution of acetic acid   (b) a mixture of acetic acid and water
    (c) pure acetic acid   (d) none of these

34. Which compound is oxalic acid?

    (a) ⬡–COOH   (b) HOOC–COOH   (c) ⬡ COOH COOH   (d) HO ⬡ COOH

35. Which compound is salicylic acid?

    (a) ⬡–COOH   (b) HOOC–COOH   (c) ⬡ COOH COOH   (d) HO ⬡ COOH

36. Carboxylic acids react with alcohols to form

    (a) acetals   (b) aldehydes   (c) esters   (d) ketones

37. Acetic acid reacts with ethyl alcohol to form

    (a) butanal   (b) butanone   (c) dimethyl ether   (d) ethyl acetate

38. What are the products of the following reaction?

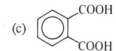

$$R'-C\overset{O}{\underset{OH}{\big|}} + R^{18}OH \underset{}{\overset{H^+}{\rightleftharpoons}} ?$$

    (a) $R'-C\overset{O}{\underset{OR}{\big|}} + H^{18}OH$ (b) $R'-C\overset{O}{\underset{^{18}OR}{\big|}} + HOH$

    (c) $R-C\overset{O}{\underset{OR'}{\big|}} + H^{18}OH$ (d) $R-C\overset{O}{\underset{^{18}OR'}{\big|}} + HOH$

39. Which of the following molecules could be used as one of the reagents necessary to prepare

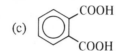

(a) CH_3CH_2OH (b) 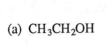 CH_3C with O and OH   (c) benzene ring $-CH_2OH$ (d) benzene ring $-C$ with O and H

40. Esters can be synthesized from carboxylic acids and alcohols in the presence of mineral acids. Which does not improve the yield of ester by this reaction?

   (a) addition of a large excess of alcohol
   (b) addition of a large excess of the carboxylic acid
   (c) addition of a large excess of water
   (d) removal of the water as it is formed

41. Methyl acetate is

   (a) CH_3-C with O and OCH_2CH_3 (b) CH_3-C with O and OCH_3 (c) CH_3CH_2-C with O and OCH_3 (d) $H-C$ with O and OCH_2CH_3

42. The ester of ethanol and benzoic acid is called

   (a) ethyl benzoic acid   (b) ethanol benzoate   (c) ethyl benzoate   (d) benzyl ethanoate

43. The IUPAC name for $CH_3CH_2CH_2C$ with O and OCH_2CH_3 is

   (a) butyl propanoate   (b) butyl ethanoate   (c) ethanol propanoate   (d) ethyl butanoate

44. The common name of 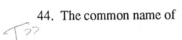 $C-CH_3$ (with O), CH_3CH_2O is

   (a) ethyl acetate   (c) ethyl formate   (b) methyl acetate   (d) ethyl methyl ester

45. Ethyl propanoate is

   (a) $C_2H_5COOCH_3$ (b) $CH_3COOC_2H_5$ (c) CH_3COOCH_3 (d) $C_2H_5COOC_2H_5$

46. The IUPAC name for CH_3CH_2C with O and OCH_3 is

   (a) propyl acetate   (b) methyl acetate   (c) methyl propanoate   (d) methyl ethanoate

47. The ester of ethanol and pentanoic acid is called

   (a) ethyl pentanoate   (b) ethyl pentanoic acid   (c) ethanol pentanoate   (d) pentyl ethanoate

48. The common name of 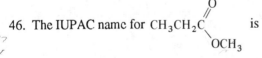 CH_3-C with O and $O-$benzene ring is

   (a) methyl benzoate   (b) phenyl acetate   (c) benzyl acetate   (d) ethyl phenoate

49. The odors associated with many of the members of this class of compounds are extremely fragrant.

    (a) aldehydes   (b) carboxylic acids   (c) esters   (d) ketones

50. Which of the following substances would you expect to be the most soluble in water?

    (a) $CH_3CH_2\overset{\text{O}}{\overset{\|}{C}}CH_3$ (b) CH_3CH_2COOH (c) $CH_3\overset{\text{O}}{\overset{\diagup\!\!\diagdown}{C}}{\diagdown}{OCH_3}$ (d) $CH_3\overset{\text{O}}{\overset{\diagup\!\!\diagdown}{C}}{\diagdown}{OCH_2CH_3}$

51. For which pure compound is hydrogen bonding not possible?

    (a) $CH_3\overset{\text{O}}{\overset{\diagup\!\!\diagdown}{C}}{\diagdown}{OCH_3}$ (b) $CH_3\overset{\text{O}}{\overset{\diagup\!\!\diagdown}{C}}{\diagdown}{OH}$ (c) CH_3CH_2OH (d) none of these

52. Which compound has the highest boiling point?

    (a) $CH_3\overset{\text{O}}{\overset{\diagup\!\!\diagdown}{C}}{\diagdown}{OCH_3}$ (b) $CH_3\overset{\text{O}}{\overset{\|}{C}}CH_2CH_3$ (c) $CH_3CH_2CH_2OH$ (d) $CH_3CH_2\overset{\text{O}}{\overset{\diagup\!\!\diagdown}{C}}{\diagdown}{OH}$

53. The compound with the lowest boiling point is:

    (a) $HOCH_2CH_2CH_2OH$ (b) $CH_3CH_2\overset{\text{O}}{\overset{\diagup\!\!\diagdown}{C}}{\diagdown}{OH}$ (c) $CH_3\overset{\text{O}}{\overset{\diagup\!\!\diagdown}{C}}{\diagdown}{OCH_3}$ (d) they are all the same

54. Four compounds of similar molecular weights belong to the following classes of organic compounds. The compound with the highest melting point is a(n)

    (a) ester   (b) aldehyde   (c) carboxylic acid   (d) ketone

55. The hydrolysis of $CH_3CH_2\overset{\text{O}}{\overset{\diagup\!\!\diagdown}{C}}{\diagdown}{OCH_3}$ in an acid solution yields

    (a) methyl alcohol and acetic acid   (b) propyl alcohol and propionic acid
    (c) methyl alcohol and propionic acid   (d) ethyl alcohol and acetic acid

56. A soap is a salt of a(n)

    (a) ester   (b) aldehyde   (c) carboxylic acid   (d) alcohol

57. Aspirin is

    (a) diacetylphenol   (b) acetylsalicylic acid   (c) acetaminophen   (d) acetylphenol

58. Dacron is a

    (a) polyacid   (b) polyether   (c) polyester   (d) polyalcohol

59. Phosphate esters are prepared from

    (a) phosphoric acid and an alcohol   (b) phosphoric acid and a carboxylic acid
    (c) phosphoric acid and an ester   (d) an ester and a phosphate salt

60. A nerve gas developed for warfare is ✱T A nerve gas is sarin

    (a) ADP   (b) DFP   (c) TCP   (d) malathion

61. Which of the following is pyrophosphoric acid?

    (a) $HO-\overset{\overset{O}{\|}}{\underset{\underset{OH}{|}}{P}}-OH$ (b) $H-\overset{\overset{O}{\|}}{\underset{\underset{H}{|}}{P}}-O-\overset{\overset{O}{\|}}{\underset{\underset{H}{|}}{P}}-H$ (c) $H-\overset{\overset{O}{\|}}{\underset{\underset{OH}{|}}{P}}-O-\overset{\overset{O}{\|}}{\underset{\underset{OH}{|}}{P}}-H$ (d) $HO-\overset{\overset{O}{\|}}{\underset{\underset{OH}{|}}{P}}-O-\overset{\overset{O}{\|}}{\underset{\underset{OH}{|}}{P}}-OH$

62. Trimethyl phosphate is

    (a) $CH_3-\underset{\underset{CH_3}{|}}{P}-CH_3$ (b) $CH_3-\overset{\overset{O}{\|}}{\underset{\underset{CH_3}{|}}{P}}-CH_3$ (c) $CH_3O-\overset{\overset{O}{\|}}{\underset{\underset{OCH_3}{|}}{P}}-OCH_3$ (d) $CH_3-\overset{\overset{O}{\|}}{P}-O-\overset{\overset{O}{\|}}{\underset{\underset{CH_3}{|}}{P}}-O-\overset{\overset{O}{\|}}{P}-CH_3$ (OH, CH₃, OH)

63. Organophosphates are used

    (a) in detergents   (b) as nerve gases   (c) as pesticides   (d) all of these

64. Botulism toxin is a deadly poison because it blocks the release of the neurotransmitter

    (a) acetylcholine   (b) choline   (c) neostigmine   (d) succinylcholine

FILL IN THE BLANKS

65. The group $-C\overset{\overset{O}{\diagup}}{\underset{\underset{OH}{\diagdown}}{}}$ is called the _____ group.

66. Give the IUPAC name for each of the following.

    (a) $CH_3CH_2CH_2CHClCOOH$ _____   (b) $CH_3CH_2COO^-$ Li^+ _____

    (c) [cyclopentane]$-COOH$ _____   (d) $CH_3CH(OCH_3)CH_2COOH$ _____

    (e) $(CH_3)_2CHCH_2C(CH_3)_2CH_2COOH$ _____   (f) $I-$[benzene ring]$-COOH$ _____

67. Give a structural formula for each of the following.

    (a) calcium butyrate _____

    (b) α-bromoacetic acid _____

    (c) cyclobutanecarboxylic acid _____

    (d) 2-hydroxy-5-methylheptanoic acid _____

    (e) succinic acid _____

    (f) m-nitrobenzoic acid _____

68. Carboxylic acids are formed by the _____ of primary alcohols.

69. Carboxylic acids have _____ boiling points than alcohols of similar molecular weight.

70. If the hydrogen ion concentration of a solution is $1.0 \times 10^{-5}\,M$, what is the pH of the solution? _____

71. The stronger the acid, the _____ the K_a and the _____ the pK_a.

72. Phenols are generally _____ acidic than carboxylic acids.

73. The carboxylate ion is stabilized by _____.

74. The hydrolysis of acetic anhydride yields _____.

75. Sodium benzoate is used as a _____.

76. The product of the reaction of a carboxylic acid with an alcohol is called an _____.

77. Give an IUPAC name for each of the following.

    (a) $(CH_3)_2CHCOOCH_2CH_3$ _____

    (b) 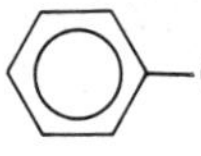 ⬡—$CH_2CH_2COOCH(CH_3)_2$ _____

    (c) $CH_3CH_2CH_2COOCH_2CH_2CH_3$ _____

    (d) CH_3COO—⬠ _____

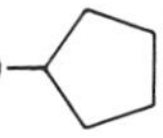

78. Give a structural formula for each of the following.

    (a) ethyl formate _____

    (b) methyl 2-ethylbutanoate _____

    (c) ethyl cyclohexanecarboxylate _____

    (d) isopropyl benzoate _____

79. Most esters are _____ in water.

80. Esters have _____ boiling points than carboxylic acids of the same molecular weight.

81. Ester hydrolysis involves the reaction of an ester with an aqueous solution of an _____.

82. Ester saponification involves the reaction of an ester with a _____.

83. The reaction of an ester with an aqueous base is called _____.

84. The acid-catalyzed hydrolysis of an ester forms an _____ and an _____ as products.

85. Esters undergo _____ to produce two alcohols.

86. Complete the following equations. (Give only the organic product(s).)

(a) $CH_3COOH + NaHCO_3 \rightarrow$ _____

(b) $CH_3COOH + SOCl_2 \rightarrow$ _____

(c) $CH_3CH_2COOH + CH_3OH \underset{\xrightarrow{H^+}}{\rightleftharpoons}$ _____

(d) $(CH_3)_2CHCH_2COOCH_2CH_2CH_3 \xrightarrow[H^+]{HOH}$ _____

(e) $CH_3COOCH_2CH_2CH_3 + KOH \rightarrow$ _____

(f) ⬡—$CHOHCH_2COOCH_2CH_3 \xrightarrow[H^+]{HOH}$ _____

(g) $CH_3OOCCH_2CH_3 + NaOH \rightarrow$ _____

(h) ⬡—$COOCH_3 \xrightarrow[ether]{LiAlH_4} \xrightarrow[H^+]{HOH}$ _____

TRUE OR FALSE

87. T   F   Carboxylic acids are also known as mineral acids.

88. T   F   The carboxyl group is another name for the carbonyl group.

89. T   F   Butyric acid is an IUPAC name.

90. T   F   Oxidation of methanol in the body yields acetic acid.

91. T   F   Lower molecular weight carboxylic acids have fragrant, perfume-like odors.

92. T   F   Carboxylic acids have high boiling points because of hydrogen bond formation between acid molecules.

93. T   F   Carboxylic acids resemble esters of similar molecular weights in their boiling points and water solubilities.

94. T   F   Formic acid is a stronger acid than hydrochloric acid.

95. T   F   The presence of an electron-withdrawing group on the α-carbon atom of a carboxylic acid increases its acid strength.

96. T   F   2-Chlorobutanoic acid and 3-chlorobutanoic acid have acidity constants (pK_a) that are practically identical.

97. T   F   Carboxylic acids may usually be distinguished from phenols by the use of a $NaHCO_3$ solution.

98. T   F   Sodium benzoate is more soluble in water than benzoic acid.

99. T   F   The formula of ammonium acetate is $CH_3COO^- \; NH_4^+$.

100. T   F   The reaction of a carboxylic acid with a base produces an ester.

101. T   F   Carboxylic acids are the only acids that form esters.

102. T   F CH_3COOCH_3 has a higher boiling point than CH_3COOH.

103. T   F CH_3OOCH is methyl formate.

104. T   F   Another name for aspirin is salicylic acid.

105. T   F   Acetylcholinesterase is an enzyme that catalyzes the hydrolysis of acetylcholine to choline and acetic acid.

CHAPTER 9

Amides and Amines

MULTIPLE CHOICE

1. The IUPAC name of $CH_3CH_2CH_2\overset{\displaystyle O}{\overset{\|}{C}}\overset{H}{\underset{\underset{CH_3}{N}}{}}$ is

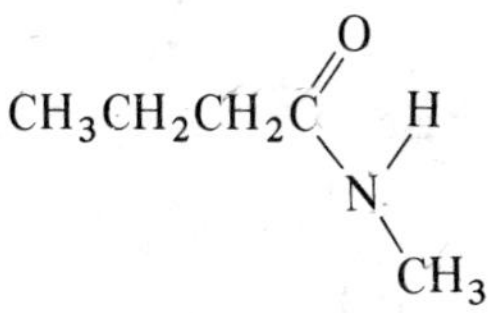

(a) *N*-methylbutanamide   (b) 1-methylbutanamide
(c) *N*-methylpentanamide   (d) 2-pentanamide

2. The IUPAC name of $CH_3CH_2\overset{\displaystyle O}{\overset{\|}{C}}\overset{H}{\underset{\underset{CH_2CH_3}{N}}{}}$ is

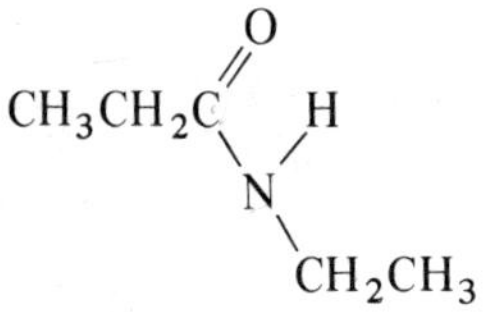

(a) *N*-propylethanamide   (b) *N*-propylacetamide
(c) *N*-ethylpropanamide   (d) 3-ethylpentanamide

3. The IUPAC name of $CH_3CH_2\overset{\displaystyle O}{\overset{\|}{C}}\overset{CH_3}{\underset{\underset{CH_3}{N}}{}}$ is

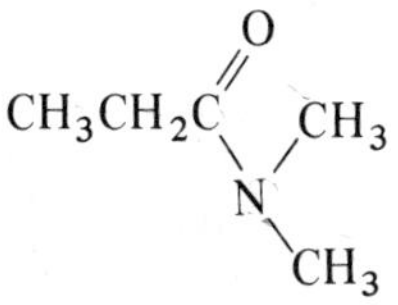

(a) *N*, *N*-dimethylpropanamide   (b) dimethylaminoproprionic acid
(c) *N*-ethylpropanamide   (d) propyldimethylamide

4. The compound $H-\overset{\displaystyle O}{\overset{\|}{C}}\overset{H}{\underset{\underset{CH_3}{N}}{}}$ is called

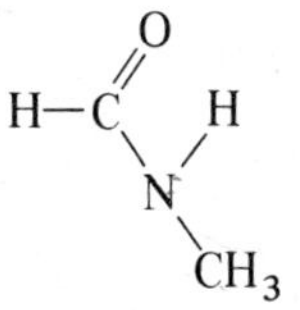

(a) ethanamide   (b) *N*-methylmethanamide
(c) *N*, *N*-dimethylformamide   (d) *N*, *N*-diethylformamide

5. The structure of 2-ethylbenzamide is

(a) 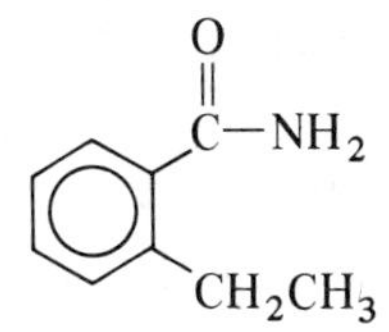   (b) 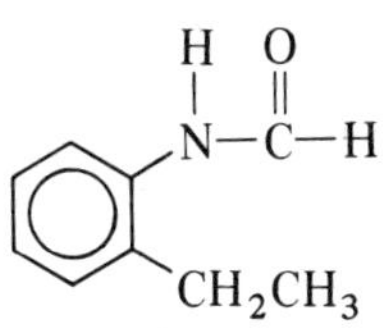   (c) 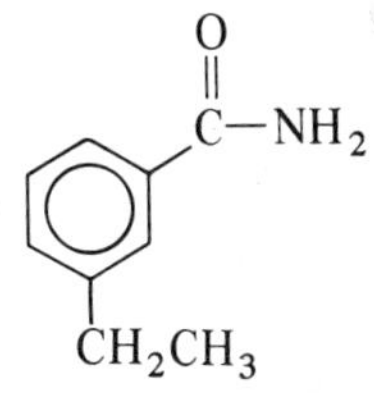   (d)

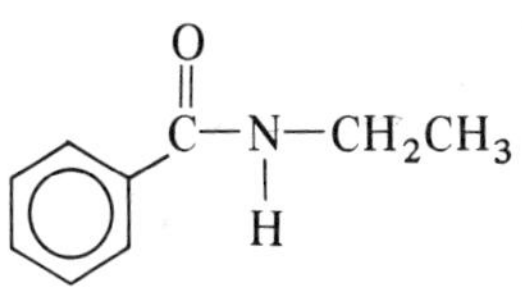

169

6. The structure of *N*-bromoacetamide is

(a) $CH_3-\underset{\underset{Br}{|}}{N}-\overset{\overset{O}{\|}}{C}\diagdown OH$ (b) $CH_3\overset{\overset{O}{\|}}{C}\diagup H$ with $\underset{\underset{Br}{|}}{N}$ (c) $CH_3\overset{\overset{O}{\|}}{C}\diagup Br$ with $\underset{\underset{Br}{|}}{N}$ (d) $\underset{\underset{Br}{|}}{\overset{\overset{H}{|}}{N}}-CH_2\overset{\overset{O}{\|}}{C}\diagdown OH$

7. When benzoic acid is heated in the presence of ammonia, the product is

(a) *p*-aminobenzoic acid   (b) aniline hydrochloride
(c) benzanilide   (d) benzamide

8. Nylon is

(a) a polyester   (b) a polyamide   (c) a natural polymer   (d) a polyamine

9. Which one of the following compounds is the weakest base?

(a) NH_3 (b) $CH_3-\overset{\overset{O}{\|}}{C}\diagdown NH_2$ (c) CH_3NH_2 (d) $(CH_3)_2NH$

10. What are the products of the following reaction?

$$CH_3\overset{\overset{O}{\|}}{C}\diagdown\overset{}{\underset{\underset{CH_3}{|}}{N}}-H \quad + \quad H_2O + HCl \rightarrow$$

(a) $CH_3\overset{\overset{O}{\|}}{C}\diagdown OH \quad + \quad CH_3NH_3{}^+Cl^-$

(b) $CH_3\overset{\overset{O}{\|}}{C}\diagdown NH_2 \quad + \quad CH_3Cl$

(c) $CH_3-\overset{\overset{O}{\|}}{C}\diagdown O^- \quad + \quad CH_3NH_3{}^+$

(d) $CH_3\overset{\overset{O}{\|}}{C}\diagdown OH \quad + \quad CH_3NH_2$

11. Ammonium chloride is a product obtained from the acid HCl hydrolysis of

(a) formamide   (b) acetamide   (c) benzamide   (d) all of these

12. The basic hydrolysis of *N*-ethylpropanamide yields

(a) $CH_3\overset{\overset{O}{\|}}{C}\diagdown OH \quad + \quad CH_3CH_2CH_2NH_2$

(b) $CH_3CH_2\overset{\overset{O}{\|}}{C}\diagdown OH \quad + \quad CH_3CH_2NH_2$

(c) $CH_3\overset{\overset{O}{\|}}{C}\diagdown O^- \quad + \quad CH_3CH_2CH_2NH_2$

(d) $CH_3CH_2\overset{\overset{O}{\|}}{C}\diagdown O^- \quad + \quad CH_3CH_2NH_2$

13. The following reaction

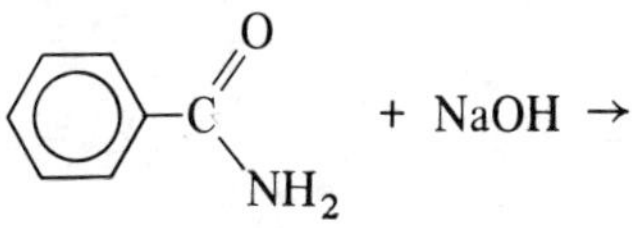

yields

(a) 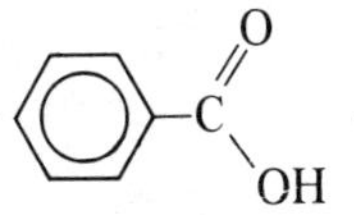 + NH_3

(b) 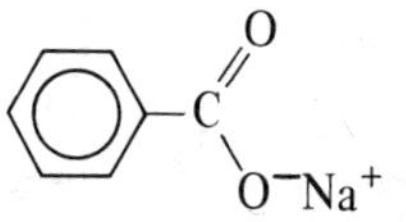 + NH_3

(c) 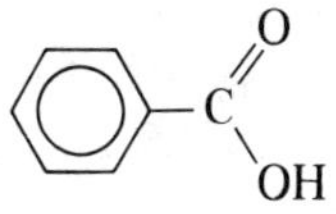 + $NaONH_2$

(d) 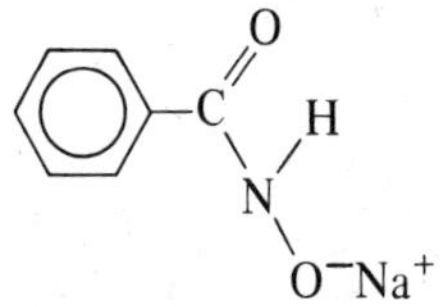 + HOH

14. Which of the following statements is not true of urea?

(a) It is a diamide of carbonic acid.
(b) It is the endproduct of protein metabolism in mammals.
(c) One of the chief uses of urea is as a fertilizer.
(d) It is a pale yellow-colored liquid.

15. An amide that is used as an antipyretic is

(a) aspirin (b) nicotinamide (c) acetaminophen (d) benzamide

16. Which compound is a secondary amine?

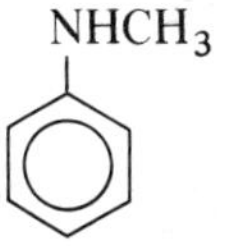

17. Which compound is a secondary amine?

(a) 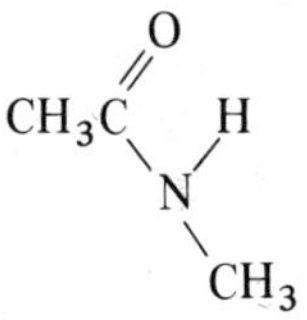 (b) $(CH_3)_4N^+ I^-$ (c) 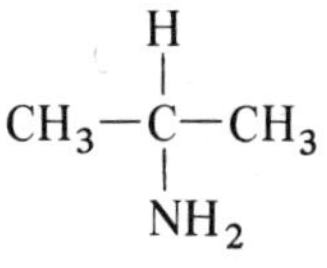 $CH_3-\overset{\overset{H}{|}}{\underset{\underset{NH_2}{|}}{C}}-CH_3$ (d) $(CH_3)_2NH$

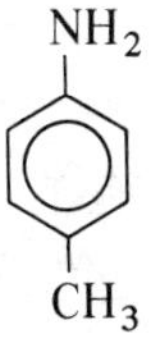

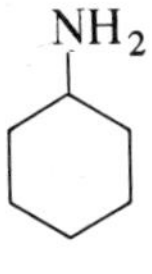

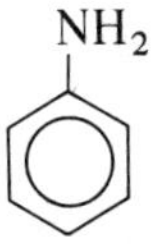

18. Which compound is a tertiary amine?

(a) $(CH_3)_3CNH_2$ (b) 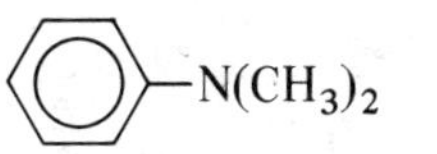 $-N(CH_3)_2$ (c) 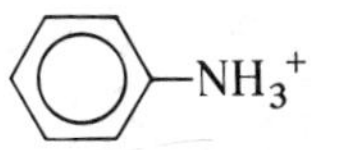 $-NH_3^+$ (d) $(CH_3)_2NH$

19. The compound 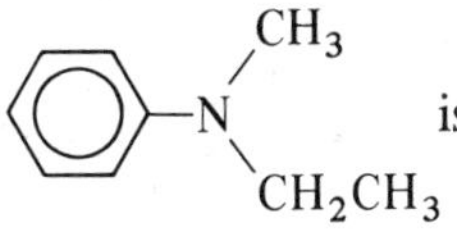 is classified as a _____ amine.

(a) primary (b) secondary (c) tertiary (d) quaternary

20. The compound $CH_3CH(NH_2)CH_2CH_3$ is classified as a _____ amine

    (a) primary   (b) secondary   (c) tertiary   (d) quaternary

21. A correct name for $CH_3-\overset{\displaystyle |}{\underset{\displaystyle CH_3}{N}}-CH_3$ is

    (a) N,N-dimethylamine   (b) trimethylamine
    (c) dimethylmethylamine   (d) tertiary methylamine

22. A correct name for [structure] is

    (a) N-methylaniline   (b) 1-methylaniline
    (c) 2-methylaniline   (d) 1-aminotoluene

23. A correct name for [structure] is

    (a) N-methylaniline   (b) 1-methylaniline
    (c) methylaniline   (d) aminomethylbenzene

24. A correct name for $CH_3CH_2-\overset{\displaystyle |}{\underset{\displaystyle H}{N}}-CH_2CH_3$ is

    (a) diethylamine   (b) diethylaniline
    (c) N,N-diethylmethylamine   (d) diethylaminobutane

25. The structure of aniline hydrochloride is

    (a) [structure] —NHCl   (b) [structure] —NH_3^+ (c) [structure] —$NH_3^+Cl^-$ (d) Cl—[structure]—NH_2

26. Which of the following is the best representation of the hydrogen bonding in methylamine?

    (a) $CH_3-\overset{\displaystyle |}{\underset{\displaystyle H}{N}}-H\cdots CH_3-\overset{\displaystyle |}{\underset{\displaystyle H}{N}}-H$

    (b) $CH_3-\overset{\displaystyle |}{\underset{\displaystyle H}{N}}-H\cdots H-\overset{\displaystyle |}{\underset{\displaystyle N}{N}}-CH_3$

    (c) $CH_3-\overset{\displaystyle \overset{H}{|}}{N}-H\cdots \overset{\displaystyle \overset{H}{|}}{\underset{\displaystyle H}{N}}-CH_3$

    (d) $CH_3-\overset{\displaystyle \overset{H}{|}}{\underset{\displaystyle H}{N}}\cdots \overset{\displaystyle \overset{H}{|}}{\underset{\displaystyle H}{N}}-CH_3$

    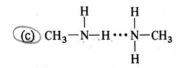

27. Which of the following amines is least soluble in water?

    (a) CH_3NH_2 (b) $CH_3CH_2NH_2$ (c) $CH_3CH_2CH_2NH_2$ (d) $CH_3CH_2CH_2CH_2NH_2$

28. Which of the following amines has the highest boiling point?

    (a) $CH_3CH_2CH_2NH_2$ (b) $CH_3CH_2NHCH_3$ (c) $(CH_3)_3N$ (d) $CH_3CH_2N(CH_3)_2$

29. In general, the boiling points of amines

   (a) are higher for primary amines than for tertiary amines (same number of carbons).
   (b) are higher for tertiary amines than for primary amines (same number of carbons).
   (c) are about the same for all classes of amines.
   (d) decrease as the molecular weight of the amine increases.

30. Amines are

   (a) weak acids   (b) weak bases   (c) heterocyclic compounds   (d) insoluble in water

31. Which compound is the weakest base?

   (a) KOH   (b) NH_3 (c) $(CH_3)_4N^+ I^-$ (d) $CH_3CH_2CH_2NH_2$

32. When methylamine is dissolved in water, one of the species in solution is

   (a) $CH_3CH_2NH_2$ (b) $(CH_3)_2NH$ (c) $CH_3NH_3{}^+OH^-$ (d) CH_3NHOH

33. A product of the following reaction is

$$CH_3CH_2Cl + NH_3 \rightarrow$$

   (a) $CH_3CH_2NH_2$ (b) $(CH_3CH_2)_2NH$ (c) $(CH_3CH_2)_3N$ (d) $CH_3NH_3{}^+ Cl^-$

34. The reaction of ethylamine with hydrochloric acid yields $CH_3CH_2-NH_2 \xrightarrow{HCl}$

   (a) $CH_3CH_2NH_2{}^+OH^-$ (b) $(CH_3CH_2)_2NH$ (c) $CH_3CH_2NHCH_3$ (d) $CH_3CH_2NH_3{}^+ Cl^-$

35. For the generalized reaction between an amine and a carboxylic acid, $RNH_2 + R'COOH$, what is a formula of the product?

   (a) $R'COO^- RNH_3{}^+$ (b) $R'COO^+ RNH_3{}^-$ (c) $R'COOH^- RNH_2{}^+$ (d) $R'COOH^+ RNH_2{}^-$

36. A characteristic of amines is that they react with acids to form

   (a) salts   (b) bases   (c) esters   (d) amino acids

37. Nitrosamines are formed by reaction of secondary amines with

   (a) HCl   (b) HNO_2 (c) NH_3 (d) $SOCl_2$

38. Which of the following is not a heterocyclic compound?

   (a)   (b)   (c)   (d)

39. Which of the following compounds is pyrrole?

   (a)   (b)   (c)   (d)

40. Which compound is a barbiturate?

   (a) halothane   (b) lidocaine   (c) phenobarbital   (d) reserpine

41. Which compound is a hallucinogen?

    (a) mescaline   (b) marijuana   (c) caffeine   (d) cocaine

42. Codeine is all of the following except

    (a) a cough suppressant   (b) a tranquilizer   (c) an analgesic   (d) a narcotic

43. Which is another name for adrenaline?

    (a) amphetamine   (b) catecholamine   (c) epinephrine   (d) norepinephrine

44. Which compound is referred to as a "downer"?

    (a) amytal   (b) cocaine   (c) psilocybin   (d) demerol

45. _____ is a neurotransmitter that is released by certain types of nerve cells.

    (a) fentanyl   (b) phencyclidine   (c) quaalude   (d) norepinephrine

FILL IN THE BLANKS

46. The common name of $CH_3-C\overset{O}{\underset{NH_2}{\big\backslash}}$ is _____.

47. The IUPAC name of $CH_3CH_2C\overset{O}{\underset{\underset{CH_3}{N}}{\big\backslash}}H$ is _____.

48. The name for $\bigcirc-C\overset{O}{\underset{NH_2}{\big\backslash}}$ is _____.

49. Draw a structural formula for *N, N*-dimethyl-*p*-nitrobenzamide. _____

50. Amides are ____less____ basic than amines.

51. Complete the following equations. (Give only the organic product(s).)

    (a) $CH_3C\overset{O}{\underset{OH}{\big\backslash}}$ + NH_3 $\xrightarrow{\text{room temperature}}$ $CH_3 C\overset{//O}{\underset{O^-NH_4^+}{}}$

    (b) $CH_3CH_2C\overset{O}{\underset{OH}{\big\backslash}}$ + NH_3 $\xrightarrow{\Delta}$ $CH_3CH_2C\overset{//O}{\underset{NH_2}{}}$

(c) 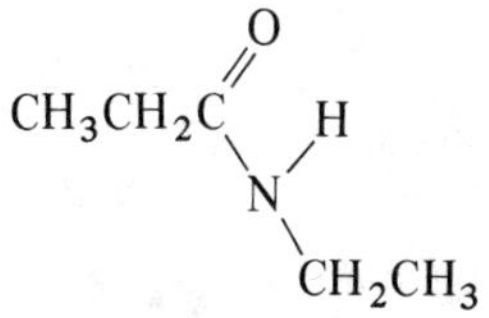 CH_3CH_2C (with $=O$ and H, N, CH_2CH_3) $\xrightarrow[HCl]{HOH}$ $CH_3CH_2C\overset{O}{\underset{OH}{}}$ + $CH_3CH_2NH_3^+Cl^-$

(d) 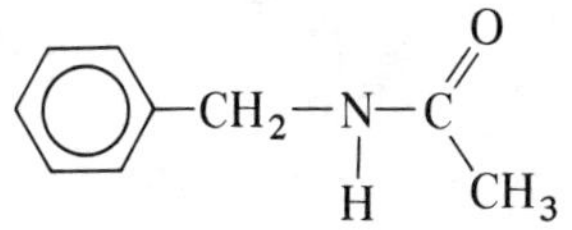 phenyl$-CH_2-N-C(=O)CH_3$ (with H) $\xrightarrow{NaOH}$ phenyl$-CH_2NH_2$ + $CH_3C(=O)O^-Na^+$

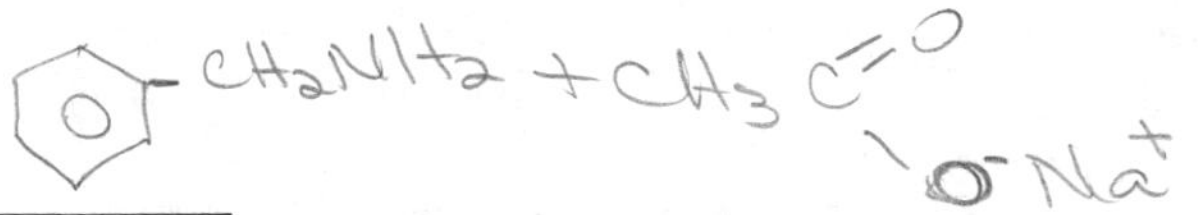

52. Give the IUPAC name for each of the following.

   (a) $(CH_3CH_2)_3N$ _____

   (b) $(CH_3)_2CHCH_2CH_2NH_2$ _____

   (c) $(CH_3)_2NCH_2CH_2CH_3$ _____

   (d) 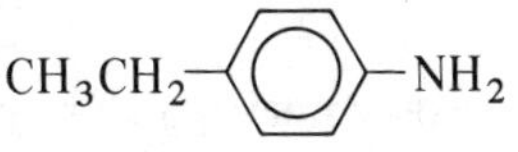 CH_3CH_2-⟨phenyl⟩$-NH_2$ _____

53. Give a structural formula for each of the following.

   (a) 2-amino-2-methylpropane _____

   (b) isopropylamine _____

   (c) diphenylamine _____

   (d) diethylmethylamine _____

54. Amines have a characteristic ___fishy___ odor.

55. In their physical properties, the low molecular weight amines resemble ___ammonia___

56. ___Aliphatic___ amines are more soluble in water than ___aromatic___ amines of comparable molecular weight.

57. Amines have ___higher___ boiling points compared to organic halides of comparable molecular weight.

58. Some drugs that contain amino groups are converted to their salts to increase their ___water solubility___.

59. Complete the following equations. (Give only the organic product(s).)

   (a) $CH_3CH_2CH_2NH_2 + HCl \rightarrow$ ___$CH_3CH_2CH_2NH_3^+Cl^-$___

   (b) $(CH_3)_3N + CH_3I \rightarrow$ ___$(CH_3)_4N^+I^-$___

   (c) 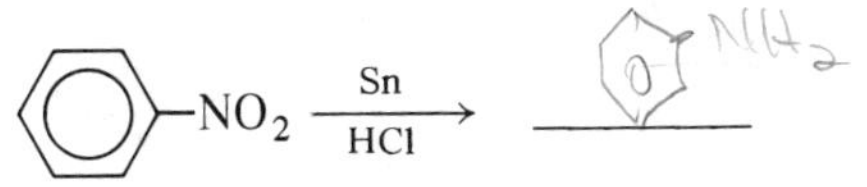 ⟨phenyl⟩$-NO_2 \xrightarrow[HCl]{Sn}$ ___⟨phenyl⟩$-NH_2$___

   (d) $(CH_3CH_2)_2NH \xrightarrow[HCl]{NaNO_2}$ ___$CH_3CH_2-N-N=O$ (with CH_2CH_3)___

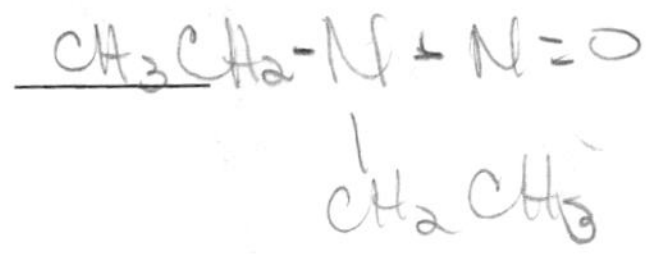

TRUE OR FALSE

60. T F Amines are classified as 1°, 2°, and 3°, in the same manner as the alcohols.

61. T F The N in the name *N*-phenylaniline locates the position of the nitrogen atom.

62. T F *N*, *N*-Dimethylaniline is a tertiary amine.

63. T F Amines react with water to give a slightly acidic solution.

64. T F Both amines and amides can exhibit hydrogen bonding.

65. T F Amines react with acids to form salts.

66. T F The most striking chemical property of amines is their ability to neutralize inorganic bases.

67. T F The alkylation of amines is an example of a substitution reaction.

68. T F Quaternary salts of amines are generally insoluble in water.

69. T F The compound $CH_3NH_2^+Cl^-$ is a quaternary ammonium salt.

70. T F [pyridine ring structure] N is called pyridine.

71. T F All alkaloids are isolated from plants and are nitrogen-containing compounds.

CHAPTER 10

Stereoisomerism

MULTIPLE CHOICE

1. A polarimeter is an instrument used to

   (a) detect structural isomers.
   (b) detect chiral compounds.
   (c) separate enantiomers.
   (d) separate diastereomers.

2. Which of the following compounds contains a chiral carbon atom?

   (a) CH_2ClBr (b) $CH_3CH_2CHICH_2CH_3$ (c) CH_3CH_2CHClF (d) $CH_3CHOHCH_3$

3. A compound is said to be optically active when it

   (a) absorbs light.
   (b) emits light.
   (c) is opaque to light.
   (d) rotates plane-polarized light.

4. The number of chiral carbon atoms in the compound $HOCH_2CHOHCHBrCHOHCH_2COOH$ is

   (a) 2   (b) 3   (c) 4   (d) 5

5. Which of the following compounds does not contain a chiral carbon atom?

   (a) $CH_3CHFCOOH$ (b) CH_3CHIBr
   (c) $CH_3CH_2CH(CH_3)CH_2CH_2CH_3$ (d) $CH_3CH_2OCH_2CH_2CH_3$ ester

6. Which of the following statements about enantiomers is false?

   (a) Most physical properties of enantiomers are different.
   (b) Most chemical properties of enantiomers are identical.
   (c) Enantiomers are mirror images that cannot be superimposed.
   (d) Enantiomers rotate plane-polarized light the same number of degrees but in opposite directions.

7. Which of the following statements about diastereomers is false?

   (a) Diastereomers have different physical properties.
   (b) Diastereomers have different chemical properties.
   (c) Diastereomers are not mirror images.
   (d) Diastereomers are mirror images that cannot be superimposed.

8. The actual number of stereoisomers of the compound $CH_3CHOHCHOHCH_2OH$ is

   (a) 2   (b) 4   (c) 6   (d) 8

177

9. The actual number of stereoisomers of the compound $CH_3CHBrCHBrCH_3$ is

   (a) 1   (b) 2   (c) 3   (d) 4

10. The number of optically active isomers of the compound $CH_3CHICHFCH_3$ is

   (a) 1   (b) 2   (c) 3   (d) 4

11. Which of the following can not exist as a pair of geometric isomers?

   (a) 1-butene   (b) 2-butene   (c) 2-pentene   (d) 3-hexene

12. The correct name for [structure] is

   (a) 3-chloro-2-butene   (b) cis-2-chloro-2-butene
   (c) trans-3-chloro-2-butene   (d) trans-2-chloro-2-butene

13. The geometric isomer of [structure] is

14. Which of the following can exhibit geometric isomerism?

   (a) 2-methyl-1-butene   (b) 2,3-diethyl-2-pentene
   (c) 2-methyl-2-butene   (d) 2,3-dichloro-2-butene

15. Which of the following statements about geometric isomers is false?

   (a) They have the same molecular formula.
   (b) They can exist as cyclic compounds.
   (c) They are alkynes.
   (d) They are alkenes.

16. Which of the following compounds could have geometric isomers?

   (a) $CH_3-C(CH_3)=CH-CH_3$ (b) $CH_3-C(F)=C(Cl)-Cl$
   (c) $CH_3-C(I)=C(CH_3)-CH_3$ (d) $CH_3-C(HO)=C(OH)-CH_3$

17. Which of the following compounds could have geometric isomers?

(c)

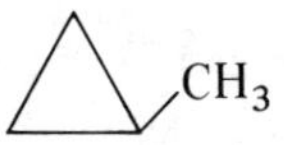

(d)

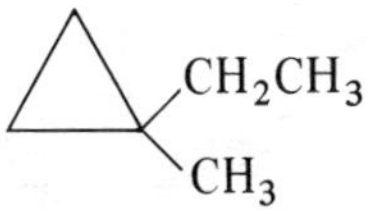

18. Which of the following compounds is *trans*-3-hexene?

(a)

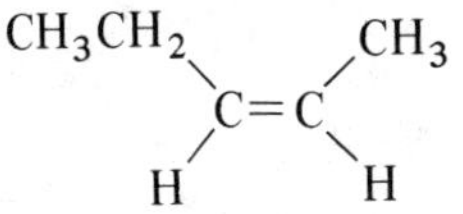

(b)

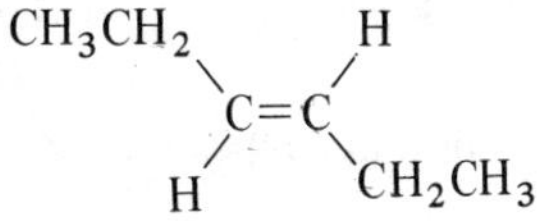

(c)

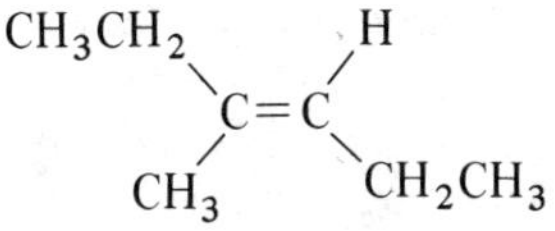

(d)

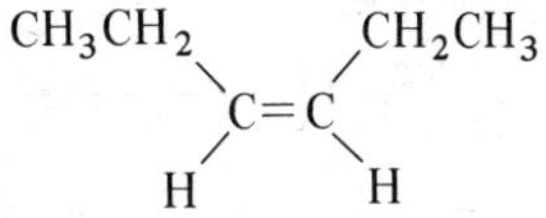

19. The structural formula for *cis*-1,2-dichlorocyclopentane is?

(a)    (b) 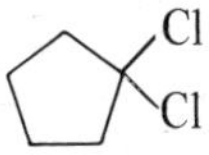   (c) 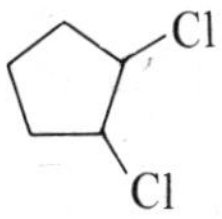   (d)

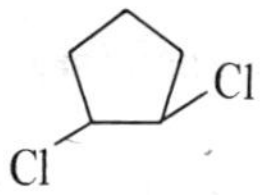

20. The structure of *cis*-3-hexene is

(a)

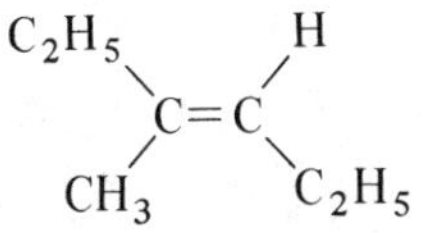

(b)

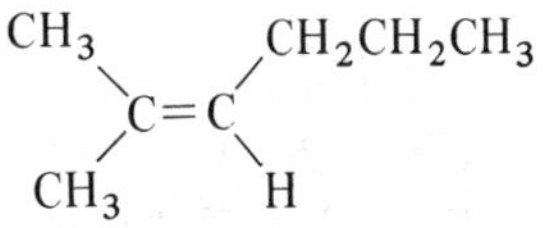

(c)

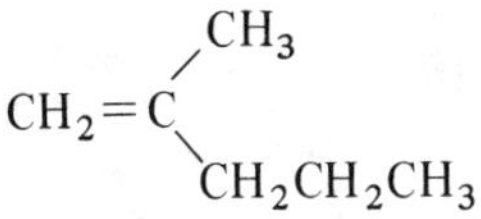

(d)

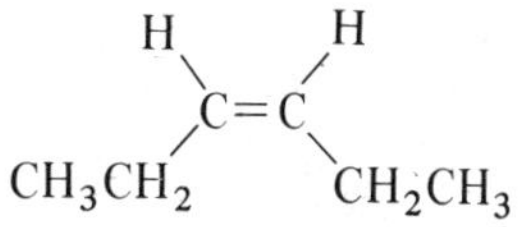

21. The *trans* isomer of 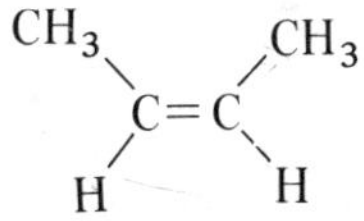 is

(a)

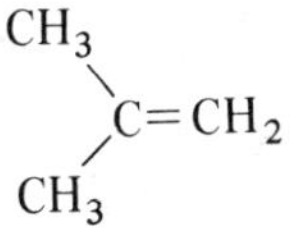

(b) CH_2=$CHCH_2CH_3$

(c) CH_3CH=$CHCH_3$

(d)

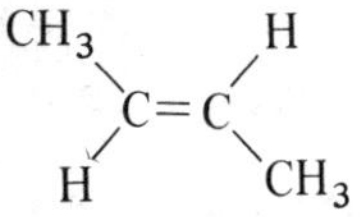

22. The correct name for 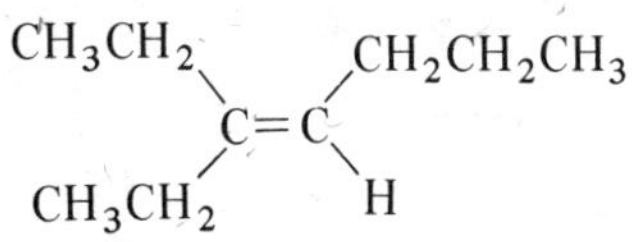 is

(a) 3-ethyl-3-heptene
(c) *trans*-3-ethyl-3-heptene

(b) *cis*-3-ethyl-3-heptene
(d) 2-ethyl-1-propyl-1-butene

FILL IN THE BLANKS

23. _Stereoisomers_ are isomers that differ with respect to the arrangement of their atoms in space.

24. Stereoisomers that are nonsuperimposable mirror images are called _Enantiomers_

25. A person's hands are examples of _Chiral_ objects.

26. A carbon atom that is bonded to four different atoms or groups of atoms is called a _Chiral_ carbon.

27. Any substance that will rotate a beam of plane-polarized light is said to be _Optically_ active.

28. The maximum number of stereoisomers of a compound is given by the formula _2^n_ where n represents the number of _Chiral_ carbons.

29. If a compound contains three chiral carbon atoms, how many stereoisomers are possible. _8_

30. Draw the enantiomers of $CH_3CHBrCH_2OH$. _$CH_3-\overset{H}{\underset{Br}{C}}-CH_2OH$_ and _$CH_3-\overset{Br}{\underset{H}{C}}-CH_2OH$_ .

31. How many chiral carbons are there in 1,2-dichloro-2-methylpropane? _none_

32. How many chiral carbons are there in 1-methoxy-2-phenylethane? _none_

33. Any two stereoisomers that are not mirror images are called _diastereomers_

34. A _meso_ compound contains chiral atoms but it can be superimposed on its mirror image.

35. Draw the meso form of 2,3-dibromobutane. _$CH_3-\overset{H}{\underset{Br}{C}}-\overset{H}{\underset{Br}{C}}-CH_3$_

36. A sample that contains equal amounts of dextrorotatory and levorotatory isomers is a _racemic_ mixture.

37. Give a name for $\overset{CH_3}{\underset{Br}{}}C=C\overset{Br}{\underset{CH_2CH_3}{}}$ _trans-2,3-dibromo-2 pentene_

38. Draw the structure of *trans*-1,3-dichlorocyclopentane. _[structure drawing]_

39. Draw the structure of *cis*-3-heptene. _[structure drawing]_

40. Draw geometric isomers for $C_2H_2I_2$. _[structure drawing]_ and _[structure drawing]_

41. Draw geometric isomers for 3-methyl-2-pentene. _____ and _____

42. Draw the structure of *trans*-2,3-dichloro-2-pentene. _____

43. Draw the meso form of 3,4-difluorohexane. _____

44. Draw geometric isomers for 1-bromo-2-phenylethene. _____ and _____

TRUE OR FALSE

45. T F   The (+)-form of any optically active substance will rotate plane-polarized light to the right (clockwise).

46. T F   The optical rotation of a compound depends upon the number of optically active molecules that are in the path of the plane-polarized light.

47. (T) F   Enantiomers differ with respect to their interaction with plane-polarized light.

48. (T) F   Enantiomers have identical melting points and boiling points.

49. (T) F   All molecules that contain only one chiral carbon atom are chiral.

50. T (F)   A molecule that contains at least two chiral carbon atoms must always be chiral.

51. (T) F   Meso forms are optically inactive because any effect that one-half of the molecule has on plane-polarized light is exactly compensated for by the effect of the other half of the molecule.

52. T (F)   The double bond between carbon atoms in a molecule permits free rotation.

53. T (F)   2-Butyne can exist in cis and trans isomeric forms.

54. (T) F   2,3-Dichloro-2-butene can exist in cis and trans isomeric forms.

55. T (F)   cis-1,4-Dimethylcyclohexene is a chiral molecule.

56. (T) F   Pheromones are chemicals that are used for communication between members of the same species of insects.

MATCHING

A. stereoisomers   E. polarimeter   I. meso compound
B. geometric isomers   F. enantiomers   J. chiral carbon
C. monochromatic   G. racemic mixture
D. plane-polarized   H. diastereomers

57. ___C___ Light of only one wavelength.

58. ___F___ Stereoisomers that are nonsuperimposable mirror images.

59. ___B___ Stereoisomers made possible by the restricted rotation about a double bond or a cyclic system.

60. ___G___ Optically inactive mixture of equal amounts of enantiomers.

61. ___E___ Instrument used to measure the optical activity of a compound.

62. ___I___ Compound with two similar chiral carbon atoms that is optically inactive because of internal compensation.

63. ___A___ Isomers that differ in the manner in which their atoms are arranged in space.

64. ___D___ Light vibrating in a single plane.

65. ___H___ Optically active stereoisomers that are not enantiomers.

66. ___J___ Carbon atom with four unlike groups attached.

CHAPTER 11

Carbohydrates

MULTIPLE CHOICE

1. The number of carbon atoms in a ketotetrose is

    (a) 3   (b) 4   (c) 5   (d) 6

2. The number of chiral carbons in an aldopentose is

    (a) 2   (b) 3   (c) 4   (d) 5

3. The chiral carbon that is farthest from the carbonyl group (in the open-chain form of a sugar) is called the _____ carbon.

    (a) anomeric   (b) epimeric   (c) optical   (d) penultimate

4. If a sugar rotates a beam of plane-polarized light to the left, we indicate this by the symbol

    (a) D   (b) L   (c) (+)   (d) (−)

5. A D-sugar is one that has the same configuration about the penultimate carbon as

    (a) dihydroxyacetone   (b) D-fructose   (c) D-glucose   (d) D-glyceraldehyde

6. The most abundant of the aldopentoses is

    (a) arabinose   (b) fructose   (c) glucose   (d) ribose

7. The one thing that glucose, mannose, galactose, and fructose have in common is that they are all

    (a) anomers   (b) epimers   (c) aldoses   (d) hexoses

8. Which of these statements is false about an aldopentose?

    (a) It has four chiral carbons.
    (b) It has four hydroxyl groups.
    (c) It has a carbonyl group.
    (d) It is a reducing sugar.

9. How many stereoisomers are possible for an aldopentose?

    (a) 6   (b) 8   (c) 12   (d) 16

10. The pentose L-xylose

    (a) rotates a beam of plane-polarized light to the left.
    (b) rotates a beam of plane-polarized light to the right.
    (c) is a mirror image of D-xylose.
    (d) is an isomer of erythrose.

11. The epimer of mannose is

    (a) fructose   (b) galactose   (c) glucose   (d) none of these

12. Ribose is a(n)

    (a) aldohexose   (b) aldopentose   (c) aldotetrose   (d) ketopentose

13. Which formula is correct for L-glucose?

    (a)   (b)   (c)   (d) none of these

14. The cyclic form of glucose results from the formation of a(n)

    (a) acetal   (b) hemiacetal   (c) hemiketal   (d) ketal

15. The anomer of α-galactose is

    (a) β-galactose   (b) α-glucose   (c) β-glucose   (d) α-mannose

16. Which of the following is a pyranose sugar?

    (a)   (b)   (c)   (d)

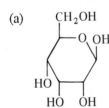 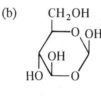 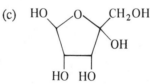 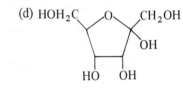

17. β-D-Fructofuranose

    (a) has the OH group on carbon-1 above the plane of the ring.
    (b) has a six-membered ring.
    (c) is an anomer of β-L-fructofuranose.
    (d) undergoes mutarotation.

18. Which of the following pairs of isomers are not epimers?

    (a) α-D-glucose and β-D-glucose   (c) D-glucose and D-mannose
    (b) D-glucose and D-galactose   (d) D-mannose and D-galactose

19. The structure of α-D-galactopyranose is

    (a)   (b)   (c)   (d)

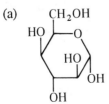 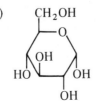 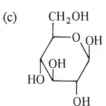 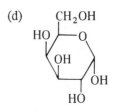

20. When pure β-D-fructofuranose is dissolved in an aqueous solution,

   (a) some of it is changed into β-L-fructofuranose.
   (b) all of it changes to the open-chain form.
   (c) some α-D-fructofuranose is formed.
   (d) it is converted to β-D-glucopyranose.

Refer to the following structures in answering questions 21–26.

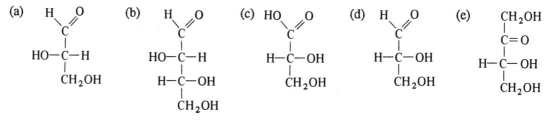

21. Which compound is the enantiomer of (a)?

22. Which compound is not a D-sugar?

23. Which compound is L-glyceraldehyde?

24. Which compound is a ketose?

25. Which compound is an aldotetrose?

26. Which compound is the product of the oxidation of D-glyceraldehyde?

27. Ribose and deoxyribose are

   (a) anomers   (b) epimers   (c) isomers   (d) none of these

28. Dextrose is

   (a) an aldose   (b) a hexose   (c) a reducing sugar   (d) all of these

29. Levulose is

   (a) an aldose   (b) a pentose   (c) a reducing sugar   (d) none of these

30. The sweetest sugar is

   (a) fructose   (b) glucose   (c) lactose   (d) sucrose

31. The majority of carbohydrates that we ingest are converted to

   (a) glucose   (b) ribose   (c) sucrose   (d) none of these

32. Lactose is also known as _____ sugar.

   (a) beet   (b) malt   (c) milk   (d) table

33. Which of the following is *not* another name for glucose?

   (a) blood sugar   (b) cane sugar   (c) dextrose   (d) grape sugar

34. A(n) _____ linkage joins the two monosaccharide units of a disaccharide.

   (a) ester   (b) ether   (c) hemiacetal   (d) glycosidic

35. Which of the following yields invert sugar upon hydrolysis?

    (a) fructose   (b) maltose   (c) lactose   (d) sucrose

36. Invert sugar is an equimolar mixture of glucose and

    (a) cellobiose   (b) fructose   (c) galactose   (d) sucrose

37. A disaccharide that contains a β-1,4-glucosidic linkage is

    (a) cellobiose   (b) lactose   (c) maltose   (d) sucrose

38. Maltose and cellobiose are

    (a) anomers   (b) enantiomers   (c) epimers   (d) isomers

39. Hydrolysis of the glycosidic bond in lactose yields D-glucose and

    (a) D-fructose   (b) D-galactose   (c) L-glucose   (d) L-galactose

40. Sucrose is different from the other common disaccharides in that it

    (a) is a nonreducing sugar.   (b) does not undergo mutarotation.
    (c) has a 1,2-glycosidic linkage.   (d) all of these

    Refer to the following structure in answering questions 41–47.

41. The monosaccharide on the left is

    (a) fructose   (b) galactose   (c) glucose   (d) mannose

42. The monosaccharide on the right is

    (a) fructose   (b) galactose   (c) glucose   (d) mannose

43. The glycosidic linkage is

    (a) alpha   (b) beta   (c) dextrorotatory   (d) levorotatory

44. The monosaccharide units are joined through a _____ bond.

    (a) 1,1   (b) 1,2   (c) 1,4   (d) 1,6

45. The disaccharide is

    (a) lactose   (b) maltose   (c) sucrose   (d) none of these

46. Upon hydrolysis, the disaccharide yields

    (a) only aldohexoses   (b) aldopentoses and aldohexoses
    (c) aldohexoses and ketohexoses   (d) aldopentoses and ketohexoses

47. The disaccharide is _____ sugar.

    (a) invert   (b) milk   (c) a nonreducing   (d) a reducing

48. Polysaccharides are characteristically

    (a) crystalline, white solids   (b) hexagonal-shaped compounds
    (c) insoluble in water   (d) sweet tasting

49. Amylose is

    (a) a polymer of mannose units   (b) highly branched
    (c) soluble in hot water   (d) all of these

50. The complete hydrolysis of starch yields

    (a) galactose   (b) glucose   (c) fructose   (d) maltose

51. Which of the following is the reserve carbohydrate in plants?

    (a) cellulose   (b) glycogen   (c) nyaluronic acid   (d) starch

52. The human body stores carbohydrates as

    (a) cellulose   (b) glycogen   (c) inulin   (d) starch

53. Which of the following is not digested by the body?

    (a) cellulose   (b) glycogen   (c) amylopectin   (d) amylose

54. Which of the following compounds contains the β-1,4-linkage?

    (a) cellulose   (b) glycogen   (c) amylopectin   (d) amylose

55. Which of the following contains an α-1,6-linkage?

    (a) glycogen   (b) cellobiose   (c) cellulose   (d) amylose

56. If cellulose contained only α-1,4-linkages, the polysaccharide would be identical to

    (a) amylose   (b) glycogen   (c) amylopectin   (d) pectin

57. Partial hydrolysis of starch gives products of intermediate size called

    (a) gums   (b) dextrose   (c) dextrins   (d) inulins

58. Upon complete hydrolysis, cellulose yields

    (a) galactose   (b) glucose   (c) maltose   (d) sucrose

59. The most branched polysaccharide is

    (a) amylose   (b) amylopectin   (c) cellulose   (d) glycogen

60. Which of the following would not yield maltose upon partial hydrolysis?

    (a) amylose   (b) amylopectin   (c) cellulose   (d) glycogen

61. Starch is a mixture of

    (a) glucose and maltose   (b) amylose and glycogen
    (c) amylopectin and amylose   (d) amylopectin and glycogen

62. A polysaccharide consisting entirely of D-glucose units joined by α-1,4 linkages is

    (a) amylose   (b) amylopectin   (c) cellulose   (d) glycogen

63. The glycosidic bonds in glycogen are

    (a) α-1,4 and α-1,6   (b) α-1,4 and β-1,4   (c) α-1,4 and β-1,6   (d) α-1,6 and β-1,4

64. Which of the following does not yield D-glucose upon complete hydrolysis?

    (a) amylopectin   (b) cellulose   (c) glycogen   (d) inulin

65. The chief energy-storage polysaccharides are

    (a) cellulose and starch   (b) chitin and starch
    (c) glycogen and cellulose   (d) starch and glycogen

66. Which of the following will not give a positive Molisch test?

    (a) cellulose   (b) fructose   (c) saccharin   (d) sucrose

67. Which of the following is not a reducing sugar?

    (a) fructose   (b) lactose   (c) maltose   (d) sucrose

68. Oxidation of a monosaccharide yields a(n)

    (a) acid   (b) alcohol   (c) disaccharide   (d) polysaccharide

69. Which of the following will give a positive Benedict's test?

    (a) amylose   (b) cellulose   (c) fructose   (d) sucrose

70. Which of the following compounds will not be oxidized by Benedict's solution?

71. If a carbohydrate gives a positive Benedict's test,

    (a) it is a reducing sugar.
    (b) it is reduced.
    (c) it will give a negative Fehling's test.
    (d) all of these

72. If a sugar undergoes _____, it is also likely to be a reducing sugar.

    (a) hydrolysis   (b) mutarotation   (c) optical rotation   (d) reduction

FILL IN THE BLANKS

73. Many carbohydrates have the general formula _____.

74. _____ is a sugar substitute that is an animal carcinogen.

75. _____ is the sweetening agent in Nutrasweet.

76. A _____ cannot be hydrolyzed into simpler carbohydrate molecules.

77. Carbohydrates containing four carbons are called _____, and those containing seven carbons are called _____.

78. The number of carbon atoms in an aldopentose is _____.

79. How many chiral carbons are there in a ketohexose?_____

80. The name of the naturally occurring ketotriose is _____.

81. What is the name of the compound that serves as the standard for the assignment of absolute configuration of sugars?_____

82. D-Glucose is the glucose enantiomer that rotates a beam of plane-polarized light in a _____ direction.

83. A pair of stereoisomers that differ only in the configuration about a single carbon atom are called _____.

84. The cyclic form of a sugar occurs as a result of an intramolecular reaction between a _____ group and a _____ group.

85. The anomer of α-glucose is _____.

86. The five-membered ring form of fructose is referred to as the _____ form.

87. The six-membered ring form of glucose is referred to as the _____ form.

88. The hydroxyl groups that are positioned to the right of the open chain form of a sugar are _____ the plane of the ring in the cyclic form.

89. The gradual change in optical rotation of a freshly prepared solution of β-D-glucose is called _____ .

90. In the furanose form of fructose, the α- and β-anomers are determined by the position of the hydroxyl group on carbon number _____ .

91. Draw the structure of α-D-glucose. _____

92. Blood sugar is another name for _____, and another name for fructose is _____.

93. How many chiral carbons are there in D-ribose? _____

94. Which of the common hexoses is not an aldose? _____

95. _____ and _____ are two pentoses that occur in nucleic acids.

96. Carbohydrates that can be hydrolyzed to yield two monosaccharide units are called _____.

97. Disaccharides generally have a _____ taste and are _____ in water.

98. The bond that joins two monosaccharides is called a _____ linkage.

99. The acetal derived from glucose is specifically called a _____ .

100. The product of the hydrolysis of sucrose is called _____ sugar.

101. The carbohydrate found in milk is _____.

102. The disaccharide obtained from the partial hydrolysis of cellulose is _____.

103. Cane sugar is another name for _____, and another name for maltose is _____ sugar.

104. Disaccharides can be hydrolyzed by _____ or _____ to their component monosaccharide units.

105. The enzyme that hydrolyzes maltose is called _____, and the enzyme that hydrolyzes lactose is called _____.

106. _____ is a disaccharide containing glucose and galactose joined by a _____ bond.

107. The hydrolysis of sucrose yields _____ and _____.

108. The storage form of carbohydrates in most plants is _____, and the reserve carbohydrate of animals is _____.

109. Complete hydrolysis of glycogen yields _____.

110. _____ and _____ are the two polysaccharides found in starch.

111. _____ comprises the skeletal material of plant cells.

112. _____ linkages join the glucose units in cellulose.

113. The plant polysaccharide that contains both α-1,4- and α-1,6- glycosidic bonds is _____.

114. Give the name of the test that can distinguish between the following pairs of sugars.

    (a) glucose and fructose _____   (b) glucose and ribose _____
    (c) galactose and lactose _____   (d) maltose and sucrose _____

TRUE OF FALSE

115. T F   Some animals can synthesize carbohydrates from carbon dioxide and water.

116. T F   In a ketotetrose, there are three pair of enantiomers.

117. T F   Most of the common monosaccharides contain three or four carbon atoms.

118. T F   D- and L-glyceraldehyde are examples of polyhydroxyketones.

119. T F   Some synthetic compounds are sweeter than sugar.

120. T F   Polyhydroxy aldehydes or ketones are called artificial sweeteners.

121. T F   Glyceraldehyde is optically active.

122. T F   Dihydroxyacetone is an isomer of glyceraldehyde.

123. T F   Dihydroxyacetone is a triose that has no chiral carbons.

124. T F   The simplest aldose is erythrose, and the simplest ketose is threose.

125. T F   The D-series of sugars is related configurationally to D-glyceraldehyde.

126. T F   A (+)-sugar is one that rotates a beam of plane-polarized light to the right.

127. T F   D-Glucose and D-mannose differ in configuration at carbon number 3.

128. T F   L-Arabinose and D-arabinose are epimers.

129. T F   D-Mannose and D-fructose are epimers.

130. T F   Ribose is an aldopentose.

131. T F   Deoxyribose is an isomer of ribose.

132. T F   Ribose and deoxyribose differ in structure only at carbon number 2.

133. T F α-D-Galactopyranose and β-D-galactopyranose are enantiomers.

134. T F   Aldohexoses can exist in both the furanose and pyranose ring forms.

135. T F   Beet sugar is another name for fructose.

136. T F   Dextrose can be administered intravenously.

137. T   F   Glucose is table sugar.

138. T   F   Glucose is sweeter than fructose.

139. T   F   The bond that joins two monosaccharides is an ester linkage.

140. T   F   The term glycoside refers to a cyclic acetal derived specifically from glucose.

141. T   F   The enzyme *zymase* catalyzes the hydrolysis of sucrose.

142. T   F   The glucose units in malt sugar are joined by a β-1,4-glucosidic bond.

143. T   F   Sucrose is a reducing disaccharide and will undergo mutarotation.

144. T   F   Sucrose can be hydrolyzed in an aqueous solution of HCl.

145. T   F   Invert sugar is another name for sucrose.

146. T   F   The hydrolysis of sucrose yields a product called "refined" sugar.

147. T   F   In sucrose, the monosaccharides are joined by an α-1,2-glucosidic bond.

148. T   F   Both starch and glycogen yield only D-glucose upon complete hydrolysis.

149. T   F   Humans can utilize cellulose for food.

150. T   F   Glycogen is stored in the liver and muscles of animals.

151. T   F   Cellulose is used to make wool and silk.

152. T   F   Amylose, cellulose, and glycogen are all straight-chain polysaccharides.

153. T   F   A polysaccharide used for the storage of energy in plants is cellulose.

154. T   F   Mannose will not give a positive Benedict's test.

155. T   F   The active ingredient in Benedict's solution is the cuprous ion.

CHAPTER 12

Lipids

MULTIPLE CHOICE

1. Lipids are classed together on the basis of their

   (a) chemical properties   (b) functional groups
   (c) solubility   (d) structure

2. Lipids are not soluble in

   (a) benzene   (b) chloroform   (c) ether   (d) water

3. The body's chief energy reserves are

   (a) alcohol   (b) carbohydrates   (c) fats   (d) proteins

4. Fats and oils are

   (a) carboxylic acids   (b) esters   (c) ethers   (d) polymers

5. Which of the following is a lipid?

   (a) dextrin   (b) glycerol   (c) stearic acid   (d) triolein

6. Oils are

   (a) hydrocarbons   (b) liquid fats   (c) solid fats   (d) steroids

7. Fats

   (a) are liquids at room temperature.
   (b) are found mainly in plants.
   (c) contain fewer than ten carbons.
   (d) are more saturated than oils.

8. Which of the following is false about fatty acids?

   (a) They are obtained from fats and oils.
   (b) They may contain one or more double bonds.
   (c) They invariably are branched chains.
   (d) They contain an even number of carbons.

9. Polyunsaturated fatty acids

   (a) are found in oils.
   (b) have low iodine numbers.
   (c) have *trans* double bonds.
   (d) none of these.

10. Which of the following is a naturally occurring fatty acid?

    (a) $CH_3(CH=CH)_3CH_2COOH$ (b) $CH_3(CH_2)_5C(CH_3)_2CH_2(CH_2)_7COOH$
    (c) $CH_3(CH_2)_{14}COOH$ (d) $CH_3(CH_2)_{17}COOH$

11. Fatty acids are joined to glycerol by _____ bonds.

    (a) acetal   (b) ester   (c) ether   (d) hemiacetal

12. Mixed triacylglycerols contain

    (a) a fat and an oil   (b) a fat and glycerol
    (c) an oil and glycerol   (d) at least two different fatty acids

13. Which is the most highly unsaturated oil?

    (a) coconut oil   (b) linseed oil   (c) olive oil   (d) peanut oil

14. Which simple triacylglycerol has the highest iodine number?

    (a) trimyristin   (b) trilinolein   (c) triolein   (d) tristearin

15. The alkaline hydrolysis of a triacylglycerol is called

    (a) esterification   (b) hydrogenation
    (c) neutralization   (d) saponification

16. When a fat is boiled with sodium hydroxide, one of the products is always

    (a) a fatty acid   (b) glycerol   (c) a glycol   (d) an oil

17. Complete saponification of an oil yields _____ and glycerol

    (a) esters of fatty acids   (b) salts of fatty acids
    (c) saturated fatty acids   (d) unsaturated fatty acids

18. Acid hydrolysis of a mixed triacylglycerol may yield

    (a) three different alcohols   (b) three different fatty acids
    (c) three different salts   (d) all of these

19. Saponification of triolein yields 3 moles of _____ and glycerol.

    (a) oleic acid   (b) sodium oleate   (c) sodium stearate   (d) stearic acid

20. If 1000 grams of tallow absorbs 400 grams of iodine, the iodine number of the tallow sample is

    (a) 4   (b) 40   (c) 400   (d) none of these

21. Liquid oils are changed into solid fats by

    (a) emulsification   (b) hydrolysis   (c) hydrogenation   (d) oxidation

22. Hydrogenation of trilinolein yields

    (a) glycerol and linoleic acid   (b) glycerol and linoleic alcohol
    (c) tristearin   (d) trilinolenin

23. Vegetable oils are not hydrogenated to produce

    (a) oleomargarine   (b) shortenings   (c) soaps   (d) all of these

24. Which reagent is used to convert oils to fats?

    (a) HCl   (b) H_2 (c) I_2 (d) NaOH

25. What are the two types of chemical reactions that cause butter to become rancid?

    (a) hydrogenation and hydrolysis   (b) hydrogenation and oxidation
    (c) hydrolysis and oxidation   (d) none of these

26. In oxidative rancidity, which portion of the triacylglycerol is attacked by oxygen?

    (a) the ester bonds   (b) the double bonds   (c) the glycerol portion   (d) the alkane portion

27. A wax is

    (a) any solid fat
    (b) a monoacylglycerol
    (c) an ester of a fatty acid and a long-chain alcohol
    (d) an ester of lanolin and a fatty acid

28. Spermaceti wax is obtained from

    (a) honeycombs   (b) lamb's wool   (c) palm trees   (d) whales

29. Sodium and potassium salts of fatty acids are called

    (a) micelles   (b) soaps   (c) syndets   (d) all of these

30. Which of the following represents a soap molecule?

    (a) ~~~~~~~⊖ K^+ (b) ~~~~~~~⊕ Cl^-
    (c) ~~~~~~~O H   (d) ~~~~~~~⊖
                                           ~~~~~~~⊖ Ca^2

31. Which fatty acid salt is soluble in water?

    (a) $RCOO^- Na^+$ (b) $(RCOO^-)_2 Ca^{2+}$ (c) $(RCOO^-)_2 Mg^{2+}$ (d) none of these

32. A synthetic detergent

    (a) is biodegradable   (b) contains a builder
    (c) is soluble in hard water   (d) all of these

33. Which of the following is not a type of syndet?

    (a) anionic   (b) cationic   (c) nonionic   (d) nonpolar

34. Sodium tripolyphosphate is added to detergents as a "builder." It functions

    (a) to tie up Ca^{2+} and Mg^{2+} ions.
    (b) to maintain the proper pH of the wash water.
    (c) to add bulk to the box of detergent.
    (d) all of these

35. Which of the following are not phospholipids?

    (a) cephalins   (b) cerebrosides   (c) lecithins   (d) sphingomyelins

36. Which of the following contains glycerol?

    (a) a cerebroside   (b) a ganglioside   (c) phosphatidic acid   (d) sphingosine

37. The chief difference between lecithin and the cephalins is

    (a) the amino alcohol that is bonded to the phosphate group.
    (b) the type of fatty acids they contain.
    (c) the presence or absence of a carbohydrate group.
    (d) the presence or absence of the glycerol backbone.

38. Which of the following does not contain glycerol?

    (a) galactocerebroside
    (b) phosphatidyl choline
    (c) phosphatidyl ethanolamine
    (d) phosphatidyl serine

39. Which of the following statements about phosphoglycerides is false?

    (a) They contain glycerol as their backbone.
    (b) They are nonsaponifiable lipids.
    (c) They contain two fatty acids
    (d) Their polar groups contain plus and minus charges at a pH of 7.

40. All glycolipids contain a carbohydrate group, a fatty acid, and

    (a) glycerol
    (b) sphingosine
    (c) either (a) or (b)
    (d) neither (a) nor (b)

41. Which of the following are the most abundant lipids found within cell membranes?

    (a) cholesterol   (b) glycolipids   (c) phospholipids   (d) triacylglycerols

42. The cell membranes

    (a) are rigid structures.
    (b) are inert barriers.
    (c) consist of a protein bilayer.
    (d) consist of a lipid bilayer.

43. Steroids are classified as lipids because they

    (a) are present in all living cells.
    (b) are soluble in nonpolar solvents.
    (c) can be saponified.
    (d) have complex organic structures.

44. Which of the following is a nonsaponifiable lipid?

    (a) cholesterol   (b) myricyl cerotate   (c) phosphatidyl choline   (d) tristearin

45. All animal steroids are derived from

    (a) bile salts   (b) cholesterol   (c) cortisone   (d) testosterone

46. Which of the following statements about cholesterol is false?

    (a) If cholesterol is present in the diet, its synthesis by the body is increased.
    (b) Fasting inhibits the biosynthesis of cholesterol.
    (c) Diets high in carbohydrate or fat increases the biosynthesis of cholesterol.
    (d) The cholesterol content of blood varies with the age, diet, and sex of an individual.

47. The bile salts are

    (a) androgens   (b) emulsifying agents   (c) glucocorticoids   (d) mineralcorticoids

48. Which of the following is not a female sex hormone?

    (a) estradiol   (b) estrone   (c) prednisolone   (d) progesterone

49. Which of the following is not an adrenocortical hormone?

    (a) aldosterone   (b) androstenedione   (c) cortisol   (d) cortisone

50. Anabolic steroids

    (a) increase muscle strength.
    (b) increase aggressiveness.
    (c) decrease fatigue.
    (d) all of these

51. All of the following function to prevent ovulation except

    (a) mestranol   (b) methandienone   (c) norethinodrone   (d) norethynodrel

52. Which of the following is not one of the hormones secreted during the normal ovarian cycle?

    (a) FSH   (b) HCG   (c) LH   (d) LTH

53. During menopause _____ secretion decreases.

    (a) DES   (b) estrogen   (c) FSH   (d) LH

54. Prostaglandins are classified as

    (a) nonsaponifiable lipids   (b) saponifiable lipids   (c) steroids   (d) terpenes

55. Prostaglandins are derivatives of

    (a) cholesterol   (b) phospholipids   (c) a saturated fatty acid   (d) an unsaturated fatty acid

56. In their physiological activity, prostaglandins resemble

    (a) carbohydrates   (b) enzymes   (c) hormones   (d) lipids

57. All prostaglandins

    (a) contribute to the inflammatory response.
    (b) induce smooth muscle contraction.
    (c) lower blood pressure.
    (d) all of these

58. Aspirin obstructs the synthesis of prostaglandins by

    (a) blocking the action of phospholipase.
    (b) inhibiting the cyclooxygenase enzyme.
    (c) both (a) and (b)
    (d) neither (a) nor (b)

59. Which vitamin is a steroid derivative?

    (a) A   (b) D   (c) E   (d) K

60. Vitamins _____ function in the body as antioxidants.

    (a) A and K   (b) A and C   (c) C and E   (d) E and K

61. Which vitamin is vital to the blood-clotting process?

    (a) A   (b) C   (c) E   (d) K

62. Vitamin A deficiency results in _____.

    (a) keratomalacia   (b) osteomalacia   (c) rickets   (d) scurvy

63. The lipids found in the highest concentration in atherosclerotic plaques are

    (a) cholesterol and cholesterol esters.
    (b) cholesterol and unsaturated fats.
    (c) cholesterol esters and saturated fats.
    (d) saturated and unsaturated fats.

64. Which molecule aids in transporting lipids in the blood?

    (a) cholesterol   (b) glycolipid   (c) glycoprotein   (d) lipoprotein

65. _____ are the major cholesterol-carrying molecules in the plasma.

    (a) chylomicrons   (b) HDLs   (c) LDLs   (d) VLDLs

FILL IN THE BLANKS

66. Lipids are soluble in _____ solvents, but insoluble in _____.

67. All fats and oils contain _____ as their alcohol component.

68. In fats and oils, fatty acids are joined to glycerol by _____ linkages.

69. Melting points of fatty acids _____ as the length of the hydrocarbon chain increases, and _____ as the degree of unsaturation of the chain increases.

70. The most abundant lipids in nature are the _____.

71. Stearic acid is a(n) _____ fatty acid, and linoleic acid is a(n) _____ fatty acid.

72. The configuration about the double bonds in most unsaturated fatty acids is _____.

73. _____ acid is the most abundant saturated fatty acid, and _____ acid is the most abundant unsaturated fatty acid.

74. Acid hydrolysis of a triacylglycerol yields _____ and _____.

75. Saponification refers to the reactions of lipids with a _____.

76. Lipids that undergo alkaline hydrolysis are referred to as _____ lipids, whereas those that do not undergo alkaline hydrolysis are referred to as _____ lipids.

77. The development of a disagreeable odor in a fat or an oil is known as _____.

78. The acid responsible for the odor of rancid butter is _____ acid.

79. Esters of long-chain acids and long-chain alcohols are called _____.

80. Sodium and potassium salts of long-chain acids are called _____.

81. The polar portion of a soap molecule is called the _____ part, and the nonpolar portion is called the _____ part.

82. The three classes of synthetic detergents are _____, _____, and _____.

83. Phosphate in detergents has been implicated as the principal cause of _____ of lakes and rivers.

84. Phosphatidic acid is the parent compound of the _____.

85. The number of ester bonds in a lecithin molecule is_____.

86. The net charge on a cephalin molecule (at pH 7) is _____.

87. Phospholipids are the major lipid components of _____.

88. In addition to lipids, what other major class of compounds is part of the composition of cell membranes? _____

89. The steroid nucleus contains _____ six-membered rings.

90. Draw the structure that is common to all steroids. _____

91. _____ results from the deposition and accumulation of plaque on the inside of arterial walls.

92. The _____ function to emulsify the lipids in our diet. They also aid in the absorption of lipids.

93. The estrogens and _____ are the female sex hormones.

94. The _____ component of birth control pills is responsible for most of the adverse side effects.

95. _____ are derivatives of arachidonic acid and function as biological regulators in the body.

96. Which vitamin is termed the sunshine vitamin? _____

97. The fat-soluble vitamins are _____, _____, _____, and _____.

TRUE OR FALSE

98. T F   A triacylglycerol contains three ester bonds.

99. T F   Fats are more saturated than oils.

100. T F   Fats and oils are solids at room temperature.

101. T F   Fats are simple triacylglycerols, whereas oils are mixed triacylglycerols.

102. T F The naturally occurring fatty acids contain an odd number of carbon atoms.

103. T F Fatty acids usually have between 12 and 14 carbon atoms.

104. T F Unsaturated fatty acids have higher melting points than saturated fatty acids.

105. T F The acid hydrolysis of oils yields mainly 18-carbon fatty acids.

106. T F Lipases are a class of saponifiable lipids.

107. T F Basic hydrolysis of a fat is called hardening.

108. T F Saponification of glyceryl tristearin with KOH will yield potassium stearate.

109. T F Vegetable oils can be converted into semisolid cooking fats by a process known as halogenation.

110. T F The odor of rancid butter is due to the release of low molecular weight acids.

111. T F In the hydrolytic rancidity of triacylglycerols, the ester bonds within the molecule are attacked.

112. T F Triacylglycerols containing only saturated fatty acids are often spoiled by oxidative rancidity.

113. T F The addition of oxidants to fats and oils retards rancidity.

114. T F Hydrolysis of a wax produces a fatty acid and a long-chain alcohol.

115. T F $CH_3(CH_2)_{12}COO(CH_2)_{14}CH_3$ is a wax.

116. T F A wax is a nonsaponifiable lipid.

117. T F $CH_3(CH_2)_{10}N^+(CH_3)_2$—⬡ Cl^- is an anionic detergent.

118. T F Lecithin and the cephalins are phospholipids.

119. T F Cerebrosides are glycolipids.

120. T F The major component (in terms of number of molecules) of most cell membranes is protein.

121. T F In the lipid bilayer formed from phospholipid molecules, the fatty acid "tails" are in the interior of the membrane.

122. T F Cholesterol occurs in just about all plants.

123. T F Bile acids are fatty acid derivatives.

124. T F Cortisone is a pituitary steroid.

125. T F The sex hormones are steroids.

126. T F Progesterone is an androgen.

127. T F Testosterone is produced only by males.

128. T   F   Anabolic steroids have been used to increase muscle mass in athletes.

129. T   F   Oral contraceptives contain a carbon–carbon triple bond.

130. T   F   Ethinyl estradiol is a synthetic analog of the estrogens.

131. T   F   Prostaglandins are steroid derivatives.

132. T   F   Prostaglandins resemble hormones in their physiological actions.

133. T   F   Prostaglandins have been used to induce abortions.

134. T   F   Vitamin E promotes the absorption of calcium into the blood and its deposition into the bones and teeth.

135. T   F   LDLs are the major cholesterol-transporting lipoproteins.

CHAPTER 13

Proteins

MULTIPLE CHOICE

1. Proteins function as

   (a) antibodies

   (b) hormones

   (c) oxygen carriers

   (d) all of these

2. Which of the following is not an α-amino acid?

   (a) $CH_2(NH_2)COOH$

   (b) $CH_3CH(NH_2)COOH$

   (c) $CH_2OHCH(NH_2)COOH$

   (d) $CH_2NH_2(CH_2)_3COOH$

3. Lysine is a _____ amino acid.

   (a) acidic   (b) basic   (c) neutral   (d) nonpolar

4. The compound whose structure is $(CH_3)_2CHCH_2CH(NH_2)COOH$ would be classified as a(n) _____ amino acid.

   (a) acidic   (b) basic   (c) polar but neutral   (d) nonpolar

5. Of the twenty amino acids that occur in proteins

   (a) one contains phosphorus.

   (b) two contain sulfur.

   (c) five have two carboxyl groups.

   (d) twenty have a primary amino group alpha to the carboxyl group.

6. An α-amino acid always contains

   (a) two amino groups

   (b) two carboxyl groups

   (c) an amino group and a carboxyl group

   (d) none of these

7. Which one of the following amino acids does not contain a chiral carbon?

   (a) alanine   (b) glycine   (c) serine   (d) valine

8. The chief distinguishing feature between the various amino acids is

   (a) their sign of optical rotation

   (b) their solubility in water

   (c) the number of amino or carboxyl groups

   (d) none of these

9. The side chains of amino acids do not include _____ groups.

    (a) polar   (b) nonpolar   (c) phosphate   (d) sulfhydryl

10. Which of the following amino acids contains a polar side chain?

    (a) alanine   (b) leucine   (c) phenylalanine   (d) serine

11. Which one of the following amino acids contains a sulfur atom?

    (a) arginine   (b) glutamine   (c) methionine   (d) threonine

12. The amino acids that occur in plants and animals are invariably

    (a) D-amino acids   (b) L-amino acids   (c) γ-amino acids   (d) racemic mixtures

13. The pH at which an amino acid has no net charge is called its _____ pH.

    (a) isoelectric   (b) nonpolar   (c) polar   (d) zwitterionic

14. Which of the following terms can be used to describe the structure of histidine at its isoelectric pH?

    (a) dipolar ion   (b) electrically neutral   (c) zwitterion   (d) all of these

15. In a solution of high pH, isoleucine would be

    (a) negatively charged   (b) positively charged   (c) a neutral molecule   (d) a zwitterion

16. At a pH that is lower than its isoelectric pH, an amino acid is

    (a) insoluble   (c) negatively charged   (b) electrically neutral   (d) positively charged

17. The zwitterionic form of glycine is

    (a) H_2NCH_2COOH (b) $H_2NCH_2COO^-$ (c) $H_3N^+ CH_2COOH$ (d) $H_3N^+ CH_2COO^-$

18. What is the form of alanine at a pH of about 2?

    (a) $CH_3\underset{\underset{NH_2}{|}}{C}HCOOH$ (b) $CH_3\underset{\underset{^+NH_3}{|}}{C}HCOOH$ (c) $CH_3\underset{\underset{NH_2}{|}}{C}HCOO^-$ (d) $CH_3\underset{\underset{^+NH_3}{|}}{C}HCOO^-$

19. What is the form of aspartic acid at a pH of about 12?

    (a) $HOOCCH_2\underset{\underset{NH_2}{|}}{C}HCOO^-$ (b) $HOOCCH_2\underset{\underset{^+NH_3}{|}}{C}HCOOH$

    (c) $^-OOCCH_2\underset{\underset{^+NH_3}{|}}{C}HCOO^-$ (d) $^-OOCCH_2\underset{\underset{NH_2}{|}}{C}HCOO^-$

20. Proteins can be considered to be

    (a) polyacetals   (b) polyamides   (c) polyamines   (d) polyesters

21. The primary structure of proteins is maintained by

    (a) disulfide bonds   (b) hydrophobic interactions
    (c) ionic bonds   (d) peptide bonds

22. The primary structure of a polypeptide is

(a) −N−C−CH−C−N−C−
         | ‖   ‖
         H O R O H O

(b) −CH−N−C−N−CH−C−
          | ‖ | ‖
          R H O H R O

(c) −N−CH−C−N−CH−C−
        | | ‖ | | ‖
        H   R O H   R O

(d) None of these

23. What is the abbreviation for the following tetrapeptide?

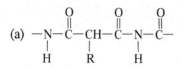

$\langle\bigcirc\rangle$−CH₂−CH−C−N−CH−C−N−CH₂−C−N−CH−COO⁻
                  | ‖ | | ‖ | ‖ | |
                ⁺NH₃ O   H CH₃ O   H   O   H CH₂OH

(a) Ser-Gly-Ala-Phe
(c) Phe-Ala-Ser-Gly

(b) Phe-Cys-Ser-Gly
(d) Phe-Ala-Gly-Ser

24. How many different tripeptides are possible from the combination of 2 molecules of valine and 1 molecule of leucine?

(a) 2   (b) 3   (c) 4   (d) 6

25. The peptide Ile-Trp-Glu-Asn-Tyr

(a) contains tyrosine at the C-terminal end.
(b) contains five peptide bonds.
(c) is the same peptide as Tyr-Asn-Glu-Trp-Ile.
(d) all of these.

26. $H_3\overset{+}{N}$−CHC−N−CH₂−C−N−CHCOO⁻ is a
        | ‖ | ‖ | |
        CH₃ O H   O   H CH₂SH

(a) dipeptide   (b) tripeptide   (c) tetrapeptide   (d) polypeptide

27. Which functional groups of amino acids interact in the formation of peptides?

(a) two amino groups
(c) an amino group and a carboxyl group

(b) two carboxyl groups
(d) a hydroxyl group and a carboxyl group

28. The hydrolysis of a heptapeptide molecule would require _____ molecules of water.

(a) 5   (b) 6   (c) 7   (d) 8

29. Five different amino acids are combined to make a pentapeptide. How many different pentapeptide sequences are possible?

(a) 10   (b) 20   (c) 60   (d) 120

30. Oxytocin and vasopressin are nonapeptides that function as _____ in the body.

(a) enzymes   (b) hormones   (c) protective agents   (d) all of these

31. Which type of bonding maintains the α-helix structure of some proteins?

    (a) covalent bonds   (b) hydrogen bonds   (c) hydrophobic interactions   (d) ionic bonds

32. The two most commonly occurring conformations of proteins are the α-helix and the

    (a) β-helix   (b) α-sheet   (c) β-pleated sheet   (d) γ-helix

33. The polypeptide chains in wool are wound in the α-helical conformation. Stretching wool fiber causes intrachain _____ bonds to break.

    (a) covalent   (b) disulfide   (c) hydrogen   (d) ionic

34. The α-helix represents one of the _____ structures of a protein.

    (a) primary   (b) secondary   (c) ternary   (d) tertiary

35. The interaction of groups in the side chains of proteins helps to maintain the _____ structure.

    (a) primary   (b) secondary   (c) ternary   (d) tertiary

36. The presence of _____ units in a polypeptide accounts for the formation of disulfide bonds.

    (a) asparagine   (b) cysteine   (c) methionine   (d) lysine

37. Interchain disulfide bonds arise from a(n) _____ reaction.

    (a) hydrolysis   (b) oxidation   (c) polymerization   (d) reduction

38. _____ involve nonpolar interactions between the side chains of certain amino acids.

    (a) disulfide bonds   (b) hydrogen bonds   (c) hydrophobic bonds   (d) salt linkages

39. Which of the following is not one of the interactions that maintains the tertiary structure of proteins?

    (a) disulfide bonds   (b) hydrophobic bonds
    (c) ionic bonds   (d) peptide bonds

40. When we specify the manner in which the four polypeptide chains of hemoglobin fit together, we are describing its _____ structure.

    (a) primary   (b) quaternary   (c) secondary   (d) tertiary

41. The structure of collagen can be described as a(n)

    (a) α-helix   (b) β-pleated sheet   (c) rope   (d) triple helix

42. Which amino acid disrupts the α-helical structure of proteins?

    (a) arginine   (b) glutamine   (c) glycine   (d) proline

43. The hemoglobin molecule contains

    (a) two protein chains and two heme groups.
    (b) two protein chains and four heme groups.
    (c) four protein chains and two heme groups.
    (d) four protein chains and four heme groups.

44. Proteins that tend to be soluble in water and have spherical, compact shapes are termed _____ proteins.

    (a) fibrous   (b) globular   (c) simple   (d) conjugated

45. The nonamino acid portion of a conjugated protein is called a(n) _____ group.

    (a) albuminoid   (b) globular   (c) historic   (d) prosthetic

46. The heme group in a hemoglobin molecule is

    (a) defective in sickle cell anemia.
    (b) different from the heme group in myoglobin.
    (c) called a prosthetic group.
    (d) none of these

47. Myoglobin is a _____ protein.

    (a) conjugated   (b) globular   (c) transport   (d) all of these

48. _____ is a structural protein.

    (a) casein   (b) collagen   (c) fibrinogen   (d) myosin

49. _____ is not a protein.

    (a) actin   (b) elastin   (c) heme   (d) insulin

50. The most abundant protein in mammals is

    (a) collagen   (b) hemoglobin   (c) α-keratin   (d) β-keratin

51. Which contains the least amount of protein?

    (a) blood   (b) bone   (c) egg white   (d) muscle

52. The isoelectric pH of a protein is

    (a) the pH at which the net charge on the protein is zero.
    (b) equal to the charge on the protein.
    (c) the sum of the positive charges on the protein.
    (d) the sum of the negative charges on the protein.

53. A protein has a negative net charge when it is in an aqueous solution at a pH that is

    (a) less than its isoelectric pH.   (b) greater than its isoelectric pH.
    (c) equal to its isoelectric pH.   (d) neither acidic nor basic.

54. At the isoelectric pH of proteins

    (a) the proteins have no charged groups.
    (b) the proteins are least soluble.
    (c) the proteins are most soluble.
    (d) none of these.

55. Protein denaturation results in the

    (a) formation of disulfide bonds.
    (b) hydrolysis of amide linkages.
    (c) disruption of the three-dimensional structure.
    (d) change from a globular to a fibrous protein.

56. Which of the following processes is least likely to occur when a protein becomes denatured?

    (a) the breaking of hydrogen bonds.   (b) the breaking of hydrophobic bonds.
    (c) the breaking of peptide bonds.   (d) the breaking of salt linkages.

57. _____ denature the enzymes of bacteria, and thus destroy the bacteria.

    (a) high temperatures   (c) acids and bases   (b) organic solvents   (d) all of these

58. Ions of heavy metals denature proteins by combining with _____.

    (a) $-\overset{+}{N}H_3$ groups and $-OH$ groups
    (b) $-COO^-$ groups and $-SH$ groups

    (c) —⬡ groups and —⬡—OH groups

    (d) $-COOH$ groups and $-NH_2$ groups

59. Picric acid and tannic acid denature proteins by combining with _____ groups.

    (a) $-OH$ (b) $-SH$ (c) $-COO^-$ (d) $-\overset{+}{N}H_3$

60. Which of the following would be least likely to denature a protein?

    (a) heating   (b) changes in pH   (c) addition of alcohol   (d) addition of a NaCl solution

61. Disulfide bonds can be disrupted by _____ agents.

    (a) dehydrating   (b) hydrating   (c) oxidizing   (d) reducing

62. Which of the following amino acids does not give a positive xanthoproteic test?

    (a) arginine   (b) phenylalanine   (c) tryptophan   (d) tyrosine

63. Which amino acid gives a positive Millon test?

    (a) $HSCH_2CHCOOH$
          NH_2

    (b) ⬡—$CH_2CHCOOH$
                      NH_2

    (c) [indole]—$CH_2CHCOOH$
                        NH_2

    (d) HO—⬡—$CH_2CHCOOH$
                          NH_2

64. Which amino acid gives a positive Hopkins-Cole test?

    (a) [indole]—$CH_2CHCOOH$
                        NH_2

    (b) ⬡—$CH_2CHCOOH$
                      NH_2

    (c) [imidazole]—$CH_2CHCOOH$
                          NH_2

    (d) HO—⬡—$CH_2CHCOOH$
                          NH_2

FILL IN THE BLANKS

65. The number of different amino acids commonly found in proteins is _____.

66. _____ is the only naturally occurring amino acid that is not optically active.

67. _____ is an amino acid that contains two chiral carbons.

68. Amino acids are insoluble in _____ solvents.

69. The dipolar ion structure of an amino acid is called a _____.

70. The pH at which an amino acid has no net charge is called the _____ pH.

71. The structure of glycine at a pH of 12.0 is _____.

72. The structure of alanine at a pH of 2.0 is _____.

73. The zwitterion form of phenylalanine is _____.

74. In the dipolar ion form of an amino acid, the amino group has a _____ charge, and the carboxylate group has a _____ charge.

75. For the amino acid serine, the pK_a for the –COOH group and the $-NH_3^+$ group are 2.21 and 9.15, respectively. What is the isoelectric pH of serine? _____

76. At pH values above its isoelectric pH, an amino acid has a net _____ charge.

77. At a pH of 6.0, a solution of glutamic acid, lysine, and valine was placed in an electrophoresis apparatus. When the current was turned on, _____ migrated to the anode, _____ migrated to the cathode, and _____ remained stationary.

78. Two of the basic amino acids are _____ and _____.

79. One of the acidic amino acids is _____.

80. The amide linkage that joins two amino acids is called the _____ bond.

81. The sequence of amino acids in a protein is referred to as the _____ structure.

82. _____ bonds maintain the primary structure of proteins.

83. The hormone insulin contains _____ polypeptide chains.

84. In the pentapeptide Leu-Thr-Glu-His-Cys, which amino acid has a free amino group? _____

85. Human insulin differs from sheep insulin at positions _____, _____, _____ in the A chain.

86. The hydrolysis of the tripeptide asparagylisoleucyltryptophan yields _____, _____, and _____.

87. One of the secondary structures of proteins is the _____.

88. The protein in silk exists in the _____ conformation.

89. The polypeptide chains in wool are wound in the _____ conformation.

90. The stretching of wool causes lengthening of the intrachain _____ bonds.

91. The precise folding and bending of the α-helices or β-pleated sheets is referred to as the _____ structure of a protein.

92. Two functional groups found in the side chains of proteins are _____ and _____.

93. Two types of noncovalent interactions found in proteins are _____ and _____.

94. The manner in which polypeptide subunits are joined together is called the _____ structure of a protein.

95. Myoglobin stores _____ in heart muscle.

96. Which amino acid disrupts the α-helical conformation of a protein?_____

97. What is the most abundant amino acid found in collagen? _____

98. One method of classifying proteins is by their shape. The two protein types are _____ and _____.

99. The complete hydrolysis of a simple protein yields only _____.

100. The hydrolysis of a conjugated protein yields a nonprotein group called a _____ group.

101. At pH values below its isoelectric pH, a protein has a net _____ charge.

102. A disruption of the three-dimensional structure of a protein is called _____.

103. Give three methods used to denature proteins. _____, _____, and _____.

TRUE OR FALSE

104. T   F   Proteins, like fats, serve as energy reserves for the body.

105. T   F   Glutamine is a basic amino acid.

106. T   F   β-amino acids are the most common amino acids found in naturally occurring proteins.

107. T   F   The proteins in all living species are constructed from just 18 of the 20 common amino acids.

108. T   F   The naturally occurring amino acids belong to the D-series.

109. T   F   Several amino acids are colorless liquids.

110. T   F   The value of the pK_a of the carboxyl group of an amino acid is always less than the value of the pK_a of the protonated amino group.

111. T   F   At its isoelectric pH, an amino acid is negatively charged.

112. T   F   A zwitterion is an ion that possesses two positive charges.

113. T   F   Polypeptides are distinguished from proteins on the basis of molecular weight.

114. T   F   Each protein has a very specific sequence of amino acids.

115. T   F   A tetrapeptide contains four peptide bonds.

116. T   F   Arginine is at the C-terminal end of the tripeptide Arg-Asp-Asn.

117. T   F   Peptide bonds in proteins are amine linkages.

118. T   F   A change in just one amino acid could alter the function of a protein.

119. T   F   Insulin is a protein hormone that shows no variation with species.

120. T   F   In sickle cell hemoglobin an Asp substitutes for a Glu.

121. T   F   Oxytocin and vasopressin are polypeptide hormones that are secreted by the pituitary gland.

122. T   F   The bonds that maintain the α-helical conformation are intermolecular hydrogen bonds.

123. T   F   Silk fibers are more elastic than wool because the secondary structure of their α-keratins is more helical.

124. T   F   Intramolecular and intermolecular hydrogen bonding between groups in the side chains of proteins play a major role in maintaining the secondary structure.

125. T   F   The β-pleated sheet is an example of a protein conformation.

126. T   F   Salt linkages are the most important bonds for maintaining the tertiary structure of proteins.

127. T   F   A disulfide bond occurs when two methionine amino acids, on adjacent polypeptide chains, are in proximity.

128. T   F   In aqueous solution most of the polar groups are on the outer surface of the protein molecule.

129. T   F   Hydrophobic interactions maintain the triple helix structure of collagen.

130. T   F   Myoglobin is used for energy storage in muscle cells.

131. T   F   Proline and hydroxyproline disrupt the α-helical conformation of proteins.

132. T   F   The most abundant amino acid in silk fibroin is cysteine.

133. T   F   All proteins possess quaternary structure.

134. T   F   Albumins do not contain heme as a prosthetic group.

135. T   F   Conjugated proteins may contain metal ions.

136. T   F   Fibrous proteins are insoluble in aqueous solution.

137. T   F   The peptide linkage joins amino acids to the prosthetic group in a conjugated protein.

138. T   F   Heme is the prosthetic group of hemoglobin and myoglobin.

139. T   F   Simple proteins contain only nonpolar amino acids.

140. T   F   Globular proteins have no tertiary structure.

141. T   F   Globular proteins are soluble in water because they contain numerous disulfide bonds.

142. T   F   Prosthetic groups are found in all proteins.

143. T   F   A nucleoprotein is a conjugated protein.

144. T   F   At the isoelectric pH of a protein, its solubility in water is maximum.

145. T   F   At its isoelectric pH, the tripeptide Glu-His-Tyr would migrate to the cathode.

146. T   F   Electrophoresis is a method that is used to separate proteins.

147. T   F   Denatured proteins lose their biological activity.

148. T   F   Protein denaturation is always reversible.

149. T   F   Denaturation does not disrupt the primary structure of proteins.

150. T   F   An example of denatured protein is fried egg.

151. T   F   A 95% alcohol solution is a better disinfectant than a 70% alcohol solution.

152. T   F   Salts of heavy metals are used internally as disinfectants.

153. T   F   Alkaloidal reagents are used to denature proteins.

154. T   F   Reducing agents denature proteins by disrupting the disulfide linkages.

155. T   F   A positive ninhydrin test is given by amino acids but not by proteins.

156. T   F   A positive biuret test is given by proteins but not by amino acids.

CHAPTER 14

Enzymes

MULTIPLE CHOICE

1. Enzymes are a class of _____ proteins.

   (a) catalytic   (b) protective   (c) storage   (d) transport

2. Which of the following statements about enzymes is correct?

   (a) They are unchanged in a biochemical reaction.
   (b) They increase the rate of a biochemical reaction.
   (c) They are proteins.
   (d) all of these

3. The substance that is acted upon by an enzyme is called a(n)

   (a) activator   (b) coenzyme   (c) cofactor   (d) substrate

4. An enzyme that catalyzes the hydrolysis of peanut oil could be called a(n)

   (a) invertase   (b) lipase   (c) peptidase   (d) transferase

5. All lipases are

   (a) esterases   (b) hydrogenases   (c) oxidases   (d) none of these

6. Enzymes that catalyze the hydrolysis of the amide linkages in polypeptides are called

   (a) aminoases   (b) hydrolases   (c) reductases   (d) transferases

7. The enzyme that catalyzes the addition of water to a compound is called

   (a) hydrogenase   (b) hydrolyase   (c) ligase   (d) lyase

8. A transferase enzyme catalyzes a(n) _____ reaction.

   (a) addition   (b) elimination   (c) isomerization   (d) group transfer

9. An apoenzyme is

   (a) a polypeptide   (b) a nonprotein organic group
   (c) a metal ion   (d) none of these

10. If a metal ion is necessary for the activity of an enzyme, the metal ion is termed a(n)

    (a) activator   (b) apoenzyme   (c) coenzyme   (d) proenzyme

11. An enzyme that is a conjugated protein is called a(n)

    (a) apoenzyme   (b) coenzyme   (c) holoenzyme   (d) zymogen

12. The coenzyme is

    (a) the same as the apoenzyme.
    (b) the protein portion of the enzyme.
    (c) derived from an activator.
    (d) a prosthetic group.

13. The inactive form of an enzyme is termed a(n)

    (a) cofactor   (b) inactivator   (c) modulator   (d) zymogen

14. Pepsinogen is activated by hydrogen ions in the stomach to

    (a) amylase   (b) chymotrypsin   (c) pepsin   (d) trypsin

15. Which of the following is not a zymogen?

    (a) chymotrypsinogen   (b) pepsinogen   (c) trypsinogen   (d) all are zymogens

16. The vitamin niacin serves as a precursor for the coenzyme

    (a) biotin   (b) FAD   (c) folic acid   (d) NAD^+

17. In an enzyme-catalyzed reaction, the enzyme and substrate interact at a small region of the enzyme's surface called the _____ site.

    (a) active   (b) allosteric   (c) induced   (d) key

18. The theory that proposes a flexible, dynamic interaction between enzyme and substrate is the _____ theory.

    (a) allosteric   (b) lock-and-key   (c) induced fit   (d) nonspecific

19. The active site of an enzyme

    (a) is a rigid structural feature.
    (b) is bound to the enzyme by hydrophobic interactions.
    (c) is the entire surface of the enzyme.
    (d) possesses a unique conformation that is complementary to the substrate.

20. In the first step of an enzyme-catalyzed reaction

    (a) bonds in the substrate are broken.
    (b) the substrate combines with the enzyme.
    (c) the substrate combines with a water molecule.
    (d) two substrate molecules interact.

21. When *trypsin* catalyzes the hydrolysis of proteins, peptide bonds involving the amino acids lysine and arginine are broken. *Trypsin* exhibits _____ specificity.

    (a) absolute   (b) group   (c) linkage   (d) stereochemical

22. The enzyme *arginase* catalyzes the conversion of L-arginine to ornithine and urea. It does not act upon D-arginine. *Arginase* exhibits _____ specificity.

    (a) absolute   (b) group   (c) linkage   (d) stereochemical

23. Which linkage is hydrolyzed by a peptidase?

(a) $R-O-R'$ (b) $R-O-\underset{\underset{O^-}{|}}{\overset{\overset{O}{\parallel}}{P}}-O^-$ (c) $R-\overset{\overset{O}{\parallel}}{C}-O-\overset{\overset{O}{\parallel}}{C}-R'$ (d) $R-\overset{\overset{O}{\parallel}}{C}-\underset{\underset{H}{|}}{N}-R'$

24. The lock-and-key theory helps to explain the _____ of enzymes

(a) high turnover number (b) inducibility (c) protein nature (d) specificity

25. The enzyme *lactase* shows specificity for only

(a) carbohydrates (b) disaccharides (c) lactose (d) polysaccharides

26. Which of the following influences the rate of an enzyme-catalyzed reaction?

(a) amount of substrate (b) pH (c) temperature (d) all of these

27. If the temperature of an enzymatic reaction is decreased

(a) the enzyme becomes denatured.
(b) the activity of the enzyme decreases.
(c) the enzyme precipitates out of solution.
(d) the rate of the reaction is unaffected.

28. An enzyme that functions in the blood has an optimum pH of

(a) 5.4 (b) 6.4 (c) 7.4 (d) 8.4

29. An enzyme that functions in the stomach has an optimum pH of

(a) 2.0 (b) 3.0 (c) 4.0 (d) 5.0

30. The optimum temperature for most enzymes that function in the human body is

(a) 0°C (b) 37°C (c) 98.6°C (d) 100°C

31. When an enzyme becomes saturated with substrate, an increase in _____ will not increase the rate of the enzymatic reaction.

(a) enzyme concentration (b) substrate concentration
(c) temperature (d) pH

32. Which of the following does not inhibit the activity of enzymes?

(a) Ag^+ (b) Hg^{2+} (c) K^+ (d) Pb^{2+}

33. The inhibition of succinic acid dehydrogenase by malonic acid is an example of _____ inhibition.

(a) competitive (b) end-product (c) irreversible (d) noncompetitive

34. _____ inhibition refers to the process by which the end product of a sequence of enzyme-catalyzed reactions inhibits the enzyme that catalyzed an early step of the sequence.

(a) allosteric (b) end-product (c) noncompetitive (d) reversible

35. In the lock-and-key model of enzyme action, a competitive inhibitor is analogous to

(a) a key that won't fit the lock.
(b) a lock that can't accept the key.

    (c) a wrong key for a wrong lock.

    (d) a key that fits the lock, but won't turn it.

36. A(n) _____ is an enzyme whose activity can be controlled by regulatory molecules.

    (a) allosteric enzyme   (b) coenzyme   (c) isoenzyme   (d) proenzyme

37. Nerve gases are toxic because they inhibit the enzyme

    (a) acetylcholinesterase   (b) succinic acid dehydrogenase

    (c) threonine dehydratase   (d) thromboplastin

38. The allosteric site on an allosteric enzyme binds the

    (a) end-product inhibitor   (b) enzyme

    (c) proenzyme   (d) substrate

39. Activation of an enzyme may occur through

    (a) conversion of an enzyme to a proenzyme.

    (b) separation of the apoenzyme from the enzyme.

    (c) binding of a positive modulator to the allosteric site of an allosteric enzyme.

    (d) all of these.

40. Penicillin acts by

    (a) forming complexes with metal ion activators.

    (b) mimicking the action of a coenzyme.

    (c) inhibiting folic acid synthesis in bacteria.

    (d) none of these.

41. Which of the following enzymes is not important for the clinical diagnoses of myocardial infarctions?

    (a) CK   (b) GOT   (c) GPT   (d) LDH

FILL IN THE BLANKS

42. Enzymes _____ the rates of biochemical reactions.

43. All enzymes are _____.

44. The name of an enzyme usually ends with the three letters _____.

45. The substrate for the enzyme urease is _____.

46. The protein portion of a holoenzyme is called the _____.

47. The nonprotein (often organic) portion of a holoenzyme is called the _____.

48. A _____ is the catalytically inactive enzyme precursor.

49. Many of the B vitamins are used by the body to synthesize _____.

50. The initials of two coenzymes are _____ and _____.

51. All catalysts lower the _____ of a chemical reaction.

52. Enzymes function as catalysts by combining with a _____ to form an _____ complex.

53. The substrate combines with the enzyme at a specific region on the surface of the enzyme called the _____.

54. The two theories that explain the mode of enzyme action are the _____ and the _____.

55. Name two types of interactions that maintain the enzyme–substrate complex. _____ and _____.

56. According to Koshland, there are at least two unique groups at the active site of enzymes. The _____ groups are responsible for substrate specificity, and the _____ groups are responsible for catalysis.

57. An enzyme that acts only on a single substrate exhibits _____ specificity.

58. The enzymes that catalyze the hydrolysis of the peptide bonds in proteins are called _____.

59. Enzymes that catalyze the transfer of a group from one molecule to another are generally called _____.

60. When the rate of an enzyme-catalyzed reaction is plotted against pH, the shape of the curve is approximately _____.

61. Most body enzymes have optimal activity within the pH range _____ to _____.

62. _____ is an example of an enzyme whose optimal pH is about 2.0.

63. Most enzymes are inactivated at temperatures above _____ °C.

64. The catalytic activity of most enzymes is a maximum at a certain temperature that is referred to as the _____ temperature.

65. _____ inhibitors compete with the substrate for binding at the active sites of enzymes.

66. Many poisons exert their effects by _____ key enzymes in living organisms.

67. A _____ inhibitor binds to the enzyme at a site remote from the active site.

68. Nerve gases inactivate the enzyme _____.

69. An _____ enzyme is an enzyme whose activity can be controlled by regulatory substances.

70. Two types of enzyme inhibitors are _____ and _____.

71. _____ is the use of chemicals to destroy infectious microorganisms without destroying the host.

72. An _____ is a compound produced by one microorganism that is toxic to another microorganism.

73. An _____ is a substance that behaves as a competitive inhibitor and kills or prevents the growth of microorganisms.

74. _____ inhibits the enzyme that catalyzes the synthesis of the bacterial cell wall.

75. Sulfa drugs have structures similar to _____.

76. What are two antibiotics (other than penicillin)? _____ and _____

TRUE OR FALSE

77. T   F   About 50–60% of all the reactions that occur in living cells are catalyzed by enzymes.

78. T   F   Enzymes are broken down during the biochemical reactions in which they participate.

79. T   F   Enzymes permit biochemical reactions to occur at lower temperatures and at faster rates compared to uncatalyzed reactions.

80. T   F   In the living cell, relatively high concentrations of each of the enzymes are required.

81. T   F   Lipases digest lipids, and amylases digest amylids.

82. T   F   The enzyme that catalyzes the hydrolysis of lactose is galactase.

83. T   F   A hydrogenase is classified as a hydrolase.

84. T   F   Oxidases are enzymes that catalyze the conversion of fats to fatty acids and glycerol.

85. T   F   Some enzymes are carbohydrates and/or lipids.

86. T   F   The active site on the substrate is the region that binds to the enzyme.

87. T   F   An apoenzyme is the active form of the enzyme.

88. T   F   Coenzymes are organic compounds.

89. T   F   Pepsinogen is the zymogen of pepsin.

90. T   F   Frequently a small peptide is removed from a zymogen to convert it into a proenzyme.

91. T   F   Many enzymes are obtained from vitamins.

92. T   F   The formation of an enzyme–substrate complex occurs simultaneously with the release of products from the enzyme's surface.

93. T   F   The lock-and-key model suggests that the substrate must fit precisely the active site of the enzyme for catalysis to occur.

94. T   F   The amino acids that make up the active site must be adjacent to each other in the primary structure of the enzyme.

95. T   F   The induced-fit theory assumes that enzymes have a rigid conformation.

96. T F The formation of the enzyme-substrate complex is consistent with the lock and key model of enzyme action.

97. T F All enzymes act only on a single substance.

98. T F If an enzyme exhibits linkage specificity, it catalyzes a single reaction.

99. T F An enzyme that catalyzes the oxidation of many different alcohols exhibits stereochemical specificity.

100. T F If a substrate has this structure,

$$-\overset{\displaystyle |}{\underset{\displaystyle R}{CH}}-\overset{\displaystyle O}{\overset{\displaystyle \|}{C}}-\overset{\displaystyle |}{\underset{\displaystyle H}{N}}-\overset{\displaystyle |}{\underset{\displaystyle R}{CH}}-\overset{\displaystyle O}{\overset{\displaystyle \|}{C}}-\overset{\displaystyle |}{\underset{\displaystyle H}{N}}-$$

its hydrolysis will be catalyzed by a peptidase.

101. T F The optimum pH of all body enzymes is about 6.

102. T F A plot of reaction rate versus substrate concentration for an enzyme-catalyzed reaction most frequently yields a bell-shaped curve.

103. T F An increase in substrate concentration will always increase the rate of an enzyme-catalyzed reaction.

104. T F For all practical purposes, an increase in enzyme concentration always increases the rate of an enzyme-catalyzed reaction.

105. T F Enzymes are heat stable catalysts.

106. T F Enzymes can be denatured.

107. T F Competitive inhibitors and irreversible inhibitors both bind to the active sites of enzymes.

108. T F The substrate binds to the allosteric site of an allosteric enzyme.

109. T F The inhibitor of an allosteric enzyme is usually the product of the first reaction in the sequence.

110. T F Antibiotics are synthesized by the body to combat pathogenic microorganisms.

111. T F Penicillin functions as an antibiotic by competing with folic acid in bacterial reactions.

112. T F Sulfa drugs are antimetabolites that are similar in structure to p-aminobenzoic acid.

113. T F GOT and GTP are abbreviations for enzymes that are found in both saliva and meat tenderizers.

CHAPTER 15

The Nucleic Acids

MULTIPLE CHOICE

1. Hydrolysis of a nucleic acid would not yield the following type of compound.

   (a) an amino acid   (b) a carbohydrate   (c) a heterocyclic base   (d) phosphoric acid

2. Which of the following is a purine base?

   (a) cytosine   (b) guanine   (c) thymine   (d) uracil

3. A nucleoside consists of

   (a) a base and a sugar.
   (b) a phosphate and sugar.
   (c) a purine base and a pyrimidine base.
   (d) ribose and deoxyribose.

4. Which of the following is a nucleoside?

   (a) cytosine   (b) cytidine   (c) cytidine monophosphate   (d) deoxycytidine monophosphate

5. A nucleotide consists of

   (a) a base and a phosphate.   (b) a base and a sugar.
   (c) a sugar and a phosphate.   (d) a sugar, a base and a phosphate.

6. Which is a component of AMP?

   (a) adenine   (b) phosphate   (c) ribose   (d) all of these

7. Which group is not part of the backbone of a nucleic acid?

   (a) base   (b) phosphate   (c) sugar   (d) All of these are part of the backbone.

8. Which base is not found in DNA?

   (a) cytosine   (b) guanine   (c) thymine   (d) uracil

9. Which nucleotide would be incorporated into DNA?

   (a) adenosine   (b) guanine
   (c) guanosine monophosphate   (d) deoxyadenosine monophosphate

10. Nucleic acids, like proteins, have two ends, one called the _____ end and the other the _____ end.

    (a) $1'$; $4'$ (b) $2'$; $6'$ (c) $3'$; $5'$ (d) C ; P

11. The base sequence in a nucleic acid is considered to be its _____ structure.

    (a) primary   (b) secondary   (c) tertiary   (d) quaternary

12. In DNA, the phosphate is bonded directly to

    (a) adenine   (b) deoxyribose   (c) guanine   (d) ribose

13. Base pairing in DNA is accomplished through

    (a) disulfide bonds   (b) hydrogen bonds
    (c) hydrophobic interactions   (d) salt linkages

14. Which of the following statements is incorrect concerning the structure of DNA?

    (a) The structure is a double-stranded helix. .
    (b) The two strands are antiparallel.
    (c) The bases are aligned on the outside of each strand.
    (d) The two strands are held together by hydrogen bonds.

15. According to the Watson–Crick model, DNA consists of two strands coiled in a _____ configuration, and the two strands are aligned _____.

    (a) β-pleated sheet; parallel   (b) β-pleated sheet; antiparallel
    (c) helical; parallel   (d) helical; antiparallel

16. In comparing the DNA of different species, we find that

    (a) the percentage of adenine in each is constant.
    (b) the length of the DNA strands are the same.
    (c) the cytosine to guanine ratios are the same.
    (d) none of these

17. Usually in a DNA molecule the content of _____ equals that of _____.

    (a) cytosine; thymine   (b) adenine; uracil
    (c) guanine; thymine   (d) adenine; thymine

18. Where does DNA biosynthesis occur?

    (a) in the cytosol   (b) at the cell membrane
    (c) in the nucleus   (d) at a ribosome

19. The biosynthesis of DNA is called

    (a) inscription   (b) replication   (c) transcription   (d) translation

20. Which base is not usually found in RNA?

    (a) adenine   (b) guanine   (c) thymine   (d) uracil

21. A base pair found in RNA, but not in DNA is

    (a) adenine–thymine   (b) adenine–uracil
    (c) guanine–cytosine   (d) guanine–uracil

22. Which is the most abundant of the RNAs?

    (a) mRNA   (b) rRNA   (c) sRNA   (d) tRNA

23. The biosynthesis of RNA from a strand of DNA is called

    (a) duplication   (b) replication   (c) transcription   (d) translation

24. A segment of a DNA molecule that directs the synthesis of a particular polypeptide molecule is called

    (a) a chromosome   (b) a gene   (c) an intron   (d) an operon

25. The triplet of bases on the mRNA that specifies a particular amino acid is called

    (a) an anticodon   (b) a codon   (c) an exon   (d) a gene

26. The most significant feature of a tRNA mole is the

    (a) anticodon region   (b) codon region   (c) 3′ end   (d) 5′ end

27. When a polypeptide is to be synthesized,

    (a) a chromosome divides and migrates to the ribosomes.
    (b) a gene is copied and the new DNA migrates to the ribosomes.
    (c) a gene is transcribed and the mRNA migrates to the ribosomes.
    (d) a mRNA is translated and the tRNA migrates to the ribosomes.

28. Activation of an amino acid requires

    (a) ATP   (b) tRNA   (c) aminoacyl–tRNA synthetase   (d) all of these

29. Which process occurs at a ribosome?

    (a) replication   (b) reverse transcription
    (c) transcription   (d) translation

30. Which of the following binds to amino acids during the synthesis of a polypeptide?

    (a) DNA   (b) mRNA   (c) rRNA   (d) tRNA

31. The codon ____ initiates the synthesis of a polypeptide chain.

    (a) CAU   (b) AGC   (c) UAC   (d) AUG

32. The activated tRNA molecule binds to the ribosome at the _____site.

    (a) aminoacyl   (b) inducer   (c) peptidyl   (d) promoter

33. The movement of a ribosome along a mRNA molecule is called

    (a) transduction   (b) transcription   (c) translation   (d) translocation

34. Cysteine is represented by the codons UGU and UGC. This is an example of the ____ of the genetic code.

    (a) degeneracy   (b) essentiality   (c) nonpolarity   (d) universality

35. If the codon is GUA, the anticodon is

    (a) UAC   (b) CAU   (c) GUA   (d) CAT

36. Three of the codons

    (a) do not code for any amino acid.
    (b) are termination codons.

(c) signal the end of the polypeptide chain.
(d) all of these

37. A mutation is a change in the base sequence in the

   (a) DNA   (b) RNA   (c) protein   (d) all of these

38. Which is not a point mutation?

   (a) deletion   (b) elongation   (c) insertion   (d) substitution

39. Antibiotics and anticancer drugs exert their effects by blocking the synthesis of

   (a) DNA   (b) RNA   (c) proteins   (d) all of these

40. In recombinant DNA research

   (a) plasmids are used as vectors.
   (b) restriction enzymes are used to cleave DNA molecues.
   (c) DNA ligase splices the strands of DNA.
   (d) all of these

FILL IN THE BLANKS

41. A nucleoside is composed of a _____ and a _____ joined by a _____ bond.

42. A _____ consists of a phosphate group, a sugar, and a purine or pyrimidine base.

43. The _____ group can be considered to be the connecting bridge between adjacent nucleotides.

44. Since a phosphate group is ionized at physiological pHs, nucleotides have a _____ charge.

45. The _____ of nucleotides in DNA accounts for its primary structure.

46. The two bases adenine and thymine are called _____ bases.

47. _____ bonds hold the two strands together in DNA.

48. The two major classes of nucleic acids are _____ and _____.

49. Nucleic acids, like proteins, have two ends; one end is designated _____ and the other end is _____.

50. DNA replication occurs in such a manner that each new molecule consists of one parental strand and one new strand. This mechanism is termed _____.

51. _____ is the enzyme that joins nucleotides together, whereas _____ is the enzyme that joins strands of nucleotides.

52. Both the leading strand and the lagging strand are synthesized in the _____ direction.

53. What sugar is a component of RNA? _____

54. Synthesis of RNA occurs in the _____.

55. In RNA molecules adenine binds to _____, and guanine binds to _____.

56. Genetic information is passed from the nucleus to the ribosomes in the form of _____ molecules.

57. A gene is a segment of a molecule of _____.

58. Protein synthesis occurs at the _____.

59. The dicodon CAUGGU codes for the dipeptide _____.

60. The anticodon is found on _____.

61. The number of bases in a codon is _____.

62. What is a possible codon sequence for the tripeptide Arg-His-Leu? _____

63. If one strand of DNA has the base sequence TCTTTCGTAAATGGC, what will be the base sequence on the complementary strand? _____

64. If the sequence on a mRNA strand is GCAAGUAGCUUC, what is the corresponding sequence of the DNA strand? _____

65. If a portion of the DNA has the base sequence CAACTACTTGGT, what is the tetrapeptide that will be synthesized? _____

TRUE OR FALSE

66. T   F   Purine and pyrimidine bases are hydrolysis products of DNA.

67. T   F   Guanine and adenine are isomers.

68. T   F   Nucleosides consist of nucleotides plus phosphate.

69. T   F   The average molecular weight of a nucleotide is about 300 daltons.

70. T   F   The average molecular weight of a nucleic acid is about 4×10^3 daltons.

71. T   F   Nucleic acids are located throughout the cell membranes.

72. T   F   Nucleic acids are polymers.

73. T   F   DNA is located at various regions throughout the cell.

74. T   F   DNA melting occurs when DNA is heated and the two strands separate.

75. T   F   Complete hydrolysis of DNA yields phosphates and nucleosides.

76. T   F   The hydrolysis of deoxycytidine yields deoxyribose and cytidine.

77. T   F   DNA always incorporates equal amounts of purines and pyrimidines.

78. T F   The pairing of adenine with thymine and guanine with cytosine permits the best hydrogen-bonded interactions between base pairs.

79. T F   The backbone of a DNA is held together by hydrogen bonds.

80. T F   In any organism the ratio of A+G to T+C is always very close to unity.

81. T F   DNA contains only purine bases, whereas RNA contains only pyrimidine bases.

82. T F   RNA and DNA are structurally the same except guanine is found in place of adenine in RNA.

83. T F   DNA and RNA are termed acids because the sugar groups can donate protons.

84. T F   Ribosomal RNA usually has a higher molecular weight than DNA.

85. T F   An RNA may fold back on itself and undergo base pairing.

86. T F   DNA and RNA contain identical nucleic acid backbones.

87. T F   Codons are represented by pairs of bases.

88. T F   Some codons call for more than one amino acid.

89. T F   More than one codon can represent a particular amino acid.

90. T F   A particular codon can represent only one amino acid.

91. T F   The order of bases on the tRNA constitutes the genetic message for the synthesis of proteins.

92. T F   The anticodon is a three-base sequence on mRNA.

93. T F   A codon sequence for the dipeptide Phe-Met is UUUAUG.

94. T F   The codons that do not call for a particular amino acid, are signals that terminate the synthesis of a polypeptide.

95. T F   The drug 5-bromouracil is used to treat cancer.

96. T F   Some genetic mutations could aid an organism.

97. T F   If a mutation occurs to change the codon UGG to UGC, then a different amino acid will be incorporated into the protein.

98. T F   Nitrous acid is a mutagen.

99. T F   Viruses contain nucleic acids surrounded by proteins.

100. T F   Viruses can independently replicate themselves.

CHAPTER 16

Carbohydrate Metabolism

MULTIPLE CHOICE

1. The energy required by all living organisms is ultimately derived from

   (a) ATP   (b) catabolism   (c) glucose   (d) the sun

2. In photosynthesis

   (a) solar energy is converted into chemical energy.
   (b) carbon dioxide is converted into glucose.
   (c) oxygen is a by-product.
   (d) all of these

3. The digestion of carbohydrates is initiated in the

   (a) mouth   (b) stomach   (c) small intestine   (d) large intestine

4. Most products of carbohydrate digestion are absorbed from the

   (a) stomach   (b) small intestine   (c) large intestine   (d) none of these

5. Which of the following is not readily absorbed into the bloodstream?

   (a) fructose   (b) galactose   (c) glucose   (d) sucrose

6. Which of the following enzymes is not involved in carbohydrate digestion?

   (a) lactase   (b) maltase   (c) ptyalin   (d) rennin

7. The type of reaction that is fundamental to the digestion of carbohydrates is

   (a) anabolism   (b) decarboxylation   (c) hydrolysis   (d) oxidation

8. In a normal individual the blood-sugar level is _____ that of a diabetic.

   (a) the same as   (b) lower than   (c) slightly higher than   (d) much greater than

9. If a glucose tolerance test shows an abnormally high level of blood glucose, the person is probably

   (a) anemic   (b) anorexic   (c) hypoglycemic   (d) none of these

10. A person whose blood-sugar level is 40 mg/100mL is

    (a) a diabetic   (b) hyperglycemic   (c) hypoglycemic   (d) normal

11. Insulin promotes the entry of _____ into most types of cells.

    (a) fructose   (b) glucose   (c) glycogen   (d) sucrose

12. Insulin enhances

(a) gluconeogenesis   (b) glycogenesis   (c) glycogenolysis   (d) none of these

13. A person with _____ diabetes requires daily injections of insulin.

(a) Type I   (b) Type II   (c) adult-onset   (d) all of these

14. Insulin cannot be taken orally because it

(a) is toxic.
(b) has a bitter, unpleasant taste.
(c) is a protein and would be broken down by digestive enzymes.
(d) would be absorbed from the stomach rather than the intestine.

15. An overdose of insulin produces

(a) high blood pressure   (b) juvenile-onset diabetes
(c) hyperglycemia   (d) hypoglycemia

16. Which of the following is not secreted by the pancreas?

(a) epinephrine   (b) glucagon   (c) insulin   (d) All are secreted.

17. Which hormone is secreted in response to hypoglycemia?

(a) glucagon   (b) FHS   (c) LH   (d) vasopressin

18. Which of the following acts to lower the blood-sugar level?

(a) epinephrine   (b) glucagon   (c) insulin   (d) norepinephrine

19. The conversion of glucose to pyruvic acid and then to lactic acid is called

(a) glycogenesis   (b) glycogenolysis   (c) glycolysis   (d) gluconeogenesis

20. Anaerobic cells do not require

(a) nutrients   (b) oxygen   (c) water   (d) metal ions

21. In the first step in glycolysis, glucose is converted into

(a) glucose 1-phosphate   (c) fructose 1-phosphate
(b) glucose 6-phosphate   (d) fructose 6-phosphate

22. When *phosphorylase* catalyzes the cleavage of a glucose unit from a glycogen molecule, the first product is

(a) fructose   (b) glucose 1-phosphate   (c) glucose 6-phosphate   (d) none of these

23. A product of fermentation is

(a) acetyl-CoA   (b) carbon dioxide
(c) glucose   (d) water

24. The number of moles of pyruvic acid formed from one mole of glucose is _____.

(a) 1   (b) 2   (c) 3   (d) 4

25. The enzyme that catalyzes the reaction in which a six-carbon compound is split into two three-carbon compounds is

    (a) aldolase   (b) enolase   (c) kinase   (d) dehydrogenase

26. When lactic acid is utilized as the starting material for the biosynthesis of glucose in the liver, the process is termed

    (a) anaerobic glycolysis   (b) cellular respiration
    (c) gluconeogenesis   (d) glycogenesis

27. During strenuous exercise, muscle cells respire anaerobically. In doing so they

    (a) incur an oxygen debt.
    (b) operate more efficiently.
    (c) produce CO_2 and H_2O.
    (d) produce creatine phosphate.

28. During high muscle activity, ATP is formed in muscle cells by the reaction of ADP and

    (a) AMP   (b) creatine
    (c) creatine phosphate   (d) inorganic phosphate

29. The main entry of carbohydrates into the Krebs cycle is by means of

    (a) acetyl-CoA   (b) glucose 6-phosphate   (c) lactic acid   (d) pyruvic acid

30. The first intermediate formed in the Krebs cycle is

    (a) acetyl-CoA   (b) citric acid
    (c) isocitric acid   (d) oxaloacetic acid

31. Which of the following compounds is not an intermediate formed within the Krebs cycle?

    (a) fumaric acid   (b) malic acid   (c) pyruvic acid   (d) succinic acid

32. Which of the following metabolites is oxidized by FAD?

    (a) fumaric acid   (b) isocitric acid
    (c) α-ketoglutaric acid   (d) succinic acid

33. Which oxidizing agent is used in both the Embden–Meyerhof pathway and the Krebs cycle?

    (a) coenzyme A   (b) FAD   (c) NAD^+ (d) O_2

34. Which of the following conversions requires the oxidation of a secondary alcohol to a ketone?

    (a) acetaldyhyde to ethanol   (b) pyruvic acid to lactic acid
    (c) pyruvic acid to acetyl-CoA   (d) malic acid to oxaloacetic acid

35. The conversion of fumaric acid to malic acid is an example of a(n) _____ reaction.

    (a) hydration   (b) hydrolysis   (c) oxidation   (d) reduction

36. Both electron transport and oxidative phosphorylation occur in the

    (a) cytosol   (b) mitochondria   (c) nucleus   (d) ribosome

37. Which of the following occurs in the cytosol?

    (a) conversion of FAD to $FADH_2$ (b) conversion of GDP to GTP

(c) conversion of NAD^+ to NADH

(d) conversion of H_2O to O_2

38. Which of the following initials represents the reduced form of a coenzyme?

(a) FADH   (b) FMNH   (c) NADH   (d) Fe·S

39. _____ are strictly electron carriers in the respiratory chain.

(a) cytochromes   (b) flavins   (c) ubiquinones   (d) all of these

40. In the iron–sulfur proteins, the iron

(a) is always in the +2 oxidation state.
(b) is always in the +3 oxidation state.
(c) can exist either as Fe(II) or Fe(III).
(d) is not involved in electron transport.

41. The coenzyme that contains a quinone group is

(a) cyt aa_3 (b) CoQ   (c) NAD^+ (d) FMN

42. Which is not true of the cytochromes?

(a) They are inhibited by carbon monoxide.
(b) They are heme-containing compounds.
(c) Their iron atoms can exist as either Fe(II) or Fe(III).
(d) They have different reduction potentials.

43. In which metabolic sequence is H_2O produced?

(a) electron transport   (b) fermentation
(c) glycolysis   (d) Krebs cycle

44. The name given to the ATP-synthesizing reactions that occur in the mitochondria is

(a) oxidative phosphorylation   (b) photophosphorylation
(c) substrate-level phosphorylation   (d) all of these

45. The oxidation of each mole of NADH in the respiratory chain is coupled with the phosphorylation of _____ moles of ATP.

(a) 2   (b) 3   (c) 4   (d) 6

46. The oxidation of each mole of $FMNH_2$ in the respiratory chain is coupled with the phosphorylation of _____ moles of ATP.

(a) 2   (b) 3   (c) 4   (d) 6

47. A total of _____ moles of ATP are produced for each mole of glucose that is metabolized in skeletal muscle cells.

(a) 15   (b) 18   (c) 36   (d) 38

48. Which of the following statements is generally true?

(a) Anabolic reactions involve oxidation reactions.
(b) Anabolic reactions are exothermic.
(c) Catabolic reactions are exothermic.
(d) Catabolic reactions are endothermic.

49. An example of an energy-releasing biochemical pathway is

   (a) gluconeogenesis   (b) glycogenesis   (c) the Cori cycle   (d) the Krebs cycle

50. The efficiency of energy conversion in most body cells is about _____ percent.

   (a) 15   (b) 40   (c) 60   (d) 75

FILL IN THE BLANKS

51. The series of reactions by which solar radiation is transferred into molecules of glucose is termed _____.

52. Glycogen in liver cells serves as a reservoir for _____.

53. The blood-sugar level is controlled by the hormones _____, _____, and _____.

54. What is the range of the normal blood-sugar level? _____

55. _____ occurs when the blood-sugar level is below 40 mg/100 mL.

56. The first step in the utilization of glycogen to provide energy is catalyzed by the enzyme _____.

57. The conversion of glucose to ethanol is called _____.

58. The general name of the enzyme that catalyzes the reaction between ATP and glucose to produce glucose 6-phosphate is _____.

59. The cleavage of fructose 1,6-diphosphate is catalyzed by the enzyme _____.

60. The conversion of dihydroxyacetone phosphate to glyceraldehyde 3-phosphate is catalyzed by an enzyme whose general name is _____.

61. Fructose 1,6-diphosphate is converted into _____ and _____.

62. The conversion 3-phosphoglyceric acid to 2-phosphoglyceric acid is catalyzed by an enzyme whose general name is _____.

63. The dehydration of 2-phosphoglyceric acid is catalyzed by the enzyme _____.

64. An enzyme whose general name is dehydrogenase catalyzes a(n) _____ reaction.

65. The end-product of glycolysis in muscle cells is _____.

66. The formation of ethanol and carbon dioxide from glucose occurs under _____ conditions.

67. Pyruvic acid is one of the "crossroads" compounds in metabolism. Depending upon the type of cell and the amount of oxygen available, pyruvic acid can be converted into _____, _____, or _____.

68. Anaerobic glycolysis is another name for the _____ pathway.

69. _____ is the term for the formation of glycogen from glucose.

70. The soreness and fatigue of muscle cells that are undergoing strenuous exercise are caused by the accumulation of _____.

71. _____ serves as the storage form of phosphate groups in muscle cells.

72. How many high-energy phosphate bonds are there in a molecule of ATP? _____

73. Two other names for the Krebs cycle are _____ and _____.

74. The conversion of pyruvic acid to acetyl-CoA occurs in the _____.

75. To be incorporated into the Krebs cycle, acetyl-CoA combines with oxaloacetic acid to form _____ acid.

76. The initials that represent the reduced form of NAD^+ are _____.

77. The hydration of fumaric acid yields _____ acid.

78. What type of reaction occurs when malic acid is converted into oxaloacetic acid? _____

79. The majority of ATP is produced in the mitochondria by a process referred to as _____ phosphorylation.

80. What is the name of the shuttle that operates in liver cells to transport cytosolic NADH into the mitochondria? _____

TRUE OR FALSE

81. T   F   Anabolic reactions are degradative reactions.

82. T   F   Catabolic pathways are reductive processes.

83. T   F   Villi are finger-like projections that line the digestive tract.

84. T   F   Glycogen is referred to as blood sugar.

85. T   F   When glucose appears in the urine, the renal threshold has been exceeded.

86. T   F   The blood-sugar level in a hyperglycemic individual is higher than the normal fasting level.

87. T   F   Insulin decreases the blood-sugar level by enhancing glycogenesis.

88. T   F   Oral administration of insulin is effective for most Type I diabetics.

89. T   F   Glucose is stored in the liver in the form of glycogen.

90. T F   A diabetic has a high level of blood glucose, yet experiences hunger and weight loss.

91. T F   Gluconeogenesis is a catabolic process.

92. T F   The conversion of glycogen to glucose is an anabolic process.

93. T F   In glycogenolysis, glucose is converted into pyruvic acid.

94. T F   During periods of normal activity, muscle cells respire aerobically.

95. T F   In glycolysis, lactic acid is converted into glucose.

96. T F   McArdle's disease and von Gierke's disease occur as a result of abnormalities in carbohydrate metabolism.

97. T F   Catabolic reactions and anabolic reactions occur at the same time within living cells.

98. T F   When glucose is metabolized in yeast cells, ethanol and ATP are produced.

99. T F   The hydrolysis of ADP to ATP releases energy.

100. T F   The ATP yield of anaerobic respiration is about the same as the ATP yield of aerobic respiration.

101. T F   Some ATP is formed during glycolysis.

102. T F   The NADH formed in glycolysis is transferred directly to the respiratory chain.

103. T F   A person's energy requirements could be met from stored glycogen for approximately one week.

104. T F   The enzyme lactic acid dehydrogenase catalyzes both the formation of lactic acid from pyruvic acid, and pyruvic acid from lactic acid.

105. T F   The majority of the lactic acid produced in the body is eventually converted into ethanol and carbon dioxide.

106. T F   Muscle cells that respire anaerobically incur an oxygen debt.

107. T F   Creatine phosphate is being converted into creatine during periods of rest.

108. T F   Creatine phosphate is a storage form of energy in the liver.

109. T F   The high-energy bond in creatine phosphate is formed during photosynthetic phosphorylation.

110. T F   Pyruvic acid enters the Krebs cycle as lactic acid.

111. T F   The Krebs cycle occurs in the mitochondria.

112. T F   Cellular oxidation is primarily a two-step process.

113. T F   The Krebs cycle is an anaerobic process.

114. T F   Cyanide is a direct inhibitor of the Krebs cycle.

115. T F NAD^+ is a coenzyme that accepts electrons in both glycolysis and the Krebs cycle.

116. T F After transporting electrons to the respiratory chain, the coenzymes FMN and CoQ are reused in the Krebs cycle.

117. T F The cytochromes are involved in electron transport.

118. T F ATP synthesis occurs primarily in the Krebs cycle.

119. T F Every step that occurs in the Krebs cycle involves the release of either CO_2 or energy.

120. T F Mitochondria are called the "power plants" of the cell because they are the major sites of ATP synthesis.

121. T F ATP synthesis is said to be tightly coupled to oxidative phosphorylation because one cannot take place without the other.

122. T F When most animal cells are deprived of oxygen, oxidative phosphorylation ceases and the cell dies.

123. T F The combustion of fuels is similar to cellular oxidation because both processes produce CO_2 and H_2O and energy.

CHAPTER 17

Lipid Metabolism

MULTIPLE CHOICE

1. The oxidation of 1 gram of a lipid releases about _____ kilocalories.

   (a) 1   (b) 4   (c) 7   (d) 9

2. Which of the following is an essential fatty acid?

   (a) arachidonic acid   (b) linoleic acid   (c) oleic acid   (d) palmitic acid

3. Fat deposits

   (a) are good insulators.   (b) cushion the vital organs.
   (c) serve as storage forms of energy.   (d) all of these

4. Most of the stored triacylglycerols are found in the

   (a) adipose tissue   (b) kidney   (c) liver   (d) muscles

5. Most often, obesity is due to

   (a) eating more food than the body requires   (b) glandular malfunctions
   (c) lipid storage diseases   (d) all of these

6. The lymphatic system transports lipids from the

   (a) blood to the tissues   (b) intestine to the blood
   (c) intestine to the tissues   (d) none of these

7. Fat is stored in adipose tissue primarily as

   (a) free triacylglycerols   (b) lipids bound to proteins
   (c) a bilayer structure   (d) all of these

8. Bile functions to

   (a) emulsify lipids   (b) hydrolyze fats   (c) hydrolyze lipids   (d) lubricate the food

9. Triacylglycerols are transported in the blood bound to

   (a) cholesterol esters   (b) glycolipids   (c) phospholipids   (d) proteins

10. The recommended daily allowance of caloric intake, for the average individual, is about

    (a) 1500 Cal   (b) 2500 Cal   (c) 3500 Cal   (d) 4500 Cal

11. A hormone that stimulates the mobilization of fats to provide energy is

    (a) epinephrine   (b) insulin   (c) oxytocin   (d) vasopressin

12. The fatty acids obtained from the hydrolysis of triacylglycerols in adipose tissue are transported through the bloodstream

    (a) bound to serum albumins   (b) bound to cholesterol
    (c) as free fatty acids   (d) as glycolipids

13. The major fuel reserve for the immediate needs of active skeletal muscles is

    (a) fatty acids   (b) glycogen   (c) ketone bodies   (d) triacylglycerols

14. The mobilization of fatty acids from adipose tissue requires a(n) _____ reaction.

    (a) dehydration   (b) hydrolysis   (c) oxidation   (d) reduction

15. The intracellular messenger that activates the mobilization of fatty acids in adipose tissue is

    (a) ADP   (b) AMP   (c) cyclic AMP   (d) adenosine

16. In order to enter the fatty acid spiral, a fatty acid must first be activated by its conversion to

    (a) acetyl-CoA   (b) coenzyme A   (c) fatty acyl phosphate   (d) fatty acyl-CoA

17. The major portion of fatty acid oxidation occurs in the

    (a) cytosol   (b) mitochondria   (c) nucleus   (d) endoplasmic reticulum

18. When lipids are metabolized, fatty acids enter the reactions of the

    (a) citric acid cycle   (b) electron transport chain
    (c) glycolytic pathway   (d) β-oxidation sequence

19. The oxidizing agent in the first step of the fatty acid spiral is

    (a) ATP   (b) coenzyme A   (c) FAD   (d) NAD$^+$

20. In one step of the fatty acid spiral, an α-unsaturated compound is converted to a β-hydroxy acid. This is an example of a(n) _____ reaction.

    (a) hydration   (b) hydrolysis   (c) oxidation   (d) reduction

21. In one step of the fatty acid spiral, a β-hydroxy acid is converted into a β-keto acid. This is an example of a(n) _____ reaction.

    (a) hydration   (b) hydrolysis   (c) oxidation   (d) reduction

22. In the first oxidation step of the fatty acid spiral

    (a) an aldehyde is oxidized to an acid.
    (b) an alkane is oxidized to an alkene.
    (c) an alcohol is oxidized to an acid.
    (d) an alcohol is oxidized to a ketone.

23. Which of the following is the end product obtained from the fatty acid spiral?

    (a) $RCCH_2C$ (with O groups, SCoA)   (b) $RCHOHCH_2C$ (with O group, SCoA)   (c) $RCH=CHC$ (with O group, SCoA)   (d) none of these

24. Which of the following is not a reactant in the fatty acid spiral?

    (a) CoASH   (b) H_2O (c) NAD^+ (d) O_2

25. How many steps are there in each spiral of the β-oxidation sequence?

    (a) 3   (b) 4   (c) 5   (d) 6

26. How many carbon atoms are removed from a fatty acid each turn around the spiral?

    (a) 1   (b) 2   (c) 3   (d) 4

27. A twelve-carbon fatty acid would require _____ trips through the fatty acid spiral to be completely metabolized.

    (a) 4   (b) 5   (c) 6   (d) 12

28. The NADH formed in the fatty acid spiral delivers two electrons to the

    (a) glycolytic pathway   (b) Krebs cycle   (c) lipogenesis cycle   (d) respiratory chain

29. The complete degradation of a 16-carbon fatty acid via the fatty acid spiral requires _____ turns of the spiral and yields _____ molecules of acetyl-CoA.

    (a) 8; 8   (b) 8; 7   (c) 7; 8   (d) 9; 8

30. After three turns around the fatty acid spiral, stearic acid is converted to a _____ carbon fatty acyl-CoA.

    (a) 6   (b) 9   (c) 12   (d) 15

31. Each molecule of acetyl-CoA that enters the Krebs cycle furnishes sufficient energy to synthesize _____ molecules of ATP.

    (a) 2   (b) 3   (c) 12   (d) 36

32. How many molecules of ATP can be synthesized when six molecules of NADH and six molecules of $FADH_2$ enter the respiratory chain?

    (a) 12   (b) 24   (c) 30   (d) 36

33. What is the net yield of ATP from the complete oxidation of 1 mole of butyric acid?

    (a) 10 moles   (b) 27 moles   (c) 30 moles   (d) 32 moles

34. A majority of the energy produced from the catabolism of fatty acids is obtained

    (a) directly from the Krebs cycle.
    (b) directly from the fatty acid spiral.
    (c) from the regeneration of coenzymes in the respiratory chain.
    (d) none of these

35. What is the percentage of energy that is conserved by fatty acid metabolism?

    (a) 25%   (b) 40%   (c) 60%   (d) 75%

36. Which does not occur in the mitochondria?

    (a) β-oxidation   (b) fatty acid synthesis
    (c) the Krebs cycle   (d) oxidative phosphorylation

37. In humans,

    (a) fatty acids provide little energy.   (b) acetyl-CoA is easily converted to glycogen.
    (c) acetyl-CoA is not converted to glucose.   (d) brain cells use fatty acids as their major energy source.

38. Which of the following, if eaten in excessive amounts, could be converted to stored fat?

    (a) fructose   (b) starch   (c) sucrose   (d) all of these

39. The accumulation of ketone bodies in the blood and tissues is called

    (a) acidosis   (b) alkalosis   (c) ketonuria   (d) ketosis

40. High concentrations of ketone bodies in the blood

    (a) cause diabetes mellitus   (b) are a symptom of starvation
    (c) increase the pH of the blood   (d) none of these

41. The starting material for the synthesis of ketone bodies is

    (a) acetyl-CoA   (b) acetone   (c) carbon dioxide   (d) lipoic acid

42. Ketosis results from

    (a) consumption of too much sugar   (b) overproduction of acetyl-CoA
    (c) underproduction of acetyl-CoA   (d) underproduction of ketone bodies

43. Which of the following is not one of the ketone bodies?

    (a) acetoacetic acid   (b) acetone
    (c) butanone   (d) β-hydroxybutyric acid

44. The following often occurs in a diabetic.

    (a) The kidneys discharge large quantities of water.
    (b) There is an impairment of oxygen transport.
    (c) Ketone bodies appear in the urine.
    (d) all of these

45. Acidosis causes

    (a) hypoglycemia
    (b) increased fatty acid metabolism
    (c) storage of fats in adipose tissue
    (d) none of these

46. An untreated diabetic could have the following symptom.

    (a) acetone breath   (b) glucosuria   (c) ketonemia   (d) all of these

47. When fats are metabolized to produce energy, glycerol enters the

    (a) β-oxidation sequence   (b) electron-transport chain
    (c) glycolytic pathway   (d) Krebs cycle

48. The glycerol released from the hydrolysis of fatty acids in adipose tissue eventually enters the Embden–Meyerhof pathway as

    (a) acetyl-CoA   (b) dihydroxyacetone phosphate
    (c) glyceraldehyde 3-phosphate   (d) pyruvic acid

FILL IN THE BLANKS

49. _____ are enzymes that catalyze the hydrolysis of triacylglycerols.

50. _____ are required for the emulsification of dietary lipids.

51. _____ are absorbed primarily into the lymphatic system rather than the bloodstream.

52. Triacylglycerols are stored in _____ tissue.

53. The release of fatty acids from adipose tissue is called _____.

54. Lipids are transported in the blood, bound to _____.

55. _____ cells can not use fatty acids to provide energy.

56. When fatty acids enter a cell to be metabolized, they are first converted to _____.

57. The four types of reactions that take place in successive steps in the fatty acid spiral are _____, _____, _____, and _____.

58. The coenzymes that are reduced in the fatty acid spiral are reoxidized by the _____ chain.

59. The acetyl-CoA that is produced in the fatty acid spiral is further broken down by the _____.

60. The complete oxidation of a molecule of a 14-carbon fatty acid yields _____ molecules of acetyl-CoA.

61. The fatty acid spiral occurs in the _____, whereas the synthesis of fatty acids takes place in the _____.

62. A key metabolite in the metabolism of carbohydrates and fats is _____.

63. Fatty acids, ketone bodies, phospholipids, cholesterol, and other steroids are all synthesized from molecules of _____.

64. The accumulation of excess acetyl-CoA molecules leads to the formation of _____.

65. During starvation, ketone bodies are used as a primary energy source for the _____.

66. The odor of _____ is often apparent on the breath of diabetics.

67. The ketone body that is not a ketone is _____.

TRUE OR FALSE

68. T   F   Fats have a caloric value more than twice that of carbohydrates.

69. T   F   Glycogen is the most efficient form of stored energy in the body.

70. T   F   Triacylglycerols are stored in specialized fat cells.

71. T   F   Glucagon stimulates the hydrolysis of triacylglycerols in adipose tissue.

72. T   F   Stored lipids are mobilized faster than stored carbohydrates.

73. T   F   Most obesity is due to glandular malfunction.

74. T   F   Fats are transported in the blood primarily as lipoproteins.

75. T   F   The concentration of lipids in the blood remains constant, even immediately following a meal.

76. T   F   To be useful as a fuel, all nutrients must be converted into glucose or pyruvic acid.

77. T   F   The oxidation of one mole of palmitic acid via the fatty acid spiral produces 8 moles of $FADH_2$ and 8 moles of NADH.

78. T   F   Physical exercise causes the mobilization of fatty acids.

79. T   F   Once fatty acids enter the liver, the products of their metabolism are only used to supply energy to the organism.

80. T   F   ATP is used as a source of energy in the first steps of both the catabolism of fatty acids and glucose.

81. T   F   A 12-carbon fatty acid yields more energy than a disaccharide.

82. T   F   Unsaturated fatty acids can not be metabolized.

83. T   F   Both NADH and $FADH_2$ transport electrons and hydrogen from the fatty acid spiral to the respiratory chain.

84. T   F   If the demand for ATP is low, and excess carbohydrate is ingested, the body responds by making and storing fat molecules.

85. T   F   Fatty acid synthesis is carried out by the same enzymes that are involved in β-oxidation.

86. T   F   Glucose metabolism has no effect on the formation of ketone bodies.

87. T   F   The presence of some ketone bodies in the blood leads to an increase in the pH of the blood.

88. T   F   Starvation can lead to ketosis.

89. T   F   Diabetes mellitus results from an abnormality of carbohydrate digestion.

90. T   F   In a diabetic, the metabolism of fatty acids occurs to a much greater extent than in a normal individual.

91. T   F   β-Hydroxybutyric acid is one of the ketone bodies.

92. T   F   In acidosis, the ability of the blood to transport oxygen decreases.

93. T   F   The glycerol resulting from lipid hydrolysis can be converted into acetyl-CoA.

CHAPTER 18

Protein Metabolism

MULTIPLE CHOICE

1. Enzymes that hydrolyze proteins are called

    (a) catalases   (b) oxidases   (c) peptidases   (d) transaminases

2. The end products of protein digestion are

    (a) amino acids   (b) denatured proteins   (c) peptides   (d) proteoses

3. Which is not secreted in the pancreatic juice?

    (a) aminopeptidase   (b) chymotrypsinogen   (c) procarboxypeptidasec   (d) trypsinogen

4. Which of the following types of proteins tends to have the longest turnover rate?

    (a) enzymes   (b) connective tissue (collagen)   (c) liver proteins   (d) muscle proteins

5. _____ is characterized by the ingestion of more nitrogen than the amount of nitrogen excreted.

    (a) nitrogen balance
    (c) positive nitrogen balance
    (b) negative nitrogen balance
    (d) all of these

6. Which of the following is a source of a complete protein?

    (a) corn   (b) milk   (c) wheat   (d) none of these

7. Essential amino acids are those that

    (a) are required for the amino acid pool.
    (b) are not required for tissue proteins.
    (c) can be synthesized in the body.
    (d) must be supplied in the diet.

8. The disease kwashiorkor results from a diet

    (a) high in proteins   (b) lacking in fats   (c) lacking in proteins   (d) lacking in vitamins

9. Which of the following is an essential amino acid?

    (a) alanine   (b) glycine   (c) lysine   (d) serine

10. A complete protein is

    (a) better tasting
    (c) rich in the essential amino acids
    (b) especially high in calories
    (d) usually a plant protein

238

11. Proteins in our diet can be converted to

    (a) carbohydrates   (b) fats   (c) tissue protein   (d) all of these

12. During starvation, tissue proteins are broken down to supply amino acids for the synthesis of

    (a) enzymes   (b) glucose   (c) lipids   (d) new proteins

13. Which of the following can be formed from amino acids of the amino acid pool?

    (a) creatine   (b) heme   (c) nucleic acids   (d) all of these

14. Amino acids are metabolized by _____ reactions.

    (a) hydration   (b) isomerization   (c) reduction   (d) transamination

15. _____ is the only purely ketogenic amino acid.

    (a) arginine   (b) histidine   (c) leucine   (d) lysine

16. Which of the following is formed in one step by the transamination of the appropriate amino acid?

    (a) α-ketoglutaric acid   (b) oxaloacetic acid   (c) pyruvic acid   (d) all of these

17. An individual suffering from a liver disease would be likely to have high serum levels of

    (a) glutamic-oxaloacetic transaminase (GOT)   (b) glutamic-pyruvic transaminase (GPT)
    (c) creatine kinase (CK)   (d) all of these

18. The coenzyme required for all transamination reactions is

    (a) FAD   (b) NAD^+ (c) pyridoxal phosphate   (d) thiamine pyrophosphate

19. The α-keto acid that most frequently serves as the amino group acceptor in transamination reactions is

    (a) α-ketoglutaric acid   (b) oxaloacetic acid   (c) pyruvic acid   (d) oxalosuccinic acid

20. The enzyme GOT catalyzes the transamination reaction involving the reactants α-ketoglutaric acid and

    (a) alanine   (b) aspartic acid   (c) glutamic acid   (d) pyruvic acid

21. Glutamic acid is formed in the transamination reaction from

    (a) aspartic acid   (b) α-ketoglutaric acid   (c) oxaloacetic acid   (d) pyruvic acid

22. The product of a transamination reaction involving alanine is

    (a) $CH_3CH(NH_2)COOH$ (b) H_2NCH_2COOH (c) CH_3CH_2COOH (d) $CH_3COCOOH$

23. Which pair of compounds is most closely related?

    (a) alanine and pyruvic acid   (b) aspartic acid and α-ketoglutaric acid
    (c) glutamic acid and oxaloacetic acid (d) all of these

24. Ammonia is released from _____ during oxidative deamination.

    (a) aspartic acid   (b) glutamic acid   (c) histidine   (d) ornithine

25. The oxidizing agent in the oxidative deamination reaction is

    (a) FAD   (b) NAD^+ (c) O_2 (d) pyridoxal phosphate

26. The two compounds that play the major roles in the conversion of the α-amino group to ammonia are

    (a) aspartic acid and pyruvic acid
    (b) glutamic acid and γ-aminobutyric acid
    (c) glutamic acid and α-ketoglutaric acid
    (d) urea and carbon dioxide

27. Which pair of compounds is related via a decarboxylation reaction?

    (a) 5-hydroxytryptophan and GABA
    (b) glutamic acid and dopamine
    (c) 3,4-dihydroxyphenylalanine and serotonin
    (d) none of these

28. Inorganic nitrogen is usually found in the body in the form of

    (a) N_2 (b) NO_3^- (c) NH_3 (d) urea

29. When nucleic acids are metabolized, the purine bases are converted to

    (a) glycine   (b) γ-aminobutyric acid   (c) urea   (d) uric acid

30. Ammonia is converted into urea by the _____ cycle.

    (a) Cori   (b) Krebs   (c) urea   (d) all of these

31. The nitrogen of amino acids appears in the urine of mammals as

    (a) ammonia   (b) N_2 (c) urea   (d) uric acid

32. A connecting link between carbohydrate, lipid, and protein metabolism is

    (a) malonyl-CoA   (b) glycolysis   (c) the Krebs cycle   (d) pyruvic acid

FILL IN THE BLANKS

33. _____ is a zymogen that is found in gastric juice.

34. The major portion of protein digestion occurs in the _____.

35. _____ and _____ are two of the zymogens found in pancreatic juice.

36. _____ is the enzyme that converts trypsinogen to trypsin.

37. _____ and _____ are exopeptidases.

38. Many plants are deficient in the amino acid _____.

39. Growing children are in a state of _____ nitrogen balance.

40. Which amino acid is converted to oxaloacetic acid through transamination? _____

41. Which α-keto acid is formed from alanine through transamination? _____

42. When glutamic acid loses ammonia via oxidative deamination, _____ is formed.

43. The decarboxylation of histidine yields _____.

44. Some nitrogen is stored in the body in the compound _____.

45. Most of the nitrogen eliminated from the body is in the form of _____.

46. Gout is characterized by high serum levels of _____.

47. The formula for urea is _____.

48. Complete the following equation.

$$\underset{N \diagdown NH}{\diagup CH_2CH(NH_2)COOH} \quad \xrightarrow{\text{decarboxylase}} \quad \underline{\hspace{2cm}}$$

49. Complete the following equation.

$$CH_3COCOOH + HOOCCH_2CH_2CH(NH_2)COOH \xrightarrow{\text{transaminase}} \underline{\hspace{1.5cm}} + \underline{\hspace{1.5cm}}$$

50. Complete the following equation.

$$HOOCCH_2CH_2CH(NH_2)COOH + H_2O + NAD^+ \xrightarrow{\text{dehydrogenase}} \underline{\hspace{1.5cm}} + \underline{\hspace{1.5cm}}$$

TRUE OR FALSE

51. T   F   Proteins are the chief dietary sources of nitrogen.

52. T   F   Digestion of proteins begins in the stomach.

53. T   F   Aminopeptidases are endopeptidases.

54. T   F   Humans can synthesize several of the essential amino acids.

55. T   F   The essential amino acids are not present in the amino acid pool.

56. T   F   Essential amino acids are required for children but not for adults.

57. T   F   The essential amino acids are synthesized by transamination reactions.

58. T   F   The half-life of collagen is relatively short.

59. T   F   Hair protein has no half-life. Only synthesis occurs.

60. T   F   Negative nitrogen balance occurs when the intake of nitrogen in food is greater than the loss of nitrogen by excretion.

61. T   F   Phenylalanine is an essential amino acid.

62. T   F   The biosynthesis of muscle protein from amino acids is an anabolic process.

63. T   F   If a protein contains only 16 of the 20 amino acids, it must be an incomplete protein.

64. T   F   A normal healthy adult should on the average excrete more nitrogen than is ingested.

65. T   F   The amino acid pool can be supplied by the breakdown of tissue protein.

66. T   F   An elevation of serum glutamic-oxaloacetic transaminase (GOT) is indicative of liver damage.

67. T   F   The carbon skeletons of glucogenic amino acids are used for gluconeogenesis.

68. T   F   The biosynthesis of aspartic acid and glutamic acid is accomplished by transamination of the corresponding α-keto acids.

69. T   F   When glutamic acid undergoes oxidative deamination to α-ketoglutaric acid, the other product is urea.

70. T   F   A transamination reaction results in the formation of uric acid.

71. T   F   Amino acids are stored in depots analogous to the storage of fats.

72. T   F   Pyruvic acid and oxaloacetic acid are α-keto acids.

73. T   F   The α-keto acids formed during the conversion of amino acids to carbohydrate are referred to as ketone bodies.

74. T   F   Transaminases catalyze all transamination reactions.

75. T   F   The reaction that transfers the α-amino group from amino acids to α-ketoglutaric acid is called an acetylation reaction.

76. T   F   A starving person would be expected to excrete abnormally low amounts of urea in the urine.

77. T   F   During periods of starvation, amino acids are utilized to provide energy for the body.

78. T   F   All animals eliminate nitrogen from the body in the form of uric acid.

79. T   F   The Lesch–Nyhan syndrome is the result of a defect in pyrimidine metabolism.

80. T   F   Glycogen may be converted into amino acids.

81. T   F   Metabolic pathways exist for the conversion of proteins into stored fat deposits.

82. T   F   Carbohydrate, lipid, and protein metabolism are closely interconnected.

CHAPTER 19

The Blood

MULTIPLE CHOICE

1. The blood helps to

    (a) control the pH of the body.
    (b) protect the organism against infection.
    (c) regulate the body temperature.
    (d) all of these

2. Which of the following is not found in blood plasma?

    (a) enzymes   (b) hemoglobin   (c) hormones   (d) vitamins

3. Which of the following is a formed element found in the blood?

    (a) erythrocytes   (b) leukocytes   (c) thrombocytes   (d) all of these

4. Which is not a plasma protein?

    (a) fibrin   (b) α-globulin   (c) β-globulin   (d) γ-globulin

5. Which of the following is a function of the albumins?

    (a) containing the antibodies   (b) role in blood clotting
    (c) maintenance of the osmotic pressure   (d) all of these

6. Anemia may result from

    (a) a decreased rate of production of erythrocytes.
    (b) an increased destruction of erythrocytes.
    (c) an increased loss of erythrocytes.
    (d) all of these

7. Erythrocytes contain

    (a) hemoglobin   (b) mitochondria   (c) a nucleus   (d) all of these

8. Erythrocytes are formed in the

    (a) bone marrow   (b) liver   (c) spleen   (d) none of these

9. Which of the following is not a variety of leukocytes?

    (a) lymphocytes   (b) phagocytes   (c) macrophages   (d) thrombocytes

10. A higher than normal leukocyte count occurs during

    (a) appendicitis   (b) chicken pox   (c) measles   (d) polio

11. Which is the most abundant type of immunoglobin?

    (a) IgA   (b) IgG   (c) IgM   (d) none of these

12. If a red blood cell is placed in a(n) _____ solution, it will swell and eventually burst.

    (a) hypertonic   (b) hypotonic   (c) isotonic   (d) none of these

13. Blood pressure can be reduced by administering

    (a) β-blockers   (b) diuretics   (c) vasodilators   (d) all of these

14. Which substance does not occur in blood serum?

    (a) antibodies   (b) calcium ions   (c) fibrinogens   (d) lipoproteins

15. In the final step of the clotting mechanism,

    (a) autoprothrombin III is converted to autoprothrombin C.
    (b) fibrinogen is converted to fibrin.
    (c) prothrombin is converted to thrombin.
    (d) Vitamin K is converted to dicumarol.

16. _____ is an enzyme that catalyzes the formation of fibrin.

    (a) coumadin   (b) prothrombin   (c) thrombin   (d) thromboplastin

17. Which of the following is not an "anticoagulant"?

    (a) dicumarol   (b) heparin   (c) sodium citrate   (d) Vitamin K

18. The heme portion of hemoglobin is degraded to

    (a) amino acids   (b) bilirubin   (c) prosthetic groups   (d) uric acid

19. Methemoglobin differs from hemoglobin in that its heme

    (a) has no iron.
    (b) has Fe^{3+} instead of Fe^{2+}.
    (c) is bound to CO.
    (d) is bound to ferritin.

20. Nitrites interfere with oxygen transport by converting Fe^{2+} of hemoglobin to

    (a) Fe   (b) Fe^+ (c) Fe^{3+} (d) all of these

21. Carbon monoxide is a poison because it

    (a) forms a complex with calcium ions.
    (b) has a greater affinity for hemoglobin than oxygen.
    (c) inhibits the enzyme carbonic anhydrase.
    (d) reacts with oxygen to form carbon dioxide.

22. Which of the following is the form of hemoglobin in which CO_2 is bound to hemoglobin?

    (a) carbaminohemoglobin   (b) carboxyhemoglobin   (c) methemoglobin   (d) oxyhemoglobin

23. Jaundice can arise from

    (a) an acceleration of erythrocyte destruction.

(b) infectious hepatitis.

(c) the obstruction of bile ducts by gallstones.

(d) all of the above

24. Most of the carbon dioxide is transported in the blood chiefly

(a) bound to hemoglobin

(b) as carbonate, CO_3^{2-}

(c) as carbonic acid, H_2CO_3

(d) as bicarbonate, HCO_3^-

25. Oxygen is transported in the blood chiefly as molecules of

(a) HCO_3^- (b) H_2O (c) dissolved O_2 (d) oxyhemoglobin

26. Bicarbonate ions that migrate from the erythrocytes into the plasma are replaced by

(a) Cl^- (b) CO_3^{2-} (c) PO_4^{3-} (d) Na^+ and K^+

27. At the lungs, the chloride shift involves the flow of Cl^- from

(a) erythrocyte to plasma

(b) interstitial fluid to plasma

(c) plasma to erythrocyte

(d) plasma to interstitial fluid

28. The reaction $H^+ + HCO_3^- \longrightarrow H_2CO_3$ occurs at

(a) the lungs   (b) the tissue cells   (c) both (a) and (b)   (d) neither (a) nor (b)

29. Which of the following reactions does not occur in erythrocytes at the lungs?

(a) $HHb + O_2 \longrightarrow HbO_2^- + H^+$

(b) $HHb-CO_2 \longrightarrow HHb + CO_2$

(c) $H_2CO_3 \longrightarrow H^+ + HCO_3^-$

(d) $H_2CO_3 \longrightarrow H_2O + CO_2$

30. Blood has a pH of

(a) 6.4   (b) 7.0   (c) 7.4   (d) 8.0

31. If the pH of the blood decreases below normal, the condition is termed

(a) acidosis   (b) alkalosis   (c) hemolysis   (d) ketosis

32. The chief buffer within the plasma is

(a) CO_2/H_2O (b) H_2CO_2/HCO_3^- (c) $H_2PO_4^-/HPO_4^{2-}$ (d) All of these

33. During vigorous physical activity, the lactic acid produced in muscle cells is neutralized by _____ when it enters the blood (to be transported to the liver).

(a) HCO_3^- (b) $H_2PO_4^-$ (c) HPO_4^{2-} (d) proteins

34. Breathing too rapidly may cause

(a) acidosis   (b) alkalosis   (c) asphyxia   (d) ketosis

35. Hyperventilation aids in controlling acidosis by

(a) increasing the concentration of carbonic acid.

(b) removing CO_2 from the blood.

(c) both (a) and (b)

(d) neither (a) nor (b)

FILL IN THE BLANKS

36. The three circulating fluids are interstitial fluid, _____, and _____.

37. The circulating blood consists of a straw-colored fluid called _____ and the blood cells, also called the _____.

38. Three of the dissolved constituents of blood plasma are _____, _____, and _____.

39. Intracellular fluids contain a relatively high concentration of K^+ and a relatively low concentration of Na^+. Which of these ions is more abundant in both the plasma and the interstitial fluid? _____

40. _____ globulins are referred to as immunoglobins or antibodies.

41. _____ is the condition arising from an abnormally high percentage of erythrocytes in the blood.

42. A major function of erythrocytes is to transport _____ to the tissues.

43. _____ is a cancer that is characterized by the uncontrollable production of leukocytes that fail to mature.

44. Substances that are foreign to our bodies contain certain proteins, termed _____, on their surfaces.

45. _____ is characterized by the accumulation of fluids within the interstitial spaces.

46. Serum consists of blood plasma minus _____.

47. _____ is an insoluble protein of the blood clot.

48. _____ and _____ are colored substances that give the bile its yellow color.

49. Ingested nitrites can lead to the production of _____.

50. Most of the oxygen transported via the blood to the various tissues is in the form of _____.

TRUE OR FALSE

51. T   F   Blood accounts for about one-eighth of the body weight of the average individual.

52. T   F   The average individual contains about 10 liters of blood.

53. T   F   The greatest component of the plasma is water.

54. T   F   Blood plasma contains a higher concentration of potassium ions than sodium ions.

55. T   F   Leukocytes have no nuclei.

56. T   F   The α- and β-globulins are involved in transporting fatty acids in the blood.

57. T F   Blood platelets function to transport triacylglycerols in the blood.

58. T F   The hemocrit value is a measure of the hemoglobin concentration in red blood cells.

59. T F   Nearly all antibodies found in the blood are lipoproteins.

60. T F   Erythrocytes are formed in the bone marrow.

61. T F   An abnormally low thrombocyte count is related to a tendency to bleed.

62. T F   B-cells and T-cells are involved in the immune response.

63. T F   The osmotic pressure is the pressure required to prevent the occurrence of osmosis.

64. T F   The concentration of protein in the interstitial fluid exceeds the concentration of protein in the plasma.

65. T F   Blood pressure measurements are reported as a ratio of the diastolic pressure to the systolic pressure (e.g., 120/80).

66. T F   The blood pressure is dependent on the total volume of blood.

67. T F   Blood serum is the same as blood plasma minus the formed elements.

68. T F   Blood clotting is a two-step process.

69. T F   Hemoglobin will bind to cyanide ions as well as oxygen.

70. T F   The function of the chloride shift is to maintain electrolyte balance within the erythrocyte.

71. T F   Alkalosis is the condition in which the blood pH is too high.

72. T F   Hemoglobin molecules are the major buffering system of the blood plasma.

73. T F   The pH of blood is increased by the slowing of breathing.

74. T F   Hypoventilation is brought on by coronary attack.

75. T F   Hyperventilation arises during anxiety.

Answers to Practice Questions

CHAPTER 2

1. b	2. d	3. b	4. c	5. a	6. d	7. a
8. d	9. c	10. d	11. c	12. b	13. b	14. a
15. d	16. d	17. a	18. c	19. c	20. c	21. d
22. d	23. b	24. d	25. c	26. b	27. d	28. a
29. c	30. a	31. d	32. b	33. c	34. b	35. d
36. c	37. d	38. c	39. a	40. b	41. hydrocarbons	

42. saturated hydrocarbons or alkanes   43. C_nH_{2n+2} 44. C_8H_{18} 45. alkyl   46. C_nH_{2n+1}

47. two   48. nonane

49. (a) 2,3-dimethylpentane   (b) 3,4-diethylhexane   (c) 3-methyl-5-propyloctane
    (d) 3-ethyl-2,2,5,6-tetramethylheptane

50. (a) $(CH_3)_2CHCH_2CH(CH_3)CH_2CH_3$ (b) $CH_3CH_2C(CH_3)_2CH(CH_3)CH_2CH_2CH_3$

    (c) $CH_3C(Cl)(CH_3)CH_2CH_3$ (d) $CH_2BrC(CH_3)_2CH_2CH_2CH_3$

    (e) $CH_3CH_2CH(CH_3)CH(CH_2CH_3)CH_2CH_2CH_2CH_3$

    (f) $(CH_3)_2CHCH(CH_3)CH(CH_3)CH(CH_2CH_3)CH(CH_2CH_3)CH(CH_3)CH_2CH_3$

51. free radical   52. oxygen   53. methane   54. propane   55. isooctane; n-heptane

56. C_nH_{2n} 57. C_7H_{14} 58. ⬠—Br   59. cyclohexane   60. 1,2-dimethylcyclopentane

61. C_6H_{12} 62. chair; boat   63. F   64. F   65. F   66. F   67. T

68. F	69. F	70. T	71. F	72. T	73. T	74. T
75. T	76. T	77. T	78. T	79. F	80. T	81. T
82. T	83. T	84. F	85. F	86. F	87. T	

CHAPTER 3

1. a	2. c	3. a	4. b	5. c	6. c	7. a
8. d	9. a	10. d	11. c	12. b	13. c	14. d
15. a	16. alkenes; alkynes		17. alkenes	18. C_7H_{10}	19. two	

20. (a) 2-methyl-3-hexene   (b) 4-chloro-2-pentene
    (c) 3-bromo-4-iodo-2-methyl-1-pentene   (d) 1-fluoro-5-methyl-2-hexene
    (e) 3,3-dimethylcyclohexene   (f) 1-methylcyclobutene
21. (a) $CH_2=CHCH_2CH_3$ (b) $CH_2=C(CH_3)CH_2CH_2CH_3$

    (c) $CH_2=CH(CH_3)CH_2CH(CH_3)CH_2CH_3$ (d) $CH_3CH_2CH=CHCHClCHICH_3$

    (e) $CH_2BrCH_2C(CH_3)=CHCH_2CH_2Br$ (f) $CH_3CH=CHCH[CH(CH_3)_2]CH_2CH_2CH_2CH_3$

    (g) $CH_3CH=C(CH_3)CH(CH_2CH_3)CH_2C(CH_3)_2CH(CH_2CH_3)C(CH_3)_3$

(h) $CH_3CH=CHCH=CHCH_3$ (i) ⬡   (j) [cycloheptene with $CH(CH_3)_2$ and CH_3 substituents]

22. cyclopentanol   23. pentane

24. (a) $CH_3CHClCH_2CH_3$ (b) $CH_2BrCHBrCH_2CH_3$ (c) $CH_3CH_2CH_2CH_3$

    (d) $CH_3CHOHCH_3$ (e) [cyclobutane with two Cl]   (f) $(CH_3)_2CBrCH_2CH_2CH_2CH_3$ (g) [cyclopentane with two CH_3 and Cl]

    (h) [cyclohexane with CH_2CH_3 and I]   (i) $CH_3CH_2CHOHCH_2OH$ (j) $CH_3\overset{O}{\overset{\|}{C}}CH_3 + CH_3\overset{O}{\overset{\|}{C}}OH$

25. polymerization   26. polymer   27. C_nH_{2n-2} 28. ethyne   29. $H-C\equiv C-H$

30. 4-methyl-2-pentyne   31. $CH_3C\equiv CCHBrCH_2CH_3$ 32. two   33. $CH_3C(Br)_2CH_3$

34. F	35. T	36. F	37. T	38. F	39. T	40. F
41. T	42. F	43. F	44. F	45. F	46. T	47. T
48. F	49. F	50. F	51. T	52. T	53. F	54. T
55. T						

CHAPTER 4

1. b	2. b	3. a	4. b	5. b	6. b	7. a
8. c	9. d	10. d	11. c	12. a	13. c	14. d
15. b	16. d	17. b	18. a	19. a	20. c	21. c
22. b	23. C_7H_8	24. ortho, meta, para				

25. (a) *p*-bromotoluene, 4-bromo-1-methylbenzene   (b) 1,4-dichloro-2-ethylbenzene
    (c) 1-bromo-1-phenylethane   (d) 1-iodo-3-phenylbutane

26. (a) ⬡—   (b) ⬡—CH_2Br (c) [naphthalene]   (d) I—⬡—CH_3

    (e) [benzene with Br and Cl]   (f) [benzene with CH_2CH_3, CH_3CH_2, OH]   (g) [benzene with two NO_2]   (h) ⬡—CH_2CHFCH_3

27. less   28. substitution   29. ⬡—Br + HBr   30. benzo[*a*]pyrene

31. T	32. F	33. F	34. T	35. F	36. F	37. T
38. T	39. T	40. T	41. F	42. F	43. F	44. F

CHAPTER 5

1. d	2. c	3. c	4. c	5. b	6. c	7. b
8. c	9. a	10. d	11. c	12. a	13. c	14. a
15. c	16. b	17. b	18. d	19. a	20. c	21. b
22. d	23. a	24. d	25. a	26. b	27. R—X	

28. (a) 1-bromo-2-chloro-1,1,2-trifluoroethane   (b) 2,4-dibromo-2,4-dimethylpentane
    (c) 1-fluoro-2-methylcyclohexane   (d) 3-iodo-1,1-dimethylcyclohexane

(e) 3-ethyl-4-fluoro-2,2-dimethylpentane (f) 2,4-dichloro-1-butene
(g) *p*-iodonitrobenzene, 4-iodo-1-nitrobenzene (h) 3-bromocyclopentene

29. (a) CH_2ClCH_2Cl

(b)
$$CH_3-CH_2-\underset{\underset{Br}{|}}{\overset{\overset{H}{|}}{C}}-\underset{\underset{H}{|}}{\overset{\overset{CH_3}{|}}{C}}-CH_2-CH_2-CH_3$$

(c) cyclobutane with CH_2CH_3, CH_2CH_3, and F substituents

(d) cyclopentane with I and CH_3 substituents

(e) CH_2Br_2

(f) CCl_2F_2

(g) benzene ring with Br, Br substituents (meta)

(h) benzene ring with I substituent

30. chloroform   31. insoluble

32. (a) cyclohexane with CH_3, Br, CH_3 substituents

(b) $CH_3CH_2CHBrCBr(CH_3)_2$

(c) cyclopentane with Br substituent

(d) $(CH_3)_3CBr$

(e) benzene ring with $-CH_2CH_2OH$ substituent

(f) CH_3CH_2CN

33. two   34. nucleophilic   35. refrigerants
36. perfluoropropylfurans   37. dichlorodiphenyltrichloroethane
38. Chlordane, Heptachlor, Endrin, Dieldrin, Aldrin, Lindane

CHAPTER 6

1. a	2. a	3. a	4. a	5. d	6. a	7. a
8. b	9. d	10. d	11. c	12. a	13. a	14. b
15. d	16. c	17. a	18. c	19. d	20. c	21. a
22. d	23. c	24. a	25. d	26. b	27. c	28. c
29. c	30. d	31. c	32. b	33. d	34. c	35. d
36. c	37. b	38. b	39. a	40. d	41. d	42. d
43. b	44. b	45. b	46. c	47. c	48. a	49. b
50. c	51. c	52. c	53. a	54. b	55. b	56. a
57. c	58. d	59. a	60. a	61. b	62. b	63. c
64. c	65. c	66. b	67. c	68. d	69. d	70. a
71. a	72. a	73. a	74. d	75. d	76. a	77. c
78. b	79. c	80. d	81. a	82. b	83. hydroxyl	

84. secondary   85. $CH_3CHOHCH_2CH_3$

86. (a) 3-methyl-1-butanol   (b) 2,3-hexanediol
(c) 5-chloro-2-methyl-2-heptanol   (d) *p*-isopropylphenol, 4-isopropylphenol
87. (a) CH_2OHCH_2OH (b) $CH_2OHCHOHCH_2OH$

(c)
$$Br \quad CH_2CH_3$$
$$CH_3CCH_2CHCH_2CH_3$$
$$OH$$

(d) (ring)—OH, H_3C, I

88. decreases 89. aldehydes, then to carboxylic acids 90. carbon dioxide and water 91. dehydration

92. (a) $CH_3CH=CH_2$ (b) $CH_3CH_2CH_2OCH_2CH_2CH_3$

(c) (ring)—$CH=CHCH_3$ (d) (ring)—O^-Na^+

(e) (ring)—Br (f) (ring)—CH_2Cl

93. denatured 94. Ethers 95. lower 96. methyl isopropyl ether 97. $CH_3CH_2OCH_2CH_2CH_3$

98. peroxides 99. sulfhydryl 100. oxygen 101. sulfides; thiols 102. more
103. (a) $CH_3CH_2S-SCH_2CH_3$ (b) $2\ CH_3SH$

104. T	105. F	106. F	107. T	108. T	109. T	110. F
111. F	112. F	113. F	114. F	115. F	116. F	117. F
118. F	119. T	120. F	121. T	122. T	123. F	124. F
125. T						

CHAPTER 7

1. a	2. b	3. d	4. b	5. c	6. d	7. b
8. b	9. a	10. c	11. c	12. c	13. a	14. d
15. c	16. d	17. d	18. c	19. d	20. d	21. c
22. b	23. d	24. b	25. b	26. b	27. c	28. c
29. a	30. d	31. a	32. a	33. d	34. c	35. c
36. b	37. b	38. a	39. d	40. b	41. carbonyl	

42. aldehydes 43. $\overset{O}{\overset{\|}{RCR'}}$
44. (a) 3-bromobutanal (b) 1-methoxy-2-butanone
(c) 4-hydroxypentanal (d) 2-iodo-4-methylcyclohexanone
(e) 3,5-dimethyl-2-hexanone (f) 3-bromo-3-methylhexanal
(g) 6-ethyl-5-methyl-3-octanone (h) 3,5-dichlorobenzaldehyde

45. (a) $CH_3CHCCHCH_2CH_2CH_3$ (O, Cl, CH_3) (b) $CH_3CCH_2CH_2CH_2CHCH_2CH_3$ (O, OCH_3)

(c) F-cyclopentanone (d) $CH_3CHCH_2CH_2CH_2C$ (Br, O, H)

(e) $CH_3CHCHCH_2CH_2CHC$ (CH_3, CH_3, O, H, ring) (f) benzaldehyde (I, O, H, CH_2CH_3)

46. secondary 47. hemiketal 48. acetal 49. alpha

50.
$$\text{(C}_6\text{H}_5)\text{C}=\text{C(CH}_2\text{CH}_3)_2 \quad \text{with } OH$$

51. acid   52. Cu_2O 53. primary

54. acetone; formaldehyde

55. (a) $CH_3\overset{OH}{\underset{H}{C}}-OCH_3$

(b) $CH_3CH_2\overset{OCH_3}{\underset{H}{C}}-OCH_3$

(c) $CH_3\overset{OH}{\underset{CH_3}{C}}-OCH_2CH_3$

(d) $\overset{OCH_3}{\underset{H}{C}}-OCH_3$

(e) $\overset{OH}{\underset{H}{C}}-CN$

(f) $CH_3\overset{OH}{\underset{CH_3}{C}}-CH_2\overset{O}{C}CH_3$

(g) $\overset{OH}{\underset{H}{C}}-CH_2\overset{O}{C}H$

(h) =O

(i) $(CH_3)_2CHCH_2\overset{O}{C}OH$

(j) $CH_3CHOHCH_2CH_2CH_3$

56. F	57. F	58. F	59. T	60. F	61. F	62. F
63. F	64. T	65. T	66. F	67. F	68. F	69. T
70. F	71. F	72. F	73. T	74. T	75. T	76. T
77. T						

CHAPTER 8

1. a	2. b	3. c	4. c	5. b	6. b	7. a
8. a	9. d	10. c	11. d	12. c	13. d	14. c
15. d	16. c	17. b	18. c	19. a	20. a	21. a
22. c	23. c	24. b	25. a	26. c	27. b	28. a
29. a	30. b	31. c	32. d	33. c	34. b	35. d
36. c	37. d	38. b	39. a	40. c	41. b	42. c
43. d	44. a	45. d	46. c	47. a	48. b	49. c
50. b	51. a	52. d	53. c	54. c	55. c	56. c
57. b	58. c	59. a	60. b	61. d	62. c	63. d
64. a	65. carboxyl					

66. (a) 2-chloropentanoic acid

(b) lithium propanoate

(c) cyclopentanecarboxylic acid

(d) 3-methoxybutanoic acid

(e) 3,3,5-trimethylhexanoic acid

(f) *p*-iodobenzoic acid, 4-iodobenzoic acid

67. (a) $(CH_3CH_2CH_2COO^-)_2Ca^{2+}$

(b) $CH_2BrCOOH$

(c) COOH

(d) $CH_3CH_2CH(CH_3)CH_2CH_2CHOHCOOH$

(e) $HOOCCH_2CH_2COOH$

(f) —COOH   with O_2N

68. oxidation 69. higher   70. 5.0   71. greater; smaller   72. less   73. resonance
74. acetic acid   75. preservative   76. ester
77. (a) ethyl 2-methylpropanoate   (b) isopropyl 3-phenylpropanoate
    (c) propylbutanoate   (d) cyclopentylethanoate

78. (a) H—C(=O)OCH$_2$CH$_3$

    (b) CH$_3$CH$_2$CHC(=O)OCH$_3$ with CH$_2$CH$_3$ branch

    (c) cyclohexyl—C(=O)OCH$_2$CH$_3$

    (d) phenyl—C(=O)OCH(CH$_3$)$_2$

79. insoluble   80. lower   81. acid   82. base   83. saponification
84. acid; alcohol   85. reduction

86. (a) CH$_3$COO$^-$Na$^+$

    (b) CH$_3$C(=O)Cl

    (c) CH$_3$CH$_2$C(=O)OCH$_3$

    (d) (CH$_3$)$_2$CHCH$_2$COOH + CH$_3$CH$_2$CH$_2$OH

    (e) CH$_3$COO$^-$K$^+$ + CH$_3$CH$_2$CH$_2$OH

    (f) phenyl—CHOHCH$_2$COOH + CH$_3$CH$_2$OH

    (g) CH$_3$CH$_2$COO$^-$Na$^+$ + CH$_3$OH

    (h) phenyl—CH$_2$OH + CH$_3$OH

87. F	88. F	89. F	90. F	91. F	92. T	93. F
94. F	95. T	96. F	97. T	98. T	99. T	100. F
101. F	102. F	103. T	104. F	105. T		

CHAPTER 9

1. a	2. c	3. a	4. b	5. a	6. b	7. d
8. b	9. b	10. a	11. d	12. d	13. b	14. d
15. c	16. d	17. d	18. b	19. c	20. a	21. b
22. c	23. a	24. a	25. c	26. c	27. d	28. a
29. a	30. b	31. c	32. c	33. a	34. d	35. a
36. a	37. b	38. a	39. b	40. c	41. a	42. b
43. c	44. a	45. d	46. acetamide	47. N-methylpropanamide		48. benzamide

49. O$_2$N—phenyl—C(=O)N(CH$_3$)CH$_3$ 50. less

51. (a) CH$_3$C(=O)O$^-$NH$_4^+$

    (b) CH$_3$CH$_2$C(=O)NH$_2$

    (c) CH$_3$CH$_2$C(=O)OH + CH$_3$CH$_2$NH$_3^+$Cl$^-$

(d) [benzene ring]—CH_2NH_2 + $CH_3\overset{O}{\underset{O^-Na^+}{C}}$

52. (a) triethylamine
    (c) dimethylpropylamine
    (b) 1-amino-3-methylbutane
    (d) p-ethylaniline, 4-ethylaniline

53. (a) $CH_3-\overset{NH_2}{\underset{CH_3}{C}}-CH_3$
    (b) $CH_3-\overset{H}{\underset{CH_3}{C}}-NH_2$

    (c) [benzene ring]—$\overset{}{\underset{H}{N}}$—[benzene ring]
    (d) $CH_3CH_2-\overset{}{\underset{CH_3}{N}}-CH_2CH_3$

54. fishy   55. ammonia   56. aliphatic; aromatic   57. higher   58. water solubility

59. (a) $CH_3CH_2CH_2NH_3^+Cl^-$ (b) $(CH_3)_4N^+I^-$ (c) [benzene ring]—NH_2 (d) $CH_3CH_2-\overset{}{\underset{CH_2CH_3}{N}}-N=O$

60. F   61. F   62. T   63. F   64. T   65. T   66. F
67. T   68. F   69. F   70. T   71. T

CHAPTER 10

1. b	2. c	3. d	4. b	5. d	6. a	7. d
8. b	9. c	10. d	11. a	12. d	13. d	14. d
15. c	16. d	17. b	18. b	19. a	20. d	21. d
22. a	23. stereoisomers		24. enantiomers		25. chiral	26. chiral

27. optically   28. 2^n; chiral   29. 8   30. $CH_3-\overset{H}{\underset{Br}{C}}-CH_2OH$ $CH_3-\overset{Br}{\underset{H}{C}}-CH_2OH$ 31. none

32. none   33. diastereomers   34. meso   35. $CH_3-\overset{H\ H}{\underset{Br\ Br}{C-C}}-CH_3$ 36. racemic

37. *trans*-2,3-dibromo-2-pentene   38. [cyclopentane ring with Cl and Cl]   39. $\overset{CH_3CH_2}{\underset{H}{}}C=C\overset{CH_2CH_2CH_3}{\underset{H}{}}$

40. $\overset{H}{\underset{I}{}}C=C\overset{H}{\underset{I}{}}$ $\overset{H}{\underset{I}{}}C=C\overset{I}{\underset{H}{}}$

41. $\overset{CH_3}{\underset{H}{}}C=C\overset{CH_3}{\underset{CH_2CH_3}{}}$ $\overset{CH_3}{\underset{H}{}}C=C\overset{CH_2CH_3}{\underset{CH_3}{}}$

42. $\overset{CH_3}{\underset{Cl}{}}C=C\overset{Cl}{\underset{CH_2CH_3}{}}$ 43. $CH_3CH_2-\overset{F\ F}{\underset{H\ H}{C-C}}-CH_2CH_3$

44.

45. T	46. T	47. T	48. T	49. T	50. F	51. T
52. F	53. F	54. T	55. F	56. T	57. C	58. F
59. B	60. G	61. E	62. I	63. A	64. D	65. H
66. J						

CHAPTER 11

1. b	2. b	3. d	4. d	5. d	6. d	7. d
8. a	9. b	10. c	11. c	12. b	13. d	14. b
15. a	16. a	17. d	18. d	19. d	20. c	21. d
22. a,c	23. a	24. e	25. b	26. c	27. d	28. d
29. c	30. a	31. a	32. c	33. b	34. d	35. d
36. b	37. a	38. d	39. b	40. d	41. b	42. c
43. a	44. a	45. d	46. a	47. c	48. c	49. c
50. b	51. d	52. b	53. a	54. a	55. a	56. a
57. c	58. b	59. d	60. c	61. c	62. a	63. a
64. d	65. d	66. c	67. d	68. a	69. c	70. d
71. a	72. b					

73. $C_x(H_2O)_y$ 74. saccharin   75. aspartame
76. monosaccharide   77. tetroses; heptoses   78. five   79. three
80. dihydroxyacetone   81. glyceraldehyde   82. clockwise 83. epimers
84. carbonyl; hydroxyl   85. β-glucose 86. furanose   87. pyranose 88. below
89. mutarotation   90. two   91.   92. glucose; levulose

93. three   94. fructose   95. ribose; deoxyribose   96. disaccharides
97. sweet; soluble   98. glycosidic 99. glucoside 100. invert   101. lactose   102. cellobiose
103. sucrose; malt   104. acids; enzymes   105. maltase; lactase
106. lactose; β-1,4-galactosidic   107. glucose; fructose   108. starch; glycogen
109. glucose 110. amylose; amylopectin   111. cellulose 112. β-1,4   113. amylopectin
114. (a) Seliwanoff's   (b) Bial's   (c) Barfoed's   (d) Benedict's

115. F	116. F	117. F	118. F	119. T	120. F	121. T
122. T	123. T	124. F	125. T	126. T	127. F	128. F
129. F	130. T	131. F	132. T	133. F	134. T	135. F
136. T	137. F	138. F	139. F	140. F	141. F	142. F
143. F	144. T	145. F	146. F	147. T	148. T	149. F
150. T	151. F	152. F	153. F	154. F	155. F	

CHAPTER 12

1. c	2. d	3. c	4. b	5. d	6. b	7. d
8. c	9. a	10. c	11. b	12. d	13. b	14. b
15. d	16. b	17. b	18. b	19. b	20. b	21. c
22. c	23. c	24. b	25. c	26. b	27. c	28. d
29. b	30. a	31. a	32. d	33. d	34. d	35. b
36. c	37. a	38. a	39. b	40. c	41. c	42. d

43. b   44. a   45. b   46. a   47. b   48. c   49. b
50. d   51. b   52. b   53. b   54. a   55. d   56. c
57. d   58. b   59. b   60. c   61. d   62. a   63. a
64. d   65. c   66. nonpolar; polar   67. glycerol   68. ester
69. increase; decrease   70. triacylglycerols   71. saturated; unsaturated
72. cis   73. palmitic, oleic   74. fatty acids; glycerol   75. base
76. saponifiable; nonsaponifiable   77. rancidity   78. butyric   79. waxes   80. soaps
81. hydrophilic; hydrophobic   82. anionic, cationic, nonionic
83. eutrophication   84. phosphoglycerides   85. four   86. zero

87. cell membranes   88. proteins   89. three   90.

91. atherosclerosis   92. bile salts   93. progesterone   94. estrogen   95. prostaglandins
96. D   97. A,D,E,K   98. T   99. T   100. F   101. F   102. F
103. F   104. F   105. T   106. F   107. F   108. T   109. F
110. T   111. T   112. F   113. F   114. T   115. T   116. F
117. F   118. T   119. T   120. F   121. T   122. F   123. F
124. F   125. T   126. F   127. F   128. T   129. T   130. T
131. F   132. T   133. T   134. F   135. T

CHAPTER 13

1. d   2. d   3. b   4. d   5. b   6. c   7. b
8. d   9. c   10. d   11. c   12. b   13. a   14. d
15. a   16. d   17. d   18. b   19. d   20. b   21. d
22. c   23. d   24. b   25. a   26. b   27. c   28. b
29. d   30. b   31. b   32. c   33. c   34. b   35. d
36. b   37. b   38. c   39. d   40. b   41. d   42. d
43. d   44. b   45. d   46. c   47. d   48. b   49. c
50. a   51. b   52. a   53. b   54. b   55. c   56. c
57. d   58. b   59. d   60. d   61. d   62. a   63. d
64. a   65. twenty   66. glycine   67. isoleucine (or threonine)   68. nonpolar   69. zwitterion

70. isoelectric   71. $CH_2NH_2COO^-$ 72. $CH_3-\underset{\overset{|}{{}^+NH_3}}{CH}-COOH$ 73.

74. positive; negative   75. 5.68   76. negative   77. glutamic acid; lysine; valine
78. arginine; lysine (or histidine)   79. glutamic acid (or aspartic acid)   80. peptide
81. primary   82. peptide   83. two   84. leucine   85. 8,9,10   86. Asn, Ile, Trp
87. α-helical (or β-pleated sheet)   88. β-pleated sheet   89. α-helical   90. hydrogen
91. tertiary   92. $-OH$; $-SH$; phenyl; phenol; $-COO^-$; $-NH_3^+$ 93. salt linkages; hydrophobic
94. quaternary   95. oxygen   96. proline   97. glycine   98. globular; fibrous
99. amino acids   100. prosthetic   101. positive   102. denaturation
103. heat; organic solvents; acids or bases; metal ions   104. F   105. F   106. F
107. F   108. F   109. F   110. T   111. F   112. F   113. T
114. T   115. F   116. F   117. F   118. T   119. F   120. F
121. T   122. F   123. F   124. T   125. T   126. F   127. F
128. T   129. F   130. F   131. T   132. F   133. F   134. T
135. T   136. T   137. F   138. T   139. F   140. F   141. F
142. F   143. T   144. F   145. F   146. T   147. T   148. F
149. T   150. T   151. F   152. F   153. T   154. T   155. F
156. T

CHAPTER 14

1. a   2. d   3. d   4. b   5. a   6. b   7. b
8. d   9. a   10. a   11. c   12. d   13. d   14. c
15. d   16. d   17. a   18. c   19. d   20. b   21. b
22. d   23. d   24. d   25. c   26. d   27. b   28. c
29. a   30. b   31. b   32. c   33. a   34. b   35. d
36. a   37. a   38. a   39. c   40. d   41. c   42. increase
43. proteins 44. ase 45. urea 46. apoenzyme   47. coenzyme
48. zymogen (or proenzyme) 49. coenzymes 50. NAD⁺; FAD 51. activation energy
52. substrate; enzyme—substrate   53. active site   54. lock-and-key; induced fit
55. hydrogen bonds; salt linkages; hydrophobic interactions 56. contact; catalytic   57. absolute
58. peptidases59. transferases   60. bell-shaped   61. 7.0–7.5 62. pepsin
63. 100°   64. optimum 65. competitive   66. inhibiting
67. noncompetitive   68. acetylcholinesterase   69. allosteric
70. competitive, noncompetitive, irreversible, or end-product71. chemotherapy   72. antibiotic
73. antimetabolite   74. penicillin 75. p-aminobenzoic acid
76. aureomycin tetracycline, streptomycin, chloramphenicol 77. F   78. F   79. T
80. F   81. F   82. F   83. F   84. F   85. F   86. F
87. F   88. T   89. T   90. F   91. F   92. F   93. T
94. F   95. F   96. T   97. F   98. F   99. F   100. T
101. F   102. F   103. F   104. T   105. F   106. T   107. T
108. F   109. F   110. F   111. F   112. T   113. F

CHAPTER 15

1. a   2. b   3. a   4. b   5. d   6. d   7. a
8. d   9. d   10. c   11. a   12. b   13. b   14. c
15. d   16. c   17. d   18. c   19. b   20. c   21. b
22. b   23. c   24. b   25. b   26. a   27. c   28. d
29. d   30. d   31. d   32. a   33. d   34. a   35. b
36. d   37. a   38. b   39. d   40. d   41. sugar; base; N-glycosyl
42. nucleotide   43. phosphate 44. negative   45. sequence 46. complementary
47. hydrogen 48. DNA; RNA 49. 3′; 5′   50. semiconservative   51. polymerase; ligase
52. 5′→3′   53. ribose   54. nucleus   55. uracil; cytosine   56. mRNA   57. DNA
58. ribosomes   59. His-Gly   60. t RNA   61. three
62. CGCCACCUC   63. AGAAAGCATTTACCG
64. CGTTCATCGAAG   65. Val-Asp-Glu-Pro   66. T   67. F   68. F
69. T   70. F   71. F   72. T   73. F   74. T   75. F
76. F   77. T   78. T   79. F   80. T   81. F   82. F
83. F   84. F   85. T   86. F   87. F   88. F   89. T
90. T   91. F   92. F   93. T   94. T   95. F   96. T
97. T   98. T   99. T   100. F

CHAPTER 16

1. d   2. d   3. a   4. b   5. d   6. d   7. c
8. b   9. d   10. c   11. b   12. b   13. a   14. c
15. d   16. a   17. a   18. c   19. c   20. b   21. b
22. b   23. b   24. b   25. a   26. c   27. a   28. c
29. a   30. b   31. c   32. d   33. c   34. d   35. a
36. b   37. c   38. c   39. a   40. c   41. b   42. a
43. a   44. a   45. b   46. a   47. c   48. c   49. d
50. b   51. photosynthesis   52. glucose   53. epinephrine; glucagon; insulin

54. 70–100 mg/100 mL   55. hypoglycemia   56. phosphorylase
57. fermentation   58. kinase   59. aldolase   60. isomerase
61. glyceraldehyde 3-phosphate; dihydroxyacetone phosphate   62. mutase   63. enolase
64. oxidation–reduction   65. lactic acid   66. anaerobic
67. ethanol; lactic acid; acetyl-CoA   68. Embden–Meyerhof   69. glycogenesis
70. lactic acid   71. creatine phosphate   72. two
73. citric acid cycle; tricarboxylic acid cycle 74. mitochondria   75. citric acid
76. NADH   77. malic   78. oxidation 79. oxidative 80. Malic acid–aspartic acid shuttle
81. F   82. F   83. F   84. F   85. T   86. T   87. T
88. F   89. T   90. T   91. F   92. F   93. F   94. T
95. F   96. T   97. T   98. T   99. F   100. F   101. T
102. F   103. F   104. T   105. F   106. T   107. F   108. F
109. F   110. F   111. T   112. F   113. F   114. F   115. T
116. F   117. T   118. F   119. F   120. T   121. T   122. T
123. T

CHAPTER 17

1. d   2. b   3. d   4. a   5. a   6. b   7. a
8. a   9. d   10. b   11. a   12. a   13. b   14. b
15. c   16. d   17. b   18. d   19. c   20. a   21. c
22. b   23. d   24. d   25. b   26. b   27. b   28. d
29. c   30. c   31. c   32. c   33. b   34. c   35. b
36. b   37. c   38. d   39. d   40. b   41. a   42. b
43. c   44. d   45. d   46. d   47. c   48. b   49. lipases
50. bile salts   51. lipids   52. adipose   53. mobilization   54. proteins
55. brain   56. fatty acyl-CoAs   57. oxidation; hydration; oxidation; cleavage
58. respiratory   59. Krebs cycle   60. seven
61. mitochondrium; cytosol   62. acetyl-CoA   63. acetyl-CoA
64. ketone bodies   65. brain   66. acetone   67. β-hydroxybutyric acid
68. T   69. F   70. T   71. T   72. F   73. F   74. T
75. F   76. F   77. F   78. T   79. F   80. T   81. T
82. F   83. T   84. T   85. F   86. F   87. F   88. T
89. F   90. T   91. T   92. T   93. T

CHAPTER 18

1. c   2. a   3. a   4. b   5. c   6. b   7. d
8. c   9. c   10. c   11. d   12. b   13. d   14. d
15. c   16. d   17. b   18. c   19. a   20. b   21. b
22. d   23. a   24. b   25. b   26. c   27. d   28. c
29. d   30. c   31. c   32. c   33. pepsinogen
34. small intestine   35. trypsinogen; chymotrypsinogen; procarboxypeptidase
36. enteropeptidase   37. carboxypeptidase; aminopeptidase   38. lysine   39. positive
40. aspartic acid   41. pyruvic acid   42. α-ketoglutaric acid

43. histamine 44. glutamine 45. urea   46. uric acid   47. H_2NCONH_2 48.

$$\begin{array}{c} CH_2CH_2NH_2 \\ \text{ring} \end{array}$$

49. $CH_3CH(NH_2)COOH + HOOCCH_2CH_2COCOOH$
50. $HOOCCH_2CH_2COCOOH + NH_3 + NADH$
51. T   52. T   53. F   54. F   55. F   56. F   57. F
58. F   59. T   60. F   61. T   62. T   63. F   64. F
65. T   66. F   67. T   68. T   69. F   70. F   71. F
72. T   73. F   74. T   75. F   76. F   77. T   78. F
79. F   80. T   81. T   82. T

CHAPTER 19

1. d	2. b	3. d	4. a	5. c	6. d	7. a
8. a	9. d	10. a	11. b	12. b	13. d	14. c
15. b	16. c	17. d	18. b	19. b	20. c	21. b
22. a	23. d	24. d	25. d	26. a	27. a	28. a
29. c	30. c	31. a	32. b	33. a	34. b	35. b

36. blood; lymph   37. plasma; formed elements

38. proteins; carbohydrates; amino acids; hormones; vitamins   39. Na^+ 40. γ-

41. polycythemia   42. oxygen   43. leukemia   44. antigens   45. edema

46. fibrinogen   47. fibrin   48. biliverdin; bilirubin   49. methemoglobinemia

50. oxyhemoglobin		51. F	52. F	53. T	54. F	55. F
56. F	57. F	58. F	59. F	60. T	61. T	62. T
63. T	64. F	65. F	66. T	67. F	68. F	69. F
70. T	71. T	72. F	73. F	74. T	75. T	